Next-Generation Batteries and Fuel Cells for Commercial, Military, and Space Applications

Books by Dr. A. R. Jha

MEMS and Nanotechnology-Based Sensors and Devices for
Communications, Medical and Aerospace Applications
ISBN 978-0-8493-8069-3

Next-Generation Batteries and Fuel Cells for Commercial,
Military, and Space Applications
ISBN 978-1-4398-5066-4

Solar Cell Technology and Applications
ISBN 978-1-4200-8177-0

Wind Turbine Technology
ISBN 978-1-4398-1506-9

Next-Generation Batteries and Fuel Cells for Commercial, Military, and Space Applications

A.R. JHA

CRC Press
Taylor & Francis Group
Boca Raton   London   New York

CRC Press is an imprint of the
Taylor & Francis Group, an **informa** business
AN AUERBACH BOOK

CRC Press
Taylor & Francis Group
6000 Broken Sound Parkway NW, Suite 300
Boca Raton, FL 33487-2742

International Standard Book Number: 978-1-4398-5066-4 (Hardback)

Library of Congress Cataloging-in-Publication Data

Jha, A. R.
  Next-generation batteries and fuel cells for commercial, military, and space applications / A.R. Jha.
    p. cm.
  Summary: "Next-generation batteries have higher power density and higher energy density and can be put into new forms with lower-cost mass production. This book focuses on technologically advanced secondary (rechargeable) batteries in both large and small format. It covers advanced technologies as replacements for NiCd and NiMH, especially advanced lithium-ion batteries that make use of new electrode materials and electrolytes. The author discusses printable batteries and thin-film battery stacks as enablers of micropower applications as well as hybrid battery/fuel cell systems, which are emerging as complements to consumer electronics batteries"-- Provided by publisher.
  Includes bibliographical references and index.
  ISBN 978-1-4398-5066-4 (hardback)
  1. Storage batteries. 2. Fuel cells. 3. Electric batteries. I. Title.

TK2941.J47 2012
621.31'2424--dc23

2012004916

Visit the Taylor & Francis Web site at
http://www.taylorandfrancis.com

and the CRC Press Web site at
http://www.crcpress.com

This book is dedicated to my parents, who always encouraged me to pursue advanced research and developmental activities in the field of science and technology for the benefits of mankind regardless of nationality, race, or religion.

Contents

Foreword

This book comes at a time during which high global demand for oil is coupled with the anticipation of a shortage in the near future. To reduce this dependency on foreign oil and eliminate the greenhouse effects associated with oil, several automobile-manufacturing companies have been engaged in the mass development and production of electric vehicles (EVs), hybrid electric vehicles (HEVs), and plug-in hybrid electric vehicles (PHEVs). To address these objectives, the author of this book, A. R. Jha, gives serious attention to cutting-edge battery technology. Advanced material technology must be given consideration in the development of next-generation batteries and fuel cells for deployment in EVs and HEVs. In addition, Jha identifies and describes next-generation primary and secondary (rechargeable) batteries for various commercial, military, spacecraft, and satellite applications for covert communications, surveillance, and reconnaissance missions. Jha emphasizes the cost, reliability, longevity, and safety of the next generation of high-capacity batteries that must be able to operate under severe thermal and mechanical environments.

This book addresses nearly every aspect of battery and fuel cell technology involving the use of the rare earth materials that are best suited for specific components and possible applications in EVs, HEVs, and PHEVs. Use of certain rare earth materials offers significant improvement in electrical performance and a reduction in the size of alternating current induction motors and generators that will yield additional space inside these vehicles. Jha proposes ultra-high-purity metallic nano-technology PVD films in the design and development of the low-power batteries best suited for implantable medical devices and diagnostic applications. This particular technology can be used in the near future in the development of noninvasive medical diagnostic equipment such as magnetic resonance imaging and computed tomography scans.

Jha continues, throughout this book, his distinguished track record of distilling complex theoretical physical concepts into an understandable technical framework that can be extended to practical applications across a wide array of modern industries. His big-picture approach, which does not compromise the basic underlying science, is particularly refreshing. This approach should help present-day students, both undergraduate and graduate, master these difficult scientific concepts with the full confidence they will need for commercial engineering applications to benefit emerging economies around the world.

This book is well organized and provides mathematical expressions to estimate the critical performance parameters of rechargeable batteries. Jha covers all of the important design aspects and potential applications of rechargeable batteries with an emphasis on portability, reliability, longevity, and cost-effective performance. This book also provides a treatment of the underlying thermodynamic aspects of cells housed in a battery pack that contains several cells. Jha identifies their adverse heating effects on the reliability and electrical performance of the battery pack. Notably, thermodynamic evaluation of the battery pack assembly is of critical importance because it can affect the reliability, safety, and longevity of the pack. Jha's background enables him to provide an authoritative account of many of the emerging application requirements for small, lightweight, high-reliability rechargeable batteries, particularly for portable and implantable medical devices and diagnostic capsules. Jha summarizes the benefits of all-solid-state lithium-ion batteries for low-power medical devices, such as cardiac pacemakers, cardioverters, and implantable cardioverter defibrillators.

Critical performance parameters and the limits of rechargeable batteries, including state of charge, depth of discharge, cycle life, discharge rate, and open-circuit voltage, are identified. The aging effects of various batteries are identified as well. Rechargeable battery requirements for EVs, HEVs, and PHEVs are summarized with an emphasis on reliability, safety, and longevity. Memory effects resulting from voltage depression are discussed in great detail. The advantages of solid polymer electrolyte technologies are briefly mentioned because the polymer electrolytes tend to increase room temperature iconic conductivity. This increase in ionic conductivity offers improved battery performance at medium to high temperatures ranging from 60° to 125°C.

Performance capabilities of long-life, low-cost, rechargeable batteries, including silver zinc and other batteries, are summarized. Such batteries are best suited for aerospace and defense applications. Batteries for unmanned underwater vehicles, unmanned air vehicles, anti-improvised explosive devices, and satellites or spacecraft capable of providing surveillance, reconnaissance, and tracking of space-borne targets are identified with an emphasis on reliability, longevity, safety, weight, and size. Cathode, anode, and electrolyte materials are summarized for several rechargeable batteries.

Jha dedicates a chapter specifically to fuel cells and describes the three distinct types of practical fuel cells, including those that use (1) aqueous electrolytes, (2) molten electrolytes, and (3) solid electrolytes. The fuel cell is an electricity generation system that combines an oxidation reaction and a reduction reaction. In a fuel cell, both the fuel and oxidant are added from an external source to react at two separate electrodes, whereas in a battery, the two separate electrodes are fuel and oxidant. Therefore, in the fuel cell in an energy conversion device, chemical energy is isothermally converted to direct current electricity. These devices are bulky and heavy and operate mostly at high temperatures (500° to 850°C). Hydrogen-oxygen fuel cells generate high power with maximum economy and are best suited for transport buses. Electrode kinetics play a key role in achieving the most efficient operation of a fuel cell. Jha identifies the basic laws of electrochemical kinetics and notes that a superior nutrient-electrolyte media is essential for generating higher electrical power in biochemical fuel cells.

A wide variety of readers will benefit from this book, in particular the advanced undergraduate and graduate students of mechanical and materials engineering who wish to pursue a career in designing next-generation batteries and fuel cells. In view of the critical interdependencies with other technical disciplines, however, this book also is of interest to a wider variety of engineering students or practicing engineers in such industries as medical equipment, defense electronics, security, and space as well as in other yet-to-be-established disciplines. This book is particularly useful for research scientists and engineers who are deeply involved in the design of the portable devices best suited for medical, military, and aerospace systems. Technical managers will also find this book useful for future applications. I strongly recommend this book to a broad audience, including students, project managers, aerospace engineers, life-science scientists, clinical scientists, and project engineers immersed in the design and development of compact, lightweight batteries best suited for industrial, commercial, military, and space applications.

Dr. A. K. Sinha
Senior Vice President
Applied Materials, Inc.
Santa Clara, California

Preface

The publication of this book comes at a time when free nations are at odds with oil-producing nations and can be threatened with an interruption of the continuous flow of oil because of political differences and prevailing conditions in these respective regions. Western and other free nations are looking for alternative energy sources to avoid the high cost of oil and to reduce greenhouse gas emissions. This book briefly summarizes the performance capabilities and limitations of existing primary and secondary (rechargeable) batteries for the benefits of readers. I address critical and vital issues affecting the performance capabilities of next-generation batteries and fuel cells for commercial, military, and aerospace applications and propose cutting-edge battery technology best suited for all-electric and hybrid electric vehicles (HEVs) in an effort to help eliminate dependency on unpredictable foreign oil sources and supplies.

I also identify the unique materials for electrolytes, cathodes, and anodes that are most cost-effective for next-generation rechargeable batteries with significant improvements in weight, size, efficiency, reliability, safety, and longevity. Likewise, I identify rechargeable batteries with minimum weight, size, and form factor that are most ideal for implantable medical devices, unmanned aerial vehicles (UAVs), and space system applications. I identify battery designs using microelectromechanical systems (MEMS) and nanotechnologies, which are best suited for applications where weight, size, reliability, and longevity are of critical importance. Integration of these technologies would lead to significant improvements in weight, size, and form factor without compromising the electrical performance and reliability of the battery.

I propose high-power battery technologies best suited for automotive-, aircraft-, and satellite-based system applications with an emphasis on reliability, safety, and consistent electrical performance over long durations. In such applications, I recommend unique battery technologies that offer exceptionally high-energy densities that exceed 500 Wh/kg. I also describe the performance capabilities of next-generation rechargeable sealed nickel-cadmium and sealed lead-acid batteries that are most ideal for satellite communications, space-based surveillance and reconnaissance systems, unmanned ground combat vehicles (UGCVs), UAVs, and other battlefield applications where high energy density, minimum weight and size, and reliability under harsh conditions are the principal performance requirements.

This book summarizes the critical performance parameters of rechargeable batteries developed for various commercial, military, and space applications backed by measured values of parameters obtained by reliable sources through actual laboratory measurements. The book is well organized and contains reliable rechargeable battery performance characteristics for a wide range of applications, including commercial, military, and aerospace disciplines. Cutting-edge battery design techniques are discussed in the book backed by mathematical expressions and derivations wherever possible. The book provides mathematical analysis capable of projecting the critical performance parameters under various temperatures. It is especially prepared for design engineers who wish to expand their knowledge of next-generation batteries.

I have made every attempt to provide well-organized materials using conventional nomenclatures, a constant set of symbols, and easy-to-understand units for rapid comprehension. The book provides state-of-the-art performance parameters of some batteries from various reference sources with due credit given to the authors or organizations involved. It comprises eight distinct chapters, each of which is dedicated to a specific application.

Chapter 1 presents the current status of various primary and secondary (rechargeable) batteries and fuel cells for various applications. The performance capabilities and limitations of batteries and fuel cells are summarized for the benefits of readers and design engineers. The current energy sources suffer from weight, size, efficiency, discharge rates, disposal issues, and recharge capacity, thus making them unsuitable for medical, battlefield, and aerospace applications. General Motors and Siemens have invested a significant amount of money in research and development of rechargeable lithium-based rechargeable batteries for possible applications in electric vehicles (EVs) and HEVs. Current fuel cells generate electrical energy by using electrochemical conversion techniques that have serious drawbacks. I discuss direct methanol fuel cells (DMFCs) for future applications that will be found most ideal for high-, portable-power sources. DMFC technology offers improved reliability, compact form factor, and significant reduction in weight and size. I identify appropriate anode, cathode, and membrane electrode assembly configurations that will yield significantly improved electrical performance over long durations with minimum cost and complexity.

Chapter 2 briefly describes the performance capabilities and limitations of current rechargeable batteries for various applications. Performance requirements and projections for next-generation primary and secondary batteries are identified with an emphasis on cost, reliability, charge rate, safety, reliability, and longevity. I discuss the performance requirements for next-generation high-power rechargeable lithium-based batteries and sealed nickel-cadmium and lead-acid batteries best suited for applications requiring high-energy and -power densities. Battery design configurations for some specific applications are identified with a particular emphasis on safety, reliability, longevity, and portability.

In Chapter 3 I discuss fuel cells that are best suited for applications where electrical power requirements vary between several kilowatts (kW) to a few megawatts (MW).

Fuel cells generate electrical power by an electrochemical conversion technique. The early fuel cells deploy this technique, and the devices using this technique suffer from excessive weight, size, and reliability problems. In past studies I have indicated that DMFC technology offers the most promising fuel cell design configuration for applications where compact form factor, enhanced reliability, and significant reduction in weight and size are the principal fuel cell design requirements. DMFC is a system that combines an oxidation reaction and reduction reaction in a most convenient way to produce electricity with minimum cost and complexity. Such fuel cells are expected to be used extensively in the future. Studies performed by C. H. J. Broers and J. A. A. Ketelaar (*Proceedings of the IEEE*, May 1963) indicate that the fuel cells developed before 1990 used high temperatures and semisolid electrolytes. Even earlier fuel cells, such as the Bacon HYDROXZ fuel cells, were designed to operate at medium temperatures and high pressures. It was reported by C. G. Peattie (*IEEE Proceedings*, May 1963) that such fuel cell operations are difficult to maintain and require constant monitoring to ensure that the fuel cell is reliable. I discuss next-generation fuel cell design configurations capable of operating with high efficiency and high power output levels over long durations.

Chapter 4 describes the high-power batteries currently used by EVs and HEVs. Performance reviews of these batteries indicate that the rechargeable batteries suffer from poor efficiency as well as excessive weight, size, and operating costs. I describe various next-generation rechargeable batteries best suited for all-electric cars, EVs, and HEVs. Some next-generation batteries might deploy rare earth materials to enhance the battery's electrical performance and reliability under harsh operating environments. I propose rechargeable battery design configurations capable of providing significant improvements in depth of discharge, state of charge, and service life or longevity.

Chapter 5 focuses on low-power battery configurations that are best suited for compact commercial, industrial, and medical applications. I identify the design aspects and performance characteristics of micro- and nanobatteries best suited for detection, sensing, and monitoring devices. These batteries offer minimum weight, size, and longevity that are highly desirable for certain applications such as perimeter security devices, temperature and humidity sensors, and health monitoring and diagnostic medical system applications. I identify compact, low-power batteries using unique packaging technology for emergency radios and security monitoring devices operating under temperatures as low as −40°C. Most batteries cannot operate under such ultra-low temperatures.

Chapter 6 describes rechargeable batteries for military and battlefield applications where sustainable performance, reliability, safety, and portability are principal operating requirements. Sustaining electrical performance, reliability, safety, and longevity are given serious considerations for rechargeable batteries operating in battlefield environments that involve severe thermal and structural parameters. I emphasize the reliable electrical performance, safety, longevity, compact packaging, advanced materials, and portability for the batteries capable of operating in military and battlefield systems such as tanks, UAVs, UGCVs, and robot-based battlefield fighting systems.

Chapter 7 is dedicated to rechargeable batteries for possible applications in aerospace equipment and space-based surveillance, reconnaissance, and tracking systems of space-based targets. Stringent performance requirements for the rechargeable batteries deployed in commercial aircraft and military aircraft—including fighter aircraft, helicopters, UAVs for offensive and defensive missions, electronic attack drones, and airborne jamming equipment—are defined to ensure sustainable electrical energy and significantly improved reliability, safety, and longevity, which are essential for carrying out successful missions. I suggest that stringent safety and reliability requirements are needed in severe vibration, shock, and thermal environments. Improved design concepts for aluminum-air batteries using alkaline electrolyte are identified for communication satellite applications, where high-energy density (>500 Wh/kg), ultra-high reliability, and high portability are the principal performance specifications. Reliable modeling and stringent test requirements are defined for the sealed nickel-cadmium and lead-acid batteries because these batteries are ideal for next-generation communications satellites, supersonic fighters, and space-based systems for precision surveillance, reconnaissance, and tracking missions.

Chapter 8 deals with low-power batteries that are widely used for various commercial, industrial, and medical devices that can operate with electrical power ranging from nanowatts to microwatts. Low-power batteries are widely used consumer electronic products such as in infrared cameras, smoke detectors, cell phones, medical devices, minicomputers, tablets, iPhones, iPads, and a host of electronic components. These low-power batteries must meet minimum weight, size, and cost requirements in addition to being exceptionally safe and long-lasting. In past studies, I have indicated that advances in materials and packaging technology can play a significant role in the performance improvements in existing batteries such as nickel-cadmium, alkaline manganese, and lithium-based batteries. I briefly summarize the performance characteristics of low-power batteries in this chapter.

I want to express my sincere gratitude to Ed Curtis (Project Editor) and Marc Johnston (Senior Project Manager) for their meaningful suggestions and assistance in incorporating last-minute changes to the text, completing the book on time, and seeing everything through to fruition—all of which they did with remarkable coherency and efficiency.

Last, but not least, I also want to thank my wife Urmila D. Jha, my daughters Sarita Jha and Vineeta Mangalani, and my son U.S. Army Captain Sanjay Jha for their support, which inspired me to complete the book on time despite the tightly prescribed production schedule.

Author

A. R. Jha received his BS in engineering (electrical) from Aligarh Muslim University in 1954, his MS (electrical and mechanical) from Johns Hopkins University, and his PhD from Lehigh University.

Dr. Jha has authored 10 high-technology books and has published more than 75 technical papers. He has worked for companies such as General Electric, Raytheon, and Northrop Grumman and has extensive and comprehensive research, development, and design experience in the fields of radars, high-power lasers, electronic warfare systems, microwaves, and MM-wave antennas for various applications, nanotechnology-based sensors and devices, photonic devices, and other electronic components for commercial, military, and space applications. Dr. Jha holds a patent for MM-wave antennas in satellite communications.

Chapter 1

Current Status of Rechargeable Batteries and Fuel Cells

1.1 Rechargeable Batteries

The need to eliminate dependency on costly foreign oil and the adverse effects of harmful gases on health has compelled energy experts to search for alternate battery technologies for possible deployment in electric vehicles (EVs) and hybrid electric vehicles (HEVs) desired for transportation. Energy experts and transportation consultants have recommended the use of high-capacity, lithium-ion rechargeable batteries for electric and hybrid vehicles. Battery designers indicate that Nissan Motor Company and Sony Corporation have deployed rechargeable Li battery packs in their electric and hybrid vehicles. General Motors and Siemens Company have invested significant money in research and development for fuel cells best suited for trucks and buses. Battery requirements for zero-emission vehicles (ZEVs) will be more stringent. Sealed nickel-cadmium (Ni-Cd) batteries are currently used in commercial (MD-80, DC-9, and Boeing 777) and military aircrafts (F-16, F-18, and E-8). To meet high longevity and reliability requirements, vented Ni-Cd batteries previously used by various commercial and military aircraft should be replaced by high-performance, maintenance-free sealed Ni-Cd rechargeable batteries. Both sealed lead-acid (Pb-acid) and vented Ni-Cd rechargeable batteries are widely deployed by the commercial and military aircraft to meet improved efficiency, reliability, longevity, and output power requirements.

Requirements of next-generation batteries will focus on low cost, light weight, compact packaging, portability, and longevity exceeding 15 years. Rechargeable batteries will be found most suitable for battlefield weapons, communication satellites, space reconnaissance and surveillance systems, underwater tracking sensors, and a host of several applications. Note that high-capacity batteries are best suited for commercial and military aircraft, helicopters, drones, hybrid vehicles, space sensors, and battlefield weapons, whereas low-power rechargeable batteries are widely deployed by the cell phones, laptops, medical devices, computers, and host of other electronic and digital devices. The integration of microminiaturization technology [1] involving the microelectrical mechanical system (MEMS) and nanotechnology techniques will be given serious consideration in the design and development of next-generation rechargeable batteries to meet the stringent performance specifications, including reliability, portability, longevity, and compact packaging.

Current aggressive research and development activities are directed toward improving the design of nickel-zinc (Ni-Zn) rechargeable batteries, which have a great promise to offer the lowest transportation costs between $0.03 and $0.44 per kilometer (km). The design and development activities must also focus on critical electrical parameters such as the state of charge (SOC), thermal runaway, charge and discharge rates, discharge cutoff detection, and correlation between the SOC and open-circuit voltage (OCV). For battlefield rechargeable batteries, weight, size, cost, reliability, and longevity are the most essential design requirements.

1.2 Fundamental Aspects of a Rechargeable Battery

A battery consists of one or more voltaic cells. Each voltaic cell consists of two half-cells. Negatively charged anions migrate to the anode electrode (negative electrode) in one half-cell, while the positively charged cations migrate to the cathode (positive electrode) in the other half-cell. The electrodes are separated by an electrolyte medium, which is ionized to create the anions and cations and to permit the movement of those ions.

In some battery design, the half-cells have different electrolytes. In such cases, a separator prevents mixing, but the ions can squeeze through in many cells, from carbon-zinc (C-Zn) through Ni-Cd and Li. The most critical performance parameters for various rechargeable batteries, namely the energy density (expressed in watt-hour per kilogram, or Wh/kg) and the power density (expressed in watt-hour per liter, or Wh/L), are shown in Figure 1.1 along the *y*-axis and *x*-axis, respectively. The electrolyte medium is merely a buffer for an ion flow between the electrodes. However, in the case of Pb-acid cells, which

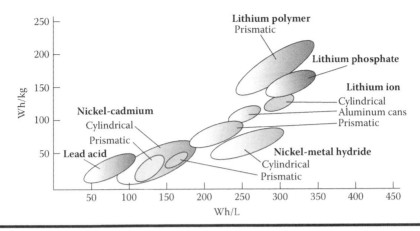

Figure 1.1 Specific power (Wh/kg) and energy density (Wh/L) for various rechargeable batteries.

are widely used in automobiles, the electrolyte is a part of the electrochemical reaction.

The OCV of a charged battery is the electromotive force (EMF) of a cell, which comes from the difference between the reduction potentials of the reactions in the half-cells. During the discharge phase, the battery converts the heat energy that would be released during the chemical reaction between the cathode and anode terminals. This heat energy is converted into electrical energy.

Real-life batteries exhibit an equivalent series resistance (ESR) internally, which drops some of the OCV when the battery is used in an external circuit or load. The ESR increases as the battery discharges or provides electrical energy to an external load, and the battery terminal voltage will drop under load conditions. Typical charge and discharge characteristics of thin-film batteries are displayed in Figure 1.2. In addition, the batteries tend to discharge during the dormant period. Some batteries tend to discharge more compared with others depending of their types. Studies performed by the author have indicated that Ni-Cd and nickel-metal-hydride (Ni-MH) [2] batteries will discharge at an approximate rate of 20% per month compared with Li batteries with 5 to 10% per month, Pb-acid batteries with 3 to 4% per month, and alkaline batteries with less than 0.3% per month. This clearly states that alkaline batteries suffer from lowest self-discharge rates during the dormant periods.

Rechargeable batteries are inherently rechargeable by reversing the chemical reactions that took place during the discharge. Charging is a classic redox chemical process. The negative material is reduced, consuming electrons, and the positive material is oxidized and the electrons are produced. Regardless of the battery type, the output energy level and the power density are the key battery characteristics, and both are expressed in per unit volume and mass, respectively.

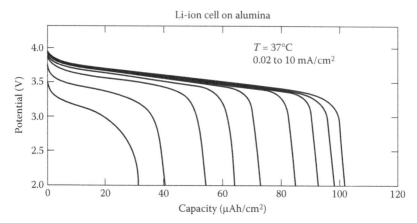

Charge/discharge characteristics of thin-film Li-ion batteries

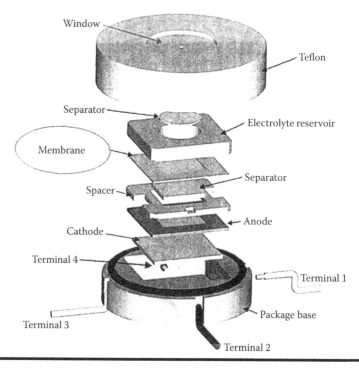

Figure 1.2 Architecture of a TF superhydrophobic nanostructured surface made from nanotubes.

1.2.1 *Critical Performance Characteristics of Rechargeable Batteries*

Two of the most critical performance characteristics of a rechargeable battery are the energy density (Wh/L) and the specific energy (Wh/kg). From these

Table 1.1 Performance Characteristics of Various Rechargeable Batteries

Battery Type and Classification	Specific Energy (Wh/kg)	Energy Density (Wh/L)
Lead-acid	25	70
Prismatic batteries		
Type 1	25	125
Type 2	40	160
Type	72	215
Nickel-cadmium	35	240
Nickel-metal-hydride	62	260
Lithium-ion	120	300
Lithium-phosphate	160	290
Lithium-polymer (prismatic)	188	294

parameters, one can estimate the energy available from 1 kg of fuel and from one volume of fuel. Studies by the author have indicated that there are several types of rechargeable batteries in the current market. One can select a right battery for his specific applications [3]. Specific energy or energy density and the power density curves for various rechargeable batteries are shown in Figure 1.1. Estimated performance characteristics of various rechargeable batteries are summarized in Table 1.1.

These estimated values are batteries developed five years ago and are accurate within ±5%. The same batteries are now improved by 5 to 10%. Reliability, efficiency, and longevity of these batteries will be specified in subsequent chapters.

1.2.2 Capabilities of Widely Used Rechargeable Batteries in Commercial Applications

Specific energy (Wh/kg), energy density (Wh/L), longevity (operating hours), and self-discharge (percent per month) and specific power (W/kg) are given serious consideration in the selection of rechargeable power batteries for commercial applications. Performance characteristics of commercial batteries are summarized in Table 1.2.

Data on the popular Duracell alkaline cell are included just to put rechargeable batteries in proper perspective. Cycle life is strictly dependent on how the

Table 1.2   Performance of Widely Used Batteries for Commercial Use

Cell Type	Nominal Voltage (V)	Specific Energy (Wh/kg)	Energy Density (Wh/L)	Self-Discharge (% per month)	Cycle Life (h)
Alkaline	1.5	150	375	0.3	1
Lead-acid	2.0	35	75	4–6	250–500
Lithium-ion	3.6	115	260	5–10	500–1,000
Lithium-polymer	3.0	100–200	150–350	−1	200–1,000
Nickel-cadmium	1.2	4–60	60–100	10–20	300–700
Nickel-metal-hydride	1.2	60	220	30	300–600
Zinc-air	1.2	146	204	−5	−200

battery is treated or used. In the case of Li cells, the anode voltage drops from 4 volt to 3 volt in a linearly fashion. Ni-MH batteries are commercially made by Toshiba and could suffer from memory loss. Zinc-air cells were initially deployed as prototype batteries for EVs and HEVs. The performance data summarized in the Table 1.2 are not obtained under identical operating conditions for all cell types.

1.2.3 Recycling of Batteries

At the end of a battery's useful life, its disposal becomes a critical issue to save the environment from the toxic effects of used batteries. Toxic effects from lead and lithium contents in batteries must be eliminated for environmental safety. In a recycling plant, as the spent batteries are dumped onto a conveyor belt, molten lead is converted into "ingots" called "pigs" and "hogs," and the operator guides the placement of recycled lead hogs into a separate container. In case of used lithium batteries, the batteries are disposed and stored in containers that can be buried underground to avoid the toxic leakage from the lithium.

Portability of electricity has become a part of daily living. Batteries are widely used by portable electric and electronic devices, such as telephones, computers, radios, compact disks, tape recorders, cordless tools, and even the electronic cars. But at the end of life, these batteries can come back to haunt us. A primary battery used for a flashlight lasts for a single life cycle, whereas in the case of a rechargeable battery, the battery can be recharged for thousands of cycles. From an environmental viewpoint, a rechargeable battery that is classified as a secondary battery is preferred over a primary battery in terms of material savings. Therefore, a single rechargeable battery is functionally equivalent to dozens of primary cells and saves recycling costs [4] involving hundreds of primary batteries. In addition, the life-cycle cost is not the slightest concern of a customer.

Worldwide, hundreds of millions of large batteries and millions of small batteries, containing tons of toxic and hazardous materials, are produced and used each year. Until recently, most of them were simply discarded. But because of the latest environmental laws, the Pb-acid and industrial Ni-Cd batteries are systematically collected for the sake of recycling their toxic and hazardous materials. Requirements for battery recycling vary from country to country, with a clear trend toward stricter controls on the requirements and disposal options. More sweeping regulations are under active consideration by the United States, Germany, France, and other European countries. A few decades back the European Union drafted a directive that would require recycling of at least 80% of all industrial and automobile rechargeable batteries. Recycling requirements are dependent on the materials used in the manufacturing of the rechargeable batteries. Only a fraction of battery materials can be recovered, which depends on three distinct factors:

- The fraction of the batteries that is returned
- The fraction of the material recoverable from each battery
- The fraction of the recoverable material actually recovered

For Pb-acid batteries, the return rate in the battery is more than 95%, the recoverable lead in the battery mass is about 60%, and the efficiency of a secondary smelter is roughly 95%. Considering all these factors, the material recycle fraction of the battery mass would be about 54%.

1.2.3.1 *Toxicity of Materials Used in Manufacturing Rechargeable Batteries*

Toxicity of some materials used in the design and development of rechargeable batteries dictates the recycling and storage requirements. Cadmium (Cd), lithium (Li), vanadium (V), praseodymium (Pr), cobalt (Co), lead (Pb), and manganese (Mn) are widely used materials in the manufacture of secondary and primary batteries. Mischmetal is produced because of the mixture of rare earth elements such as lanthanum (La), neodymium (Nd), praseodymium (Pr), and cerium (Ce). The deployment of rare earth elements in some alloys improves the characteristics of the battery electrodes, namely wide temperature range, high specific power and energy density, long cycle life, and significantly improved electrochemical activity. The valance characteristics of various rare earth and other elements widely used by the rechargeable batteries are summarized in the following table:

Valance Characteristics of Various Rare Earth and Other Elements Used in Battery Alloys

Element Used in Alloys	Symbol	Valence
Cadmium	Cd	2
Cerium (RE)	Ce	3, 4
Cobalt	Co	2, 3
Lanthanum (RE)	La	3
Lithium	Li	1
Manganese	Mn	2, 3, 4, 6, 7
Neodymium (RE)	Nd	3
Nickel	Ni	2, 3
Praseodymium (RE)	Pr	3
Sulfur	S	2, 4, 6
Vanadium	V	3, 5
Zirconium	Zr	4

RE, rare earth element.

Consumer batteries are generally smaller and are discarded with other products in the municipal solid waste. When the waste reaches a landfill, water is leached and then nickel, cadmium, and mercury are extracted from the used and broken batteries. High concentrations of the metals are sorted out from the landfill base. When the waste reaches the incinerators, the batteries contribute high levels of metal fumes to the stack emissions and ash, which escalates the cost of environmental control. Battery producers claim that used batteries accounted for close to 1.5 million metric tons of municipal from solid waste. However, this quantity is less than 1% of the total municipal solid waste generated. This solid waste contains about 67% of the lead, 90% of the mercury, and more than 50 percent of the cadmium. In countries where county and municipal regulators mandate the removal of Pb-acid batteries from the municipal solid-waste incinerators and landfills, they require safe disposal of used batteries with appropriate certification.

1.2.3.2 Safe Toxicity Limits for Workers

In many industrially advanced countries, toxicity limits are set for the workers handling various toxic and hazardous materials widely used in the manufacturing batteries. In addition, toxic limits are also set for drinking water and ambient air standards. In general, a worker's maximum allowable inhalation of the substance is measured in milligrams per cubic meter (mg/m^3) during an eight-hour working period. For nickel, it is 1 mg/m^3; for lead, it is 0.15 mg/m^3; and for cadmium, it is 0.005 mg/m^3.

Some advanced industrial nations have established strict guidelines for water supply facilities to protect their citizens from the adverse effects of toxicity and hazardous materials. Some countries have set up maximum contaminant levels (MCLs) beyond which the drinking water is considered unhealthy. The MCL established by the United States for lead is 0.05 mg/L and for cadmium is 0.01 mg/L. The United States has not defined an MCL for nickel. Specific detailed on adverse effects of lead, cadmium, mercury, nickel, and their compounds are well documented in the U.S. Environmental Protection Agency list. This list contains the Toxic Release Inventory (TRI) for chemicals, and the list is updated every five years.

1.2.4 Three Main Characteristics of a Rechargeable Battery

Regardless of the application, a secondary or rechargeable battery has the following three distinct characteristics:

- Energy performance
- Power performance
- Lifetime (both in actual time and charge-discharge cycles)

These characteristics are inextricably linked. In other words, increase one and one or both of the others must decrease. Increase the size of the current collector in the battery to boost the power density (Wh/kg), for example, and there will be less room for the active electrode material, which will decrease the energy density (Wh/L). Both the power density and the energy density are of critical importance for the rechargeable batteries deployed for electric and hybrid electric rechargeable batteries. Currently, no electric car can drive far without refueling. To overcome this serious problem, the latest technology is essential in the design and development of rechargeable batteries for possible applications in EVs and HEVs.

The following terms are used to define the performance capabilities of batteries regardless of their applications:

- *Anolyte*: A liquid anode.
- *Catholyte*: A liquid cathode.
- *DOD*: Depth of discharge.
- *Electrolyte*: A medium in which the flow of electrons is mediated by the migration of ions.

■ *Energy density*: The amount of the electrical energy a battery stores per unit volume at a specific discharge rate. This is also known as volumetric energy density and is expressed in watt-hour per liter.

■ *Power density*: The amount of power a battery can deliver per unit volume at a specified state of charge, which is typically is 20%. This quantity is also known as volumetric power density and is expressed in watts per liter.

■ *Specific energy*: The amount of electrical energy a battery is able to store per unit mass at a specified state of charge. This is also known as the gravimetric energy density and is expressed in watt-hour per kilogram.

■ *Specific power*: The amount of electric power a battery can deliver per unit mass at a specified state of charge, which is usually 20%. This quantity is also known as gravimetric power density and is measured in watt per kilogram (W/kg).

■ *SOC*: This represents the percentage of its total ampere-hour capacity stored in a battery.

1.2.5 Cost-Effective Justification for the Deployment of a Specific Rechargeable Battery for a Specified Application

A brief market survey on rechargeable batteries can provide unlimited selections of batteries for a specific application. However, cost-effective justification and selection of a rechargeable battery for a specific application requires trade-off studies in terms of cost, energy density, power density, charge and discharge rates, and life cycle.

Many types of batteries have higher specific energy (Wh/kg) ratings than Pb-acid batteries. But all cost more, many perform less well, and some entail greater safety or environmental risks. For example, the search for the "ideal" EV or HEV battery is a matter of optimization: what battery technology will offer the best combination of performance, reliability, life, and cost with adequate safety and minimum environmental risks.

The three distinct characteristics of a rechargeable battery identified in Section 1.2.4 are inextricably linked and, therefore, improvements in any one characteristic come at the expense of one or both of the others. The specific constraints of the battery vary from application to application. For example, the specific constraints are different for hybrid electric vehicles compared with all-electric vehicles, which require high-power density and lower energy density to meet optimum vehicle performance level.

It is worth emphasizing that any new fuel technology currently suffers to some extent from the same dilemma. For example, at present in United States, at least, gasoline is so cheap and the fueling infrastructure is so well established that nothing else can compete with it economically unless and until huge investments are made. Furthermore, the society in the near future is likely to impose strict safety and environmental controls on the new technologies over the existing ones. This means such concerns have to be addressed before the new technologies can be introduced with solid proof of economic advantage and independence from foreign oil.

1.2.5.1 Techniques to Improve Battery Performance in Terms of Weight and Cost

Trade-off studies are essential in terms of critical battery parameters to achieve optimum performance with minimum cost for a given application. Furthermore, the battery performance cannot be expressed by any single design parameter, and the key performance parameters tend to be interlinked. For example, a cost-effective way to improve the power performance of a battery is to use thinner electrodes, which would lower both the energy density and the life expectancy. This means thinner electrodes are not best suited for rechargeable batteries for applications in which the energy density and longevity are the principal performance requirements, such as space systems and HEVs. Note a standard gas tank in a passenger car contains approximately 60 liters or 15 gallons of fuel, which weighs around 50 kg or 33 pounds. If a Pb-acid battery pack large enough to provide the same driving mileage or distance is used, this battery pack will weigh more than the car itself (more than 2 tons) and will occupy space equivalent to passenger compartment. If an Li battery pack is selected for an electric car, the battery pack would require a minimum of 144 cells with minimum weight and size. The lithium battery pack, however, would cost more than $8,000 to replace, if the automobile is involved in a rear-end accident. In brief, to build a practical electric vehicle, the energy storage requirement, battery pack size, and cost must carefully balance what is desirable against what is possible. The battery and vehicle design processes and parameters are interdependent. An electric vehicle defines an envelope of usable battery performance, while the range of achievable battery performance at any point in time limits the electric vehicle performance in terms of mileage and miles traveled without recharging the batteries.

One disagreeable side effect of this interdependence is that the battery performance goals are always subject to change. Because the perfect battery is not achievable in actual practice, electric vehicle design processes tend to shift in response to battery performance predictions. The wide ranges shown in Table 1.3 obscure the real differences between the Pb-acid battery needs for electric and hybrid vehicles. In general, the HEV batteries will be smaller in energy content than those for EVs, but they will be required to produce more electric power per unit mass or volumes as illustrated in Table 1.3. Note batteries in hybrid vehicles can be recharged from an onboard power source or an alternator and are often operated over only a small fraction of their capacity for many thousands of cycles.

Table 1.3   Performance Goals for Electric and Hybrid Electric Vehicles

Performance Parameter	Batteries for		Typical Lead-Acid Battery
	Electric Vehicles	Hybrid Electric Vehicles	
Specific energy (Wh/kg)	85–200	8–80	25–40
Energy density (Wh/L)	130–300	10–100	30–70
Specific power (W/kg)	80–200	600–1,600	80–100
Life expectancy (cycle/year)	600–1,000/5–10	100–105/5–1	200–400/2–5
Cost ($/Wh)	100–150	175–1,000	60–100

The data presented in Table 1.3 are the estimated values of the parameters that are accurate within ±10%. The manufacturing techniques for Pb-acid batteries are highly optimized and the key electrochemical contents lead and sulfuric acid are inexpensive, whereas the non-Pb-acid-type batteries are continuously undergoing design variations because of changes in material technology and material costs.

1.2.5.1.1 Prediction of Battery Life

Prediction of battery life is not only difficulty but also expensive. Furthermore, the battery life is dependent on the number of charge and discharge cycles, materials used, and operation type (intermittent or constant). For starting, lighting, and ignition (SLI) applications, prediction of Pb-acid battery life is relatively simple, because the Pb-acid battery design is fully matured and the material cost and maintenance procedures are fully known. In case of batteries using solid electrodes and exotic solid-state materials, prediction of battery life is more complex. As stated earlier, prediction life is strictly dependent on the materials used for the cathodes and anodes, properties of the materials, and the maintenance procedures and schedules used.

Studies performed by the author on rechargeable batteries for EVs and HEVs indicate that predicting the battery system life or reliability is much more difficult as well as expensive. Note that it requires testing a full-size battery system known or battery pack over several years under various climatic and driving conditions for reasonable accuracy. The high degree of uncertainty stems from the configuration of EV batteries, which typically requires 100 or more electrochemical cells in series. In addition, with manufacturing variations between the cells and cell-to-cell temperature variations during use, it is extremely difficult if not impossible to keep so many cells in electrical performance balance over the battery life.

Varying degrees of system control and maintenance schedules are needed to achieve a balance, perhaps including some combination of thermal management systems, periodic overcharging, and active electronic systems to maintain the SOC status of the battery within its intended operating range. In brief, the practical effect of such variability means that the life of such a battery system is also highly variable and unpredictable.

1.2.5.1.2 Reliability and Failure Mechanisms of Rechargeable Batteries

The reliability of any device or system, such as a rechargeable battery, is strictly dependent on the probability of reliability with zero failures during a time and the failure rates associated with the components involved. In this case, anode and electrode are the principal components because the surfaces of anode and electrode are degraded due to the electrochemical process over a long duration. It is necessary to identify the failure mechanisms in rechargeable batteries, in particular, that are deployed for military and satellite applications. In commercial applications, failure mechanisms are easy to identify and correction steps can be taken without any interruption in the operation of the device or system. Determination of component reliability involves computing the failure rate during a specified interval of time under operating conditions and conducting a failure-mode analysis. Because the battery is composed of two essential components connected in a series, reliability aspects of both components must be considered. The reliability of a component or device can be estimated using the mathematical theory of reliability. On the basis of this theory, the battery's reliability can be computed using the exponential failure rate (λ), which can be written as follows:

$$R(t) = [e^{-\lambda t}] \qquad (1.1)$$

where $R(t)$ is the reliability with zero failures during time t and λ is the combined failure rate for anode (A) and electrode (C), which can be expressed as follows:

$$\lambda = [\lambda_A + \lambda_C] \qquad (1.2)$$

where λ_A is the failure rate for anode and λ_C is the failure rate for the electrode.

In some critical applications such as reconnaissance satellite or covert communication, satellite redundancy must be considered. This requires a system component, such as a battery charge controller, which is considered the smallest unit, to provide the redundant function necessary to achieve continuous operation aboard the satellite.

The reliability of a simple battery consisting of one anode and electrode can be expressed as follows:

$$R_B(t) = [R_A(t)\, R_C(t)] \qquad (1.3)$$

where R_A is the reliability of the anode over the time t and R_E is the reliability of the electrode over the time t.

1.2.5.2 Why Use Pb-Acid Batteries for Automobiles?

Studies performed by the author on Pb-acid batteries indicate that these batteries were used in automobiles as early as 1912. The studies further indicate that more than 90% of all electric vehicles ever manufactured in the United States, deployed Pb-acid batteries for their onboard energy storage. The Pb-acid battery technology is fully matured and all possible improvements have been implemented in their designs. In brief, unrestricted wide availability and low procurement cost are the principal advantages of this battery. Furthermore, Pb-acid batteries have been man-ufactured in present form for many decades, and they have proven to be reliable and cheap for automobile applications [3] since 1924.

1.2.5.3 Description of Flow Batteries

If we ever are to run our electric grid on the on-again, off-again power that wind and solar power systems provide, we are going to need highly reliable and well-designed batteries. A 30-year-old battery technology known as a flow battery could be best suited for such an application. In other words, flow batteries using two liq-uid electrolytes that react when pumped through a cell stack can replace the solid-state electrodes currently used in a conventional battery. Essentially, the battery is broken down into a cell stack and two large liquid electrolyte tanks. Charging and discharging cycles in the battery are generated as the electrolyte flows past a porous membrane in each cell, and ions and electrons flow back and forth. Recharging requires putting in fresh liquid electrolyte to increase the energy storage capacity. Therefore, slightly larger tanks may be required to improve the charge duration and the charging efficiency. Such batteries already have been used for backup power at factories and cell phone towers.

Now manufacturers are looking for venture capital to design and develop such batteries for grid-level systems that will be best suited for applications in conjunc-tion with wind turbines and solar power systems. The U.S. Department of Energy (DOE) has allocated $31 million in Recovery Act Funds to jump-start five utility-grade projects. Cost is critical for grid storage application, and this is where the flow batteries do the job with minimum cost and complexity. Studies performed by the author on energy storage techniques seem to indicate that zinc-bromide cells cur-rently in the works could store the electrical energy for less than $450 per kilowatt-hour (kWh), a third as much as for Li batteries and about three-quarters as much as for the sodium-sulfur batteries.

These batteries are relatively more safe and cost-effective because of the inherent architecture of flow. In addition, with a flow battery, one can store a megawatt-hour (MWh) of electrical energy in the electrolyte tanks. A recently published article in *IEEE Spectrum* reveals that a trailer-transportable zinc-bromide battery (ZBB) has demonstrated a storage capacity close to 2.8 MWh. Such batteries, when tied to utility grids, are best suited in reducing the peak loads. Flow battery technology is fully matured, but it still faces a challenge in adapting to utility scale and doing complete system integration, which involves high-efficiency power electronics and controls with fast response.

1.3 Rechargeable Batteries Irrespective of Power Capability

Various types of rechargeable batteries are available for different applications. In the past decade or so, rechargeable batteries were progressively improved in terms of cost, size, energy and power densities, shelf life, reliability, portability, and safety. According to a market survey, most of the rechargeable batteries are used by portable electronic devices and digital sensors. High capacity batteries are deployed in EVs and HEVs, commercial transports and military aircraft, remote power installations, communication satellites, and other commercial applications.

1.3.1 Rechargeable Batteries for Low- and Moderate-Power Applications

Two of the rechargeable batteries, namely the Pb-acid and Ni-Cd, have impressively long histories. The Pb-acid batteries were first manufactured as early as 1860, whereas the Ni-Cd batteries were first manufactured around 1910. The other rechargeable batteries such as zinc-manganese dioxide ($Zn-MnO_2$), Ni-MH, and Li are relatively young, but they are widely deployed in portable devices and sensors. The Li and Ni-MH batteries are best suited for portable electronic components, where reliability, longevity, and uninterrupted power sources are the principal requirements. Table 1.4 describes the performance capabilities and important characteristics of rechargeable batteries widely used to power portable electronic devices and sensors.

Furthermore, these values are for the commercially available rechargeable batteries, which were designed, developed, and tested before 2000. Obviously, one will find further improvement in the characteristics and performance capabilities of rechargeable batteries because of rapid advances of the Ni-MH and Li batteries designed and developed after 2005. One will notice significant performance improvement, particularly in the case of Li batteries, which come in four distinct categories: Li (monopolar), Li (bipolar), Li (polymer), and Li (using polymeric electrolyte).

Table 1.4　Performance Capabilities and Critical Design Characteristics of Rechargeable Batteries

Performance Capabilities	Pb-Acid	Ni-Cd	Ni-MH	Li (BP)
Power density	Good	Excellent	Average	Excellent
Specific energy (Wh/kg)	35–55	45–65	60–95	160–195
Energy density (Wh/L)	80–95	150–200	310–360	350–480
Cycle life (to 80% capacity)	200–325	1,100–1,500	600–1,100	700–1,250
Self-discharge at 20°/mo. (%)	<5	<20	<25	<3.5
Fast charging time (h)	8–16	1	1	2–3
Nominal cell voltage (V)	2.0	1.20	1.20	3.6
Operating temperature range (°C)	−20–60	−20–60	−20–60	−20–65
Peak load current at (°C)	10	20	5	<2
Continuous current at (°C)	1	1	<0.5	<0.8
Procurement cost ($/Wh)	0.5	0.5	0.5	0.75
Overcharge tolerance	High	Moderate	Low	Very low

Note: The values presented in this table are estimates and are accurate within ±5%.

None of the rechargeable batteries summarized in this table possesses the same degree of the principal characteristics that are essential in a rechargeable battery. In the case of primary batteries, the user can make select a battery to satisfy requirements for cost, recharging facilities, continuous or intermittent use, and low- or high-temperature application.

1.4　Rechargeable Batteries for Commercial and Military Applications [5]

Battery requirements are stringent for certain commercial and most military applications. For commercial aircraft, helicopters, communication satellites, and military jet fighters and bombers, sealed Pb-acid and vented Ni-Cd rechargeable batteries are highly desirable to meet system reliability and critical performance requirements under severe temperature and mechanical environment. The performance improvement of a rechargeable battery is strictly dependent on the electrochemical technology.

1.4.1 High-Power Batteries for Commercial Applications

Battery designers are focusing their efforts to research and develop activities on sealed, vented, and maintenance-free rechargeable batteries, in particular, those deployed by commercial transports and other commercial applications. Shelf lives and capacity losses for various rechargeable batteries as a function of operating temperature are shown in Figure 1.3. The capacity losses for zinc-air and lithium-based batteries are

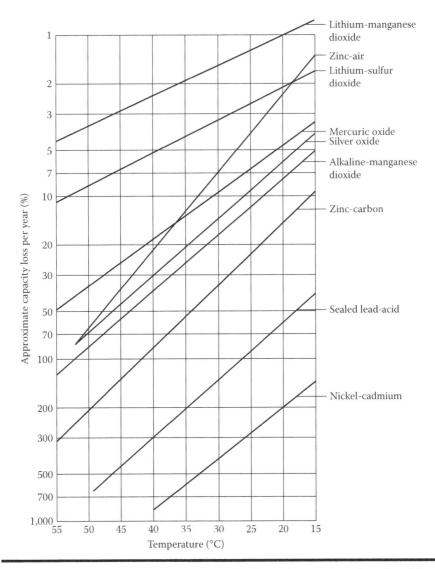

Figure 1.3   Shelf life for various rechargeable batteries as a function of ambient temperature.

the lowest as illustrated by Figure 1.3. Interestingly, advanced maintenance-free sealed Ni-Cd batteries are most attractive for medium-power military applications. Research and development efforts are directed for the development of maintenance-free sealed Ni-Cd rechargeable batteries capable of delivering high-power outputs and improved reliability. According to defense officials, the major thrust in advanced aircraft battery development is the sealed Ni-Cd rechargeable batteries. Energy density for rechargeable batteries is strictly dependent on the operating temperature. Optimum energy density for most of the batteries can be achieved if the operating temperature is maintained over 60°F to 80°F temperature ranges as illustrated in Figure 1.4.

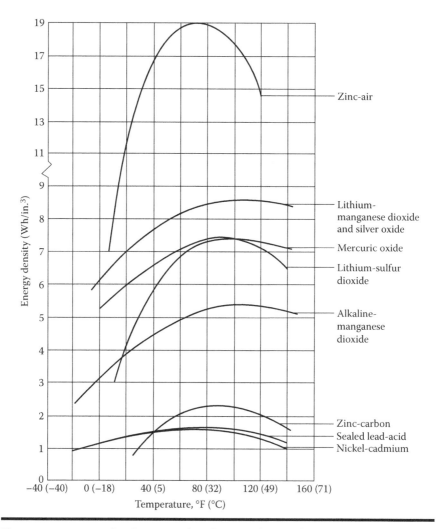

Figure 1.4 Energy density per unit volume (Wh/in.³) for various batteries as a function of temperature.

1.4.2 Critical Role of Ni-Cd in Rechargeable Batteries for Military Aircraft

Sealed Ni-Cd rechargeable batteries are being manufactured by several companies, including Acme Electronic Utah Research and Development Corp., Eagle Picher Industries, and others. Sealed Ni-Cd batteries developed by Acme are currently flying in several commercial aircraft, including MD-80, MD-90, DC-9, and Boeing 777 commercial transports. In addition, some F-16 fighter aircraft and Apache helicopters are also using these rechargeable batteries. Maintenance-free sealed Ni-Cd batteries have been approved for F-16, F-18, B-52, and E-8 advanced Airborne Warning and Control System (AWACS). Typical Ragone charts for Li, Ni-MH, and Ni-Cd batteries are shown in Figure 1.5. These charts are extremely helpful in selecting the right battery to meet specific requirements in terms of specific density (Wh/kg) and power density (W/L).

Market research indicates that nothing quite matches the high-power density of a vented Ni-Cd rechargeable battery capable of starting the aircraft engine in −45°C temperatures. It is justified to state over the next 10 to 15 years that both the

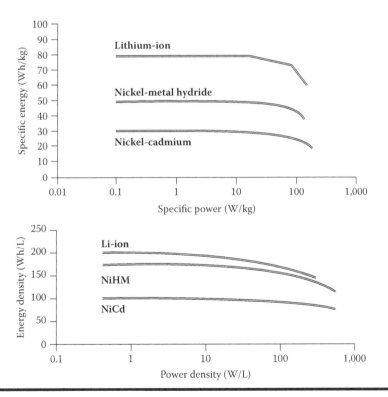

Figure 1.5 **Typical Ragone plots in terms of specific energy and energy density for various rechargeable batteries widely deployed by vehicles.**

commercial and military aircraft will depend on the high-power sealed Pb-acid and sealed Ni-Cd rechargeable batteries.

1.4.3 Benefits of Ni-MH Rechargeable Batteries for Military Aircraft

Studies performed by the author on rechargeable batteries reveal that higher performance is possible from Ni-MH batteries because of unique electrochemistry technology and prismatic bipolar design. This particular battery design offers about 20% improvement in battery performance and 25% reduction in manufacturing cost, according to the vice president of corporate development at Electro Energy Corp.

The Ni-MH battery can be subjected to overcharge and overdischarge with identified reactions on both the positive and the negative electrodes. The cell capacity is limited by the positive electrode, with a negative-to-positive ratio between 1.5 and 2.0. During overcharge, oxygen is evolved at the positive electrode and diffuses to the negative electrode to form water (H_2O). During overdischarge, hydrogen is evolved at the positive electrode and again gives rise to water at the negative terminal. Both hydrogen and oxygen recombine to form water, thereby ensuring the sealed operation of an Ni-MH battery. Interestingly, the hydrogen ion moves back and forth between the two electrodes during the charge–discharge process.

Material scientists reveal that deployment of appropriate alloys that incorporate rare earth materials such as lanthanum nickel ($LaNi_5$) or zirconium vanadium (ZrV_2) will significantly improve the Ni-MH batteries in terms of wide temperature range, long cycle life, high-power and -energy capacity, enhanced electrochemical activity, high hydrogen diffusion velocity, low cost, and environmentally friendly operation. Progressive optimization steps of the ($LaNi_5$) hydrogen storage alloy for the negative electrode of a sealed Ni-MH battery are shown in Figure 1.6. The current component used, that is, Mischmetal (Mm), is nickel-cobalt-manganese-aluminum (Ni-Co-Mn-al), which offers a discharge capacity of 330 mAh/g and is 10% higher than $LaNi_5$. This particular alloy works better than ZrV_2 alloy at low and high temperatures and demanding discharge rates. In addition, this alloy is cheaper and easier to work. Thus, $LaNi_5$ is preferred for Ni-MH rechargeable batteries.

The performance of Ni-MH batteries also depends on the cathode formulation and on the separator characteristics. Higher capacities from 700 to 1,000 Ah/kg are possible using high-performance rare earth elements. Preliminary studies performed on various rare earth materials indicate that an alloy known as V_3Ti composed of vanadium $V_{3, 4, 5}$ and titanium $Ti_{2, 3, 4}$ has twice the capacity of $LaNi_5$. Material scientists believe that addition of cobalt-hydroxide [$Co(OH)_2$] to the cathode electrode produces cobalt-oxide with much higher conductivity than nickel-hydroxide (Ni-OOH). Any oxygen evolved at the positive electrode on overcharge

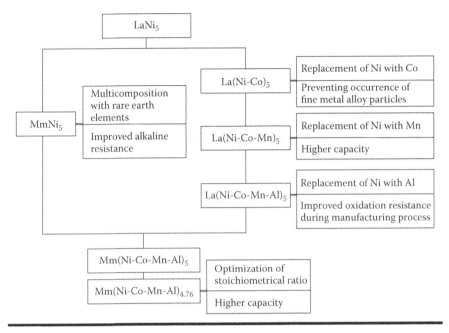

Figure 1.6 Progressive optimizing steps for La-Ni$_5$ alloy–based sealed Ni-MH rechargeable batteries. (Mm: Mischmetal.)

may oxidize the separator surface; moreover, to avoid the formation of this oxide, a chemical separator is required such as sulfonated polypropylene.

1.4.3.1 Electrode Material Cost and Characteristics for Ni-MH Batteries

Electrode material cost and characteristics affect the battery price and its life. The amount of material used is strictly dependent on the application, and its power capacity requirement. High-power and energy-dense Ni-MH batteries are best suited for heavy-duty vehicles, such as buses and trucks. Approximately 10 years ago Energy Conversion Devices Inc. manufactured Ni-MH batteries for EVs and HEVs. Battery weight and size depend on the number of cells, density of the materials used, and the power output capacity of the battery. Ni-MH batteries offer significant reduction in weight and size compared with other battery technologies. Ni-MH batteries offer increased ridership on buses and automobiles, expanded payload on trucks, and greater stealth and offensive capability on military electric vehicles because of higher energy density than Ni-Cd and Pb-acid batteries [5]. In addition, high reliability, maintenance-free operation, and ability to accept peak regenerating currents make the Ni-MH batteries an ideal solution for heavy-duty commercial and military vehicles.

Table 1.5 Approximate Price Quotations for Various Cathode Materials

Base Element	Price in $/lb. ($/kg)	Cathode Material	Price in $/lb. ($/kg)
Cobalt (Co)	18 (40)	$LiCoO_2$	30 (60)
Nickel (Ni)	3.6 (8)	$LiNi_{0.8}Co_{0.2}O_2$	35 (77)
Manganese (Mn)	0.3 (0.66)	$LiMn_2O_4$	30 (66)

Regarding cathode material cost, precise price quotation is difficult. The price per pound or kilogram of the material depends on the purity of the material, the amount of material to be purchased, and the transaction time. On the basis of a preliminary price survey of various cathode materials, Table 1.5 summarizes the price quotations.

These prices are all based on small quantities. The manganese price is based on unrefined ore. The price for manganese will increase depending on the percentage of refinement. Similarly, the price for refined nickel and cobalt would be higher.

1.4.3.2 Impact of Temperature on Discharge Capacity of Ni-MH Batteries

Ni-MH battery designers recommend that these batteries should not be used for optimum performance outside the temperature range from 0° to 40°C. Discharge capacity better than 90% can be achieved over this temperature range. Room temperature storage of this battery does not produce permanent capacity loss. Tests conducted by suppliers indicate that continuous temperature exposure to 45°C could reduce the battery cycle life by approximately 60%. Although an Ni-MH battery is capable of sustaining high discharge currents, repeated discharges at high current levels can reduce the battery life. Best life-cycle performance has been observed with rates of 0.2° to 0.5°C.

1.4.3.3 Charging Procedure for a Ni-MH Battery

Battery designers indicate that charging is the most critical step in determining the electrical performance and overall life of an Ni-MH battery because of its sensitivity to the charging conditions. Therefore, utmost attention must be given to the charging rate, temperature range, and effective techniques indicating the end of charge. Ni-MH batteries must be charged at constant current to achieve improved battery performance and long cycle life. The batteries designed about 10 years or so ago were suffering from memory loss. But the latest Ni-MH batteries do not suffer from memory loss. The charging current levels must be limited to avoid overheating and incomplete oxygen recombination. These two conditions have significant impact on battery performance and longevity. Battery tests indicate that the charge process is exothermic (forms due to evolution of heat) in Ni-MH batteries, whereas it is endothermic (forms due to absorption of heat) in Ni-Cd batteries.

1.4.3.4 Degradation Factors in Ni-MH Battery Performance

■ *Sensitivity*: The battery sensitivity is dependent on the storage temperature. The battery residual capacity loss drops at faster rate at higher temperatures and longer storage durations. This capacity drops to 20% value after 30-day storage.

■ *Voltage drop*: Voltage drop rate is higher at low temperatures.

■ *Voltage plateau*: This is applicable only to low-rate charges.

■ *Temperature cutoff*: This method stops charging when the temperature reaches a preset limit indicating overcharging.

■ *Rate of temperature increase*: This measures the rate of temperature rise as a function of time and stops the charge when a predetermined value equal to 1°C per minute is reached. This is the preferred method to stop high-rate charges, because it ensures battery longevity or longer battery cycle life. The Ni-MH battery will suffer, if subjected to repeatedly overcharge.

1.4.4 Thermal Batteries for Aerospace and Defense Applications

A distributed battery system consisting of thermal batteries plays a critical role in the use of electric brakes control actuators in an aircraft. An aircraft employing more electric technology will require enhanced backup power systems. This backup power system is composed of thermal batteries. Recent advances in the design of thermal batteries reveal significant improvement in performance, shelf life, and high-energy emergency backup power capability.

Thermal batteries developed by Eagle Pitcher have demonstrated unlimited shelf lives and reliable, continuous operation exceeding two hours without failure. These thermal batteries are best suited for aircraft, sonobuoys, cruise missiles, and other applications for which reliable operation over extended periods is the principal requirement.

Previous generations of thermal batteries have operating lives of 10 minutes or so. Such batteries developed in 1995 have demonstrated operational lives exceeding two hours. Thermal batteries designed and developed after 2005 have demonstrated an operational life close to five years. Use of improved thermal insulation of the housing and incorporating a low-wattage heater in the battery will further improve the operational life in a thermal battery. Improved electrolytes and cathode materials capable of operating over wide temperature ranges and long-term activity will significantly improve the overall performance of thermal batteries. In case of missile, aircraft, and satellite applications, battery weight, size, and longevity are the most demanding requirements.

1.4.4.1 Batteries for Space Applications

The electrical power systems for spacecrafts and communications satellites generally consist of energy-conversion devices in combination with energy storage devices, such as batteries and power conditioning components. In addition, pulse-duration modulation regulators will be found smaller, lighter, and cheaper than conventional voltage regulators. High-system reliability requires a certain amount of redundancy in the power system. As stated earlier, the spacecraft electrical power system consists of three distinct components, and the battery is one of them. Each component is made to meet a specific reliability goal by a technique that adds redundant components, which could add to the system cost and weight. Redundancy in battery, however, would slightly increase the cost and weight compared with the other components.

As far as the batteries for space applications are concerned, Ni-Cd batteries are widely deployed by the communications satellites and orbiting spacecrafts as energy storage devices. These batteries and other batteries are in use from the past three decades. In some cases, more efficient and reliable batteries have been used by the space systems since 2000. Regardless of the types of batteries used, if a dark period is experienced by the satellite-based solar cells, the onboard batteries must meet the power consumption requirements for the onboard electronic sensors and electrical devices. Different battery electrical performance specifications are required for low-earth-orbiting (LEO) and geosynchronous earth orbit (GEO) satellites. Battery types and their performance requirements for various orbiting satellites will be discussed in Chapter 7, which is dedicated to communications and surveillance satellite. Studies performed by the author indicate that both the Ni-Cd and nickel-hydrogen (Ni-H$_2$) rechargeable batteries are best suited for orbit satellites. Currently, most GEO satellites deploy Ni-H$_2$ batteries. Nickel-based electrodes yield improved cycle life and high reliability. Published literatures reveal that Ni-H$_2$ rechargeable batteries have been used in several planetary missions. In terrestrial applications, standby power for emergency or remote sites are limited due to high initial cost of these batteries and some drawbacks such as rapid discharge even at +10°C and 10% capacity loss after three days. In addition, typical but not serious weak points of this battery are low volumetric energy, high thermal dissipation at high data rates, and safety hazards. These weak points and other drawback can be eliminated by the sealed and maintenance-free Ni-Cd batteries.

1.4.5 Rechargeable Batteries for Commercial Applications

More rechargeable batteries are deployed in commercial applications, namely automobiles, telephones, cell phones, iPads, medical devices, cameras, watches, and so on. Battery power requirements are very low for these applications except for batteries for EVs and HEVs. Two important electrical performance parameters for EV batteries are the specific energy (Wh/kg) and the energy density (Wh/L). The performance specifications requirements for low-power rechargeable batteries widely used for commercial applications are not very stringent. In the case of EV and HEV applications, battery weight, size, longevity, and cost are of critical importance. For medical devices battery size, electrical noise, voltage drift, and voltage fluctuation are of prime considerations.

The aqueous secondary or rechargeable batteries are widely used for commercial and low-power sensors and devices. The Ni-MH, Ni-H$_2$, and Ni-Zn batteries have been added recently to this category of secondary batteries. Typical characteristics of various aqueous secondary batteries are summarized in Table 1.6.

Ni-MH batteries were introduced in the market as early as 1992. Since then, significant improvements in their characteristics, weight, size, and cost have been observed. These batteries have substituted Ni-Cd batteries, which were being deployed in several applications, including portable commercial devices and sensors.

Zinc-silver-oxide (Zn-AgO) batteries are widely used for commercial applications. These batteries offer high specific energy (Wh/kg) and energy density (Wh/L), proven reliability, enhanced safety, and highest power output per unit weight and volume. These batteries can be discharged even at 20°C. However, low cycle life, high procurement cost, and poor performance at low temperatures are drawbacks of Zn-AgO batteries.

Zn-AgO batteries are widely deployed in space application because of their outstanding electrical performance capabilities. These batteries have been used by the astronauts of several missions in their extravehicular activities because of their high-energy capability and portability aspects. These batteries can be used in portable applications, such as television cameras, medical equipment, communications equipment, and lighting systems.

In terms of market shares, Pb-acid batteries have and will maintain a dominant position. In case of portable applications, however, the Pb-acid battery has a marginal position. Thus far, only three other electrochemical batteries, namely Ni-Cd, Ni-MH, and Li, have been sold in recent years because of their portability, high reliability, high-energy capability per unit weight and volume, and maintenance-free operation.

Table 1.6 Characteristics of Various Aqueous Rechargeable Batteries

Battery Type	Voltage Range (V)	Temperature Range (°C)	Cycle Life (h)	Specific Energy (Wh/kg)	Energy Density (Wh/L)	Self-Discharge (%/mo.)
Ni-MH	1.4–1.2	−30–65	900–1,200	65–85	200–850	15–20
Ni-H$_2$	1.5–1.2	−10–30	>2,200	45–60	68–84	40–60
Ni-Zn	1.9–1.5	−20–50	326–650	55–65	100–140	14–18
Zn-air	1.2–1.0	0–45	20–30	150–220	160–240	5–10
Zn-AgO	1.8–1.5	−20–60	50–85	80–100	175–185	4–6

Table 1.7 Characteristics of Nonaqueous Rechargeable Batteries

Battery Type	Voltage Range (V)	Operating Temperature (°C)	Life Cycle (cycles)	Specific Energy Density (Wh/kg)	Energy (Wh/L)
Li-Al/FeS	1.7–1.2	375–500	1,000	140	225
Li-Al/FeS$_2$	2.0–1.5	375–500	1,000	185	375
Li-metal	3.0–2.0	40–60	800	140	175
Li-polymer	3.2–2.0	60–80	600	120	160

Since 2001, Ni-Cd and Ni-MH rechargeable batteries have kept at an almost-constant market share. But since 2005, among the nonaqueous batteries, lithium-based batteries have been receiving the most attention. Particularly, laminated Li batteries are in great demand for commercial applications and increasingly are being sold in the market. Characteristics of some nonaqueous batteries best suited for commercial applications are shown in Table 1.7.

Nickel-iron (Ni-Fe) batteries belong to aqueous rechargeable battery series. According to the published reports, these batteries were developed at the beginning of the twentieth century, based on the iron (Fe) as a negative electrode and NiOOH as a positive electrode. The cell voltage is 1.37 V. Even under harsh operating conditions, such as shocks and vibrations, overcharge or overdischarge, and storage in a fully charged or discharged state, the Ni-Fe battery can undergo more than 3,000 deep-discharge cycles with a calendar life exceeding 22 years. This particular battery suffers from high self-discharge, low efficiency, poor energy and power density, borderline low-temperature performance, and higher procurement cost than Pb-acid battery. Because of the battery's extreme ruggedness, ability to retain charge under severe operating environments, and long cycle life, battery designers and material scientists are looking for advanced configurations capable of yielding significant improvements in energy and power characteristics. Its potential applications include military vehicles, material handling, and other industrial operations.

Currently, several companies are actively engaged in the research and development activities to improve the electrical performance of the lithium-polymer rechargeable batteries, because there will be great demand for such batteries in the near future for various applications.

1.4.5.1 Ni-Zn Batteries for Commercial Applications

Brief studies undertaken by the author on rechargeable batteries indicate that Ni-Zn batteries offer several advantages for commercial applications. In the last decade, battery designers have initiated development activities to upgrade the performance of these batteries and also to reduce the procurement costs. These newly designed Ni-Zn rechargeable batteries will be found most ideal for consumer and mobile applications, such as electric bicycles, scooters, and medium-power EVs and HEVs. The electrical performance of these batteries has been significantly improved through the development of a reduced solubility zinc electrode, incorporating a patented sealed cell design, and improving the battery longevity. Materials scientists and the battery designers claim that Ni-Zn batteries provide a commercially viable alternative for Pb-acid, Ni-Cd, and Ni-MH batteries for many commercial applications. Beside significant performance improvements, the upgraded Ni-Zn batteries offer impressive reduction in procurement costs. Summing all the battery improvements, the Ni-Zn rechargeable batteries provide the lowest costs and highest energy density for alkaline rechargeable systems. The unique patented technology recently developed provides even greater opportunity for substantial reduction in large-volume production. The patented graphite-based composite nickel electrode offers the lowest cost, thereby providing substantial cost savings over the conventional nickel electrodes. Zinc provides the highest specific energy, lowest materials cost, and the least environmental impact for any potential anode material.

The new patented technology offers an energy density of 60 Wh/kg, which is more than sufficient to provide a vehicle range of more than 200 km. Newly designed batteries provide a total energy cost of less than $0.04 per km for a 25 kWh battery. An Ni-Zn rechargeable battery with a 25 kWh capacity could cost around $6,000 or less in large production volume. These batteries are being manufactured with capacity rating of 12.5 kWh, 25 kWh, and 50 kWh to meet driving requirements of various EVs and HEVs. For electric bicycles and scooters, the Ni-Zn capacity rating and procurement cost are substantially lower. The most significant design aspect of the improved Ni-Zn battery is that this particular battery is capable of providing sustained high-power capability while retaining a high specific energy rating, which is not possible with Ni-MH batteries.

1.4.6 Rechargeable Battery Requirements for Electric and Hybrid Electric Vehicles

Studies performed by the author on batteries for the EVs and HEVs reveal that specific energy and energy density performance parameters are of critical importance. Second to these two electrical performance parameters are the cost, size, weight, and longevity of the rechargeable battery. Performance of Pb-acid batteries in terms of these two parameters is very poor. Silver-zinc (Ag-Zn) batteries have the highest energy density and the specific energy as illustrated in Table 1.8. After the Ag-Zn battery, silver-metal-hydride (Ag-MH) batteries offer the next highest electrical performance parameters. Electrical performance parameters, namely the specific energy and the energy density of various commercial available batteries, are summarized in Table 1.8.

Table 1.8   Electrical Performance Parameters of Commercially Available Batteries

Battery Type	Specific Energy (Wh/kg)	Energy Density (Wh/L)
Lead-acid (Pb-acid)	25	85
Nickel-iron (Ni-Fe)	46	125
Nickel-cadmium (Ni-Cd)	48	122
Nickel-zinc (Ni-Zn)	75	170
Nickel-metal-hydride (Ni-MH)	52	175
Nickel-hydrogen (Ni-H$_2$)	65	118
Silver-zinc (Ag-Zn)	163	308
Silver-metal-hydride (Ag-MH)	105	242

The electrical parameter values summarized in Table 1.8 are valid for the rechargeable batteries designed, developed, and tested during the 1990–1995 timeframe. Furthermore, the parameter values are accurate within plus or minus 5%. During the past 15 years or so, the electrical parameters of these batteries have improved significantly.

Currently, various automobile manufacturers are focusing on the EVs and HEVs. The battery requirements for all-electric vehicles are different than those for HEVs. The rechargeable batteries are required to produce more electrical power per unit weight or volume to be more cost-effective. The batteries for the hybrid vehicles are recharged from an onboard power source or a charging device composed of an alternator and inverter, which converts the alternating current (AC) to the direct current (DC). Hybrids are often operated over only a small fraction of the battery capacity for many thousands of cycles.

In some hybrid cars, the battery capacity is only large enough to power the vehicle for a single-cycle operation, after which the battery needs to be recharged. On the other hand, the batteries in all-electric cars are recharged from the onboard power source, and they may be completely discharged over a period ranging from hours to days before they require recharging. This is a fundamental difference in the recharging requirement for all electric cars and hybrid cars.

1.4.6.1 Test Requirements for Rechargeable Batteries Needed for Electric and Hybrid Vehicles

Because of the unusual demands of electric vehicles, tests performed for other types of batteries are poor predictors on the chargeable batteries for electric vehicle.

The tests are often conducted on the basis of actual or synthetic driving behavior; nevertheless, the test results have been accepted for electric and hybrid vehicle battery testing. In the United States, the battery test procedures have been developed and approved by joint government and industry authorities. The tests performed have been standardized as "recommended practices" by the American Society of Automotive Engineers (ASAE) located in Warrendale, Pennsylvania. Similar test procedures have been developed by the Japanese, European, and Korean automakers, and the tests results have been approved by their respective governments and industry authorities. Recently, international standardizations for the tests have been approved. If the performance requirements of the electric or hybrid vehicle are well defined, standardized tests can reasonably predict the performance of the rechargeable battery for that specific vehicle. Furthermore, battery requirements must be carefully defined to meet the performance requirement for a compact, midsize, or full-size electric or hybrid car. Battery test results can be incorporated into the rechargeable battery models, which can be used to predict the performance of the EVS or HEVs based on a specified battery technology used to design the battery.

1.4.6.2 Predicting the Battery Life of Electric and Hybrid Vehicles

As discussed, predicting the rechargeable battery life is difficult and expensive, if not impossible. The degree of uncertainty stems from the battery configuration and the technology used to design that particular battery. An electric vehicle battery pack typically requires 100 or more electrochemical cells in a series. It is extremely difficult to keep so many cells in balance because of manufacturing variations between the cells and because of cell-to-cell temperature variations while the car is running.

Depending on the type of rechargeable battery, various control mechanisms and maintenance procedures are needed to achieve a perfect balance. In some cases, thermal management systems, periodic overcharging, and active electronic systems are needed to maintain the SOC in the battery within its intended operating range. Under these complex requirements, the battery life is highly variable and, hence, unpredictable.

1.4.6.3 Performance Capabilities of Batteries Currently Used for Electric and Hybrid Vehicles

High-power and energy-dense Ni-MH are being widely used by heavy-duty vehicles, such as buses. These rechargeable batteries have demonstrated remarkable electrical performance and economical advantages. Ovonic Battery Company has been recognized as the leader in the design, development, and commercialization.

1.4.6.3.1 Ni-MH Battery Packs for Electric and Hybrid Electric Cars

According to the published automobile articles, Ni-MH battery technology has been identified by virtually every major automobile maker as the highest performance, most reliable, and cost-effective energy storage technology for EVs and HEVs. According to the Ovonic Battery Company, Ni-MH battery technology offers high reliability, maintenance-free and cost effective operations, and an ability to accept peak regenerative braking currents an ideal solution for heavy-duty vehicles. Additional projects are being developed to achieve significant reduction in weight, size, and manufacturing costs. When these reductions are realized, the Ni-MH batteries will be recognized as the sole battery supplier for EVs and HEVs. Performance capabilities of Ni-MH batteries manufactured by Ovonic automobile applications are summarized in Table 1.9.

The dimensions shown in Table 1.9 are approximate. Ovonic has designed and developed Ni-MH battery packs to meet the performance requirements for EVs and HEVs made by General Motors Corporation and other automobile companies. In addition, Ovonic is developing advanced Ni-MH batteries with specific power ratings close to 550 W/kg best suited for both hybrid electric buses and trucks. Preliminary life-cycle testing at the module level using a simulated school bus has demonstrated a module cycle life exceeding 1,000 cycles, which is equivalent to more than 100,000 miles or three years of bus operation. Test data indicate that a 108 kWh battery pack composed of Ni-MH batteries can provide more than 100 miles of bus service per charge, which demonstrates more than 25% improvement over the Pb-acid battery. Note two, four, or more cells can be

Table 1.9 Typical Ni-MH Battery Performance Characteristics Best Suited for Electric, Hybrid, and Hybrid Electric Vehicles

Battery Characteristics	Ni-MH Battery Classification		
	A	*B*	*C*
Current capacity (Ah)	85	235	345
Cell voltage (V)	12	12	3.6
Specific energy (Wh/kg)	70	80	62
Energy density (Wh/L)	171	245	155
Specific power (W/kg)	240	200	165
Power density (W/L)	605	600	410
Weight (kg)	17.4	34	19.7
Maximum dimension (in.)	15.3 × 4.0 × 6.9	20.47 × 4.0 × 8.4	17.5 × 4.0 × 6.9

connected in series to meet the voltage and power capacity requirements for a given application.

1.4.6.3.2 Lithium-Based Battery Packs and Battery Systems for Electric and Hybrid Electric Cars

In the current decade, material scientists and battery designers have focused on the lithium-based battery technology. Material scientists claim that several lithium-based batteries have demonstrated high specific energy capabilities exceeding 400 Wh/kg as shown in Table 1.10. The specific energy values shown are adequate for small electric cars and HEVs.

There are other lithium-based batteries, but their energy densities are not suitable for the current EVs and HEVs. If aggressive research and development activities continue on other borderline batteries, such batteries will be able to join the lithium-based battery category in the near future.

1.4.6.3.3 Rechargeable Lithium-Based Thermal Batteries

Lithium-based thermal batteries have a molten-salt electrolyte that works excellently above room temperatures. These batteries use lithium- or sodium-based negative electrodes, with the exception of the lithium-metal polymer. Positive electrodes are generally fabricated using a mixture of iron sulfide (FeS) and electrolyte into the current collector or loading the material into a honeycomb matrix. Graphite, cobalt-sulfide, and nickel-sulfide are sometimes used in the matrix structure for performance enhancement. Lithium-aluminum iron sulfide (Li-Al/FeS and Li-Al/FeS$_2$) thermal batteries were investigated by the battery designers for road transport applications. Even though Li-Al/FeS$_2$ provides higher cell voltage and improved electronic conductivity, it suffers from corrosion problems. The corrosion problems can be solved using thicker electrodes, which can slightly increase the manufacturing costs. In a bipolar cell configuration, the positive and negative electrodes have

Table 1.10 Theoretical Energy Densities of Lithium-Based Rechargeable Batteries

Battery System	Symbol	Energy Density (Wh/kg)
Lithium-sulfur oxide	$LiSO_2$	1,175
Lithium-copper chloride	$Li-CuCl_2$	1,135
Lithium-vanadium oxide	$Li-V_6O_{13}$	870
Lithium-titanium sulfide	$Li-TiS_2$	562
Lithium-manganese oxide	$Li-MnO_2$	432

back-to-back electrical contact through the Mg-O separator or a conducting lane. These cells will meet specific energy requirements ranging from 80–100 Wh/kg to 200 Wh/kg. Japan has designed and developed high-capacity sodium-sulfide (Na-S) cells for stationary energy storage applications. These cells are best suited for stationary energy sources. Development for traction applications is currently pursued cautiously. Use of polymeric electrolytes is currently being investigated for possible applications in EVs and HEVs.

1.4.6.3.3.1 Sodium-Nickel Chloride Known as ZEBRA Battery for Electric Vehicles — The negative electrode and electrolyte of this particular battery are similar to those used in a Na-S battery, but a metal chloride such a $NiCl_2$ is used instead of sulfur as a positive electrode. This particular ZEBRA (zero-emission battery research activity) was developed by South Africa battery designers for electric vehicle applications. This battery offers higher cell voltage of 2.58 V. The tolerance to overcharge and overdischarge is remarkable. The battery architecture allows several cells to connect in series, without parallel connections, because the cell imbalances are leveled out by the electrochemical reactions within the battery. This battery system offers a wide operating temperature range, enhanced safety, remarkable design flexibility, high cycle life, and low corrosion. The ZEBRA batteries are most cost-effective and ideal for EVs depending on the design and development activities directed particularly toward EVs and HEVs.

The material scientists have described the evolution of polymer electrolytes in three distinct generations:

■ First generation of electrolytes based on the combination of high molecular weight of polyethylene-oxide (PEO) polymer host materials and lithium salts
■ Second generation of electrolytes based on modified PEO structures combined with lithium salts
■ Third generation of electrolytes formed by trapping a low molecular weight liquid solution of lithium salt in an organic solvent of a high molecular weight material

The electrolyte can be fabricated in the form of a thin solid film, thereby eliminating the need of a separator element requirement. The very thin electrolyte, combined with a thin electrode structures, could allow electrode high rate performance and improved lithium everlasting morphology. The possibility of greater intrinsic safety combined with improved rate capability makes the polymer electrolyte battery system a viable candidate for a high-performance battery. Major advantages of polymer electrolyte batteries can be summarized as follows:

■ Stable electrolytes materials provide nonvolatile, sold-state materials.
■ Wide electrochemical window permits cathodes with high-energy capabilities.

- Low electrode loading yields better lithium cycle life.
- Flexible shape factor offers efficient and compact packaging, which offer ideal application, for which weight and size are the most critical requirements.

1.4.6.3.4 Unique Design Features of Polymer Electrolyte Batteries

The polymer electrolyte battery design is based on thin-film technology components that incorporate very large-area electrolyte and electrode layers with thickness ranging from 20 to 200 microns. This technology minimizes the impedance of the electrolyte, thereby enhancing the electrode kinetics. Furthermore, the all solid-state structure allows for most compact sizes to be combined with the possibility of offering the most efficient electrochemical performance. The sold-state design has demonstrated high-energy efficiency and a life cycle of more than 100 cycles for small cells operating even at 100°C. The high-energy density for a lithium-based system is due to low atomic mass of the lithium element and its high reactivity with most positive materials. Materials scientists claim that that the energy density of lithium-based batteries is higher by a factor ranging from 2 to 10 relative to a Pb-acid battery. These batteries, however, suffer from problems associated with rechargeability and performance degradation. Specific power levels from 100 to 200 W/kg were achieved for the first generation of batteries, which have been improved to 400 W/kg. Modeling studies of the energy and power capabilities and experimental verifications as a function of battery design parameters are needed to project performance capabilities of polymer batteries. The rechargeable lithium-polymer electrolyte battery was not commercialized ever after 2005 because of some safety concerns. Research studies are needed for improvements in electrolytes, interfacial behavior, electrode structure, and time-dependent effects of cycling multicell batteries and individual cell reliability.

1.4.6.3.5 Lithium-Metal-Polymer Batteries for Electric Auto Applications

This particular battery was originally developed for backup power applications. The latest research and development activities seem to indicate that the battery design architecture can be modified to meet its application to EVs and HEVs. The battery uses lithium as a negative electrode, vanadium oxide as a positive electrode, and polymeric solution as electrolyte. The conductivity of the electrolyte is less than $0.0001/\Omega$-cm at temperatures greater than 40°C, which offers the most satisfactory battery performance as backup power source for telecommunications applications. The cells can be connected to a bus bar to supply a voltage of 24 V or 48 V, which is most ideal for telecommunications applications. The battery loses only 1% per year of its capacity and offers a battery life in excess of 10 years even at an ambient or storage temperature as high as 60°C. Battery tests indicate that

Table 1.11   Percentage Change in Battery Capacity as a Function of Ambient Temperature

Temperature (°F)	Battery Capacity (%)
−10 (frozen)	9
23	57
35	70
50	82
60	88
70	96
77	100

the battery can be operated with reliability for at least 12 years during the float and backup periods.

1.4.6.3.6 Impact of Ambient Temperature on Rechargeable Battery Capacity

Timing is of extreme importance when considering when to charge a battery. If you wait until the battery is completely discharged, it might take several hours in some cases to recharge the battery to its full capacity. In addition, the reduction in battery capacity is strictly dependent on the ambient temperature. The rate of reduction, however, is solely based on the electrolyte density. Percentage change in battery capacity as a function of ambient temperature is evident from the data shown in Table 1.11.

The data presented in Table 1.11 reveal that the battery capacity drops at a faster rate as one approaches to freezing temperatures.

1.5 Batteries for Low-Power Applications

As discussed, the rechargeable batteries can be used in commercial, medical, space, and military applications. Performance requirements, however, will differ from application to application. For medical and portable diagnostic devices, smallest package, ultra-low-power consumption, low offset drift, and superior noise performance are the principal design requirements. For EVs and HEVs, high-power density, high-energy density, longevity greater than 10 years, and cost are the basic requirements. For space applications, rechargeable batteries must meet stringent performance requirements such as light weight, radiation hardening, compact packaging, high efficiency, longer life cycle, and ultra-high reliability. For battlefield applications, batteries are required to meet stringent performance requirements, namely light weight, high efficiency, ultra-high reliability, and continuous operation to accomplish the mission objectives [5]. The following paragraphs describe in greater detail the performance requirements for current batteries for various commercial applications.

Low-power batteries are widely used by electronic circuits, digital sensors, and electrical devices, such as toys, electric clocks, watches, radios, computers, medical devices, electric toothbrushes, smoke detectors, parameter security devices, and a host of other commercial appliances. Older electrochemical systems such as C-Zn, zinc-air (Zn-air), Ni-Cd, and Pb-acid continue to get better in terms of performance, cost, and size. They are most ideal for medium-power applications. Most of the low-power commercial batteries are known as primary batteries. Primary cells known as D cells (or dry cells) are widely used in flashlights, smoke detectors, toys, electric clocks, radios, and other entertainment appliances. During the past decade or so, lithium-manganese oxide (L-Mn-O) batteries have dominated the commercial market. Primary batteries are growing in size as new electric and electronic devices are designed around their higher voltage, higher energy capacity, and superior shelf life.

Environmental regulations continue to affect battery use and disposal. Therefore, more interest is in secondary or rechargeable batteries that can be reused several times before disposal. Use of rechargeable secondary batteries offers the most cost-effective use of such batteries. Existing and emerging battery systems will be discussed in terms of energy content, shelf life, longevity, one-time procurement cost, discharge rate, duty cycle, operating voltage, and other relevant characteristics. Advancement in battery technology has been linked primary to electronic device and sensor applications. Advancements in materials science, design configuration, and packaging technology play a key role in improvements in energy density level, longevity, reliability, size, and weight.

1.5.1 Batteries Using Thin-Film and Nanotechnologies [6]

Material scientists and battery designers have identified interesting developments involving thin-film (TF) and nanotechnologies. The energy and power density, relative to their size, make them most attractive for energy harvesting. Fast charging capability and low ESR are the most significant characteristics of such batteries. In addition, these batteries do not self-discharge, so that they remain for normal use for a decade or longer. In other words, these batteries do not require recharging even after a very long storage. When coupled with supercapacitors, these batteries round out the energy storage picture for many mesh-network applications. Material scientists claim that it is possible to integrate a photovoltaic (PV) cell with a TF battery to retain a self-charging capability, which opens a wide door for applications in cases in which charging facilities are not readily available.

TF batteries were first designed and developed by the Oak Ridge National Laboratory (ORNL). TF batteries can be fabricated by direct deposition onto thin plastic sheet or chips. Unlike conventional batteries, TF batteries offer maximum bending capability when fabricated on a thin plastic and can be shaped into whatever form-factor needed by a particular application. These TF batteries also scale nicely in terms of size and geometric shape. Operational tests conducted by various users reveal that these batteries exhibit no deterioration in performance when operated over a wide temperature range from –30°C and +140°C. Furthermore, the battery performance remains unaffected by a heating temperature of 280°C under automated solder-reflow.

During the fabrication process, different layers can be deposited by sputtering or evaporation techniques. The stack from the current collector to anode is typically less than 5 microns. TF battery designers estimate that the total battery thickness can be anywhere from 0.35 to 0.62 mm. The charge and discharge characteristics of thin-film lithium ion batteries are shown in Figure 1.1. In this particular battery design, the voltage starts at 4.0 V because the Li cells have lower operating voltage than those of batteries with lithium anodes.

This current limitation is the most serious drawback of flexible TF batteries. Therefore, to achieve high current densities, it is necessary to heat-treat the cathode at temperatures higher than 700°C. This shortcoming tends to prevent the use of flexible polymer substrates for cathode films in some specific applications. The internal resistance of such batteries is dependent on the polyimide sheet thickness and the annealing temperature, which must not exceed 400°C. If the battery is made on a rigid ceramic substrate with a cathode of comparable thickness, annealing temperatures as high as 750°C can be used.

1.5.2 TF Microbatteries

Studies performed by the author are most ideal for applications in which ultra-compact size, portability, miniature power source, space limitation, and low weight are the principal requirements. In brief, cutting-edge, flexible nanotechnology-based TF batteries are best suited for radiofrequency identification (RFID) tags, smart cards, portable sensors, and medical embedded devices. According to the battery designers, these TF batteries can be made as thin as 0.002 inch, including the packaging.

These TF batteries use a lithium-phosphorous-oxynitride (LiPON) ceramic electrolyte that was developed by ORNL. The battery cathode is made of lithium cobalt oxide ($LiCoO_2$) and the anode is made from lithium. Both the cathode and anode contain no liquid or environmentally hazardous material. Even though the lithium material is slightly toxic, the small amount of lithium in the microbattery would not cause a fire if the hermetic seal was broken. Thus, the battery offers optimum reliability and the safest operation over its stated life.

1.5.3 Charge-Discharge Cycles and Charging Time of Low-Power Batteries

The charging time is strictly dependent on the battery capacity. Battery designers claim that a 0.25 milliampere hour (mAh) TF battery can be charged to 70% of the rated capacity in less than two minutes and to full capacity in four minutes. Every battery can be discharged at rates more than 10 C, and the batteries are good for more than 1,000 charge-discharge cycles at 100% depth discharge. Self-discharge for such batteries is less than 5% per year. During the discharge, a battery experiences a capacity loss. Capacity losses per year for various rechargeable batteries are illustrated in Figure 1.3. These batteries can be customized to meet specific size requirements. A 0.1 mAh capacity battery can be designed to a physical size not exceeding 20 millimeters (mm) by 25 mm by 0.3 mm according to the battery design engineers. Batteries can be stored in an environment temperature ranging from –40° to +85°C without any structural damage or performance degradation. In addition, the battery design is such that operating temperature does not affect the performance significantly. At high temperatures, these batteries can be charged and discharged at a higher rate and with higher capacity. At very high temperatures up to +170°C, however, the battery capacity drops at faster rate during the cycling. In colder environments down to –40°C, one can expect reduced charge and discharge rates. Drops in energy densities (Wh/in^3) for various rechargeable batteries as a function of operating temperature are shown in Figure 1.4. It is evident from this figure that the battery energy density drops at faster rate at lower operating temperatures. Furthermore, the energy drop is the lowest over a wide temperature range particularly for sealed Pb-acid and Ni-Cd batteries.

Charging these batteries requires a constant voltage of 4.2 V. Furthermore, one cannot overcharge these nanoenergy batteries according to the battery designers. When this battery is charged at 4.2 V and discharged at 1 mA to 3.0 V, the battery looses roughly 10% capacity over 1,000 charge-discharge cycles. The charging period required to attain a charge of 95% of the rated capacity is less than four minutes at the first cycle and increases to six minutes at the end of 1,000 cycles. As far as the electrical performance of these batteries is concerned, a miniaturized version of a commercial battery would provide about 80 mAh per discharge cycles with more than 400 Wh/L energy density.

Low-power batteries are identified by their power ratings, which cover the capacity ranging from few microwatts for watches to 10–20 W for notebook computers. Energy level and power output per unit volume are the most critical requirements for many portable devices. The energy delivered by a specific battery depends on the rate at which the electrical power is consumed or withdrawn. Both the energy density and the power capacity are strictly constrained by the battery structural dimensions, cell size, and the duty cycle used. The discharge rate, frequency, and the cutoff voltage are normally selected to meet electrical requirements to meet specific applications, such as a smoke detector or a camera. The primary cells are rated at a current that is one-one thousandth of the battery capacity, whereas the secondary cells or the rechargeable batteries are rated at C/20, where C denotes the full capacity rating of the battery.

1.5.4 Structural Configuration for Low-Power Batteries

Most primary or dry cells with aqueous electrolytes employ single, thick electrodes arranged in parallel or concentric configurations. Typical battery configurations are categorized as "cylindrical," "bobbins," "buttons," or "coin cells." Some primary cells are made in prismatic and thin, flat constructions to achieve the lowest volume. These form factors yield poor energy density and power capacity levels.

1.5.5 Most Popular Materials Used for Low-Power Batteries

C-Zn batteries continue to dominate the low-power household battery market on a worldwide basis. These batteries were widely used from 1920 to 1990 and demonstrated significant improvements in electrical performance, life cycle or longevity, and leakage. There are two distinct versions of this battery. A "premium" version, which uses manganese-dioxide (MnO_2) electrolyte, and the zinc-chloride (Zn-Cl) electrolyte, which offers better electrical performance and improved reliability over longer durations. The majority of the standard-size batteries come in D, C, AA, or other configurations and Zn-Cl electrolyte. According to the market survey, the unit sale ratio of alkaline Zn-Cl cell to C-Zn cell is unity, whereas in the United States, the ratio is 3.5:1 for alkaline batteries using MnO_2 electrolyte. According to an international market survey, China alone produces roughly six million C-Zn cells per year. The quality of Chinese cells is slightly poor, but the cell cost is lower. In brief, alkaline dry cells using mercury-free MnO_2 are the most option popular around the world because recycling process is easy as well as environmental friendly.

1.5.5.1 Low-Power Standard Cells

The standard cell uses MnO_2 as the cathode with ammonium chloride as the electrolyte, and this cell is known as a Zn-Cl or an alkaline battery. The C-Zn batteries are not satisfactory for electrical devices, such as tape recorders and disc players, high-resolution automatic cameras, flash units, and certain toys, because they are not able to provide electrical energy needed for satisfactory and reliable operation. Because of power limitations of C-Zn batteries, alkaline batteries are widely used by low-power devices. Typical characteristics of C-Zn and alkaline batteries are summarized in Table 1.12.

The shelf life for medium- and high-power alkaline batteries is roughly three to six years depending on the power rating. Alkaline batteries offer very long lives and

Table 1.12   Characteristics of Carbon-Zinc and Alkaline Batteries

Battery Category	Energy per Unit Volume (Wh/L)	Energy per Unit Weight (Wh/kg)	Shelf Life (yr.)
Alkaline	270	115	3–6
Carbon-zinc	155	90	3–4

retain the battery charge over long durations. In addition, alkaline batteries provide higher reliability and lower production cost. That is why alkaline rechargeable batteries are widely used in the United States.

1.5.5.2 Miniature Primary Batteries

Miniaturized cells are used for applications in which power consumption is tens to hundreds of microwatts. These batteries are best suited for watches, smoke detectors, temperature-monitoring sensors, and other low-power electronic components. Typical electrical performance parameters of such batteries are summarized in Table 1.13.

In the commercial market, cells in "button" configurations are selling large quantities. Market surveys reveal that Zn-air and silver oxide, mercuric oxide, and MnO_2 button cells are the most popular. The sale of mercuric cells is banned in industrial countries because of fear of mercury content. Zinc-air cells are preferred for hearing aid devices, whereas $Zn-MnO_2$ batteries are widely used in watches. With the exception of Zn-air cells, however, the miniature cells offer excellent shelf life and reliable service life. Most watch batteries offer high reliability and longevity at an ambient temperature of 37°C. Lifetimes of watch batteries range from five to seven years. According to miniature battery suppliers, because the new devices are designed around lithium coin cells, the zinc-anode miniature cells will decrease in unit volume. Regardless of these shortcomings, Zn-air cells enjoy the highest energy density levels.

Table 1.13   Performance Parameters of Miniaturized Cells

Battery System	Open-Circuit Voltage (V)	Operating Voltage (V)	Capacity (mAh)	Voltage versus Time Response
Zn-air	1.4	1.3–1.2	550	Slight slope
Zn-HgO	1.4	1.3–1.2	220–280	Near flat
$Zn-Ag_2O$	1.6	1.55	180	Very flat
$Zn-Mn_2O$	1.5	1.25	150	S-shaped

1.5.5.2.1 Miniature Lithium Batteries

Miniature lithium batteries have been in great demand over the past two decades because of their optimum operating life and ultra-high reliability. Lithium cells have demonstrated very high-energy density levels but low power. Lithium-iodine batteries are widely used in pacemakers. Lithium iodine is the low-conductivity solid-state electrolyte, which limits the output current to a few microamperes. Recent advances in both battery materials and pacer technology could permit the most reliable battery operation over 10 to 12 years. High-power implantable batteries using lithium-silver vanadium (Li-Ag V) oxide are being used to power the heart pacers as well as portable automatic defibrillation devices. One gram of lithium is equal to 3.86 Ah of storage energy. Disposal of large quantities of lithium batteries creates a serious environmental problem and produces hazardous waste, which must be disposed of or stored under strict environmental guidelines. Despite these problems, lithium-based rechargeable batteries are experiencing a whopping 25 percent growth rate fueled by the explosion in cellular phones, iPads, portable computers, camcorders, and entertainment devices.

1.5.6 Low-Power Batteries Using Nanotechnology

The smart battery is developed using nanotechnology. The smart battery contains a "superhydrophobic nanostructured surface" made of nanotubes as illustrated in Figure 1.2. This battery technology keeps the electrolyte separate from the anode and cathode electrodes. Upon the application of an electric field, the electrolyte experiences the "electrowetting" process. This introduces a change in the surface tension that permits it to move through the barrier flow, thereby generating a voltage across the battery electrodes. The power rating of this type of battery is not impressive, but it offers significant improvements in size, weight, and efficiency. Typical applications of a smart or nanobattery could include a mission-critical cell phone. Just as the conventional cell phone battery was about to give up power, for example, the reserve battery could be actuated to provide another 10 minutes of talk time, which may be of critical importance.

1.5.7 Paper Batteries Using Nanotechnology [7]

Research scientists and postdoctoral fellows at Stanford University, California, are trying to develop paper batteries using ordinary paper and ink that will deposit carbon nanotubes (CNTs) and silver nanowires. The scientists feel that when the paper is coated with an ink infused with nanomaterials, the paper becomes highly conductive and could be used to produce microbatteries and supercapacitors that are ultra-cheap, flexible, and light weight.

These nanomaterials [7] are one-dimensional structures with very small diameters, which helps the infused ink strongly adhere to the fibrous paper, thereby making the batteries and supercapacitors durable and cost-effective. The Stanford scientists claim that the nanomaterials make the most ideal conductors because they can move electricity more efficiently compared with ordinary conductors. The ink-coated paper is baked and then folded into an electricity-generating source to create a battery source. Potential applications of this paper battery, including small electric and hybrid cars, depend strictly on the quick transfer of electrical energy. Such batteries could be made using thin sheets of plastic. But the ink adheres more strongly to the paper surface because of its porous texture.

The paper can be crumpled, folded, even soaked in acid with no noticeable degradation in performance as a battery. The work done by the research scientists is also best suited for energy storage devices with minimum cost and complexity. The scientists are conducting more tests on these paper batteries.

1.6 Fuel Cells

The demand for long-running portable power sources has grown rapidly in the past few years. Increased electronic usage by the military systems and rapid deployment of consumer electronic devices, such as laptops, mobile phones, and camcorders, require instantly rechargeable, mobile power solutions. Li batteries are widely used to power electronic devices and sensors, but their well-known drawbacks, namely discharge rate, recharge capability, safety concerns, and disposal issues, have forced the power systems designers to investigate the potential application of fuel cells. Manufacturers of rechargeable batteries are increasingly using fuel cells to replace Li batteries. Fuel cells generate electrical power through the electrochemical conversion of a fuel that can be instantly replenished. For the portable power segment, direct methanol fuel cells (DMFCs) offer the most promising solutions.

1.6.1 Description of the Most Popular Fuel Cell Types and Their Configurations

Metal fuel cells (MFCs) can provide the electrical power in a reliable and cost-effective manner. Preliminary studies undertaken by the author indicate that Zn-air fuel cells can yield electrical energy in excess of 4 kWh/kg, which are roughly 1,000 times the energy available from the Pb-acid batteries and three times the energy provided by the gasoline.

In Thailand, DMFC fuels cells have been seriously considered to power scooters with minimum cost and complexity. In the transportation segment, there is great interest in the design and development of low-cost fuel cells to provide a clean, reliable, and safe alternate power source of electrical energy. These low-cost, low-power fuel cells can be designed to meet the energy demand ranging from 1 to 10 kWh capacity.

1.6.2 Types of Fuel Cells

Fuel cells are classified by the electrolyte used by the module. The following four distinct types of fuel cells are gaining great interest for applications as a power source:

■ The low-temperature phosphoric acid fuel cell (PAFC)
■ The proton exchange membrane (PEM) fuel cell
■ The high-temperature molten carbonate fuel cell (MCFC)
■ The solid oxide fuel cells (SOFCs)

Research and development activities pursued by various fuel cell companies indicate that PEM fuel cells offer simple design, improved reliability, reduced procurement, low operating cost, and a small footprint. The Dow Chemical Company and Ballard Power Systems, Inc. are dedicated to commercializing the PEM fuel cells for distributed power-generating markets. Regardless of who manufactures the fuel cells, the fuel cells exhibit the following unique characteristics:

■ The fuel cells can be stacked in various configurations to accommodate different capacity requirements with the same cell design.
■ Fuel cells have fairly high efficiencies that are relatively independent of the size. Fuel cells are easy to site because of extremely low environmental intrusion.
■ Fuel cells can use a variety of fuels with quick change-out provision.
■ Fuel cells offer operational advantages such as electrical energy control, quick ramp rate, remote and unattended operation, and high reliability because of inherent redundancy feature.

Because of their ultra-high reliability, as early as 1960, fuel cells were used to provide onboard electrical power for manned spacecraft, and the exhaust produced safe drinking water for the astronauts. Within the past few years, the U.S. military has provided significant research and development in fuel cell support for possible applications in battlefield applications. Chapter 3 provides specific design concepts and material requirements for various types of fuel cells.

1.7 Conclusion

The chapter provides brief descriptions of primary and secondary (rechargeable) batteries. Performance capabilities and limitations of rechargeable batteries are discussed with a particular emphasis on reliability and longevity. Battery requirements for EVs and HEVs are defined in terms of cost per kilometer of travel. Estimates of specific energy density and safe power levels are provided for specific applications.

Applications of high-, medium-, and low- rechargeable batteries are identified with an emphasis on cost and longevity. Important properties of alloys and rare earth elements widely used in the construction of rechargeable batteries are summarized. Toxicity aspects of various battery materials are highlighted for user safety. Techniques for improving the performance batteries are discussed in great details. Failure mechanisms in rechargeable batteries are identified. Battery requirements for space, commercial, and military applications are summarized with a particular emphasis on reliability, cost, weight, and size. Performance requirements of rechargeable batteries for battlefield applications are identified with an emphasis on weight, reliability, and charge-discharge rates. Critical performance capabilities of Pb-acid, Ni-Cd, Ni-MH, and Li rechargeable batteries are described. Materials best suited for electrode fabrication are listed with a particular emphasis on the electrochemical process. Reduction rates in battery capacity as a function of ambient temperature for various rechargeable batteries are mentioned for the users' benefits. Performance capabilities and limitations of TF batteries, paper batteries, microbatteries, and lithium-based batteries are summarized with an emphasis on cost and reliability. Performance parameters of microbatteries for watches, heart pacers, and hearing aids are identified. Charging and discharging rates for lithium-based and other rechargeable batteries are specified. Performance parameters of various types of fuel cells and their design configurations are summarized. The advantages of fuel cells for various commercial, spacecrafts, battlefield, unmanned air vehicles, and other critical military applications are identified with a particular emphasis on their potential applications to neutralize improvised explosive devices in the battlefield.

References

1. Courtney E. Howard, "Electronics miniaturization," *Military and Aerospace Electronics* (June 2009), p. 32.
2. Robert C. Stempel, S. R. Ovshinsky et al., "Nickel-metal hydride: Ready to serve," *IEEE Spectrum* (November 1998), pp. 29–30.
3. Gary Hunt, "The great battery search," *IEEE Spectrum* (November 1998), p. 21.
4. F. C. McMichael and C. Henderson et al., "Recycling batteries," *IEEE Spectrum* (February 1998), pp. 35–37.
5. Editor-in-Chief, "Military leads in battlefield development," *Military and Aerospace Electronics* (February 1994), p. 27.
6. A. R. Jha, *MEMS and Nanotechnology-Based Sensors and Devices for Communications, Space, and Military Applications*, Boca Raton, FL: CRC Press (2008), p. 344.
7. Christina Dairo and E. Padro, "Nanotechnology enables paper batteries," *Electronics Products* (January 2010), p. 13.

Chapter 2

Batteries for Aerospace and Communications Satellites

2.1 Introduction

Rechargeable battery requirements for spacecrafts and communications satellites are stringent. Reliability, weight, size, and longevity are the demanding performance requirements of the rechargeable batteries for their deployment in spacecrafts and communication satellites. In addition, redundancy could be another critical performance requirement. The performance requirements for low-earth orbiting (LEO) will be stringent, and the requirements for geostationary satellites will be even more stringent. Because of launch cost and system complexity, longevity requirement will be in excess of 10 to 15 years. Space batteries must be capable of providing electrical energy for the lights, spin-control stabilizing sensors, space-monitoring systems, air conditioning systems, drinking water products, and a host of other critical appliances for comfort and safety of astronauts.

The electrical systems for communication satellites, covert surveillance-reconnaissance satellites, and spacecrafts for astronaut research activities consist of energy-conversion devices, (solar cells) in combination with energy storage devices (batteries) and power-conditioning (PC) components, namely pulse-duration voltage regulators (PVRs). PVRs play a key role in maintaining a constant voltage during the pulse period to achieve optimum performance. Specific details and performance capabilities of PVRs and other regulators will be mentioned under Section 2.2.1. Weight, size, and power consumption all these components must be kept to a minimum to minimize launch costs of the satellites and spacecrafts.

Nickel-cadmium (Ni-Cd) batteries were widely deployed by earlier communication satellites and spacecrafts from 1960 to 1995. When advanced designs of nickel-hydrogen (Ni-H$_2$) batteries were available in early 2000, most of the satellites and spacecrafts preferred to use Ni-H$_2$ batteries. These batteries will be described in detail with an emphasis on reliability and improved electrical performance. For short mission satellites and LEO communication satellites, both the Ni-Cd and Ni-H$_2$ batteries are still being used.

During the dark period experienced by a satellite, the onboard batteries must meet the electrical power consumption requirements of all the operating electronic sensors and electrical devices aboard the satellite or spacecraft. The electronic and electrical components are not getting any electrical energy from the satellite-based solar cells during the dark period. As soon as the dark period is over, the onboard electronic sensors and electrical components will receive the electrical energy from the solar panels attached to the satellite or spacecraft. As mentioned, the solar arrays and batteries are interconnected with a regulator bus as illustrated by Figure 2.1.

The dark period experienced by a satellite is known as "satellite eclipse" during which the local time is a serious consideration for receiving audio video signal from the satellite. The maximum eclipse duration, when the satellite is in the earth's shadow at the season of the equinoxes, is typically 72 minutes, and its peak is passed at the apparent solar midnight for a point at the same longitude as the satellite.

The peak starts about 36 minutes before, and there is an eight-minute correction for the discrepancy between the apparent and the mean solar time. This results in an eclipse onset about 11:15 p.m. for the television coverage at the subsatellite point. As the coverage zone moves east or west of the subsatellite point, the clock time for the eclipse onset can be as much as two hours earlier or later on the local clocks. The effect on the spacecraft design is not trivial. High-power television transponders require heavy batteries to operate during the eclipse period, and the batteries must be capable of providing all of the electrical power requirements during the eclipse or dark period. This is the most critical design requirement for the satellite batteries.

2.2 Onboard Electrical Power System

The onboard electrical power system [1] has three distinct components, namely the battery, solar array, and voltage regulation circuitry. Critical devices and circuits used by these three components are clearly shown in Figure 2.1.

2.2.1 Electrical Power-Bus Design Configuration

There are two types of electrical power-bus configurations, namely a regulated power-bus configuration and an unregulated power-bus configuration. In the case

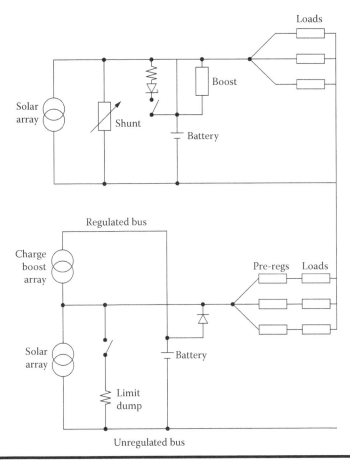

Figure 2.1 **Regulated and unregulated power-bus concepts for potential applications for satellite power systems. (Adapted from Jha, A.R.,** *Solar Cell Technology and Applications*, **CRC Press, Boca Raton, FL, 2010.)**

of a regulator bus, a booster element and a switch-controlled diode provide the voltage regulation. In the case of an unregulated bus configuration, preregulated electrical loads are deployed, thereby eliminating the need for a separate regulation element. In the case of sophisticated covert surveillance and reconnaissance satellite applications, redundant power systems may be required if ultra-high reliability and long satellite operating life exceeding 10 years are the design requirements.

2.2.2 *Solar-Array Panels*

Requirements for solar cells and solar panels are strictly dependent on continuous electrical power consumption; an additional 30 to 40% more electrical power is needed to meet power consumption requirements depending on the operating

Table 2.1 Typical Power Output Variations in Satellite Solar Panels

Operating Life (yrs.)	Power Output (W)	Reduced Power Level (%)
0 (after launch)	650	100
1	583	93.5
2	558	90.2
3	543	87.6
4	517	83.4
5	500	80.6
6	455	73.4
7	389	62.7
8	342	55.2

lifetime of the satellite. Typical solar-panel power variations in a space-based satellite are summarized in Table 2.1.

In space- and satellite-based systems, batteries generally experience power variations and degradation as a function of lifetime. In a typical space-based battery, the battery power output is reduced to 55.2% of the original value after eight years as illustrated in Figure 2.2. This is the performance degradation of a space-based maintenance-free battery that is well designed and sealed. Some batteries may last slightly more or less depending on the electrochemical reaction efficiency and operating space environments. The effects of space radiation on the battery power performance can be neglected, if the battery is inside the satellite structure.

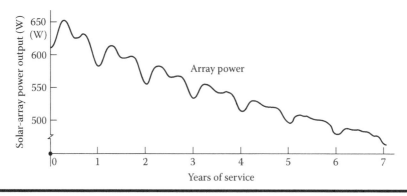

Figure 2.2 Satellite-based solar-array power variations as a function of service duration.

2.2.2.1 Solar Panel Performance Requirements to Charge the Space-Based Batteries

Installation of solar panels is required to charge the onboard batteries that are providing the electrical power to various electronics devices, stabilizing and attitude control sensors, space parameter monitoring instrument, lighting, and a host of other electrical systems that are vital in maintaining the desired performance of the satellite or spacecraft over the intended mission duration.

Schematic representation of orbital characteristics and the solar-array location for a spinning satellite is shown in Figure 2.3. Frequency of charging of batteries onboard a communication satellite or a surveillance-reconnaissance satellite is strictly dependent on the electrical lead requirements. In the case of a communication satellite, the magnitude of the electrical load is moderate compared with that for a surveillance-reconnaissance satellite. The surveillance-reconnaissance satellite has several microwave, electro-optical, optical, and infrared sensors, such as high-resolution side-looking radar, precision laser tracking sensor [2], high-resolution infrared camera, fast digital computer with ultra-high computational capability, and other climate-parameter monitoring instruments. In addition, a surveillance-reconnaissance satellite will have large number of solar panels to meet the high electrical load requirements of the primary sensors and to power the redundant systems needed to meet the reliability and intended mission objectives.

2.3 Battery Power Requirements and Associated Critical Components

Battery requirements for satellites traveling in different orbits will be discussed with a particular emphasis on the associated components needed for reliable and sustained electrical performance. Battery performance specifications are quite different for LEO and geostationary earth orbit (GEO) satellites. Typical orbit characteristics of the first-generation spinning satellite are shown in Figure 2.3. The on-board batteries are designed to meet the electrical load requirements for mission parameter-monitoring instruments as well as to provide electrical power to the redundant systems selected by the satellite design engineers. Both the satellite-based solar arrays and the onboard batteries must meet the electrical energy requirements of electronic sensors, stabilization components, and control mechanisms. Typical electrical consumption requirements for the early communications satellites launched between 1970 and 1980 vary from 10 kilowatt (kW) to 20 kW, which subsequently increased more than 25 kW as the powerful communications satellites entered into the communications area. The electrical power requirements for the latest surveillance and reconnaissance spacecraft can be at least 25 kW or higher because of deployment of high-resolution mm-wave side-looking radars,

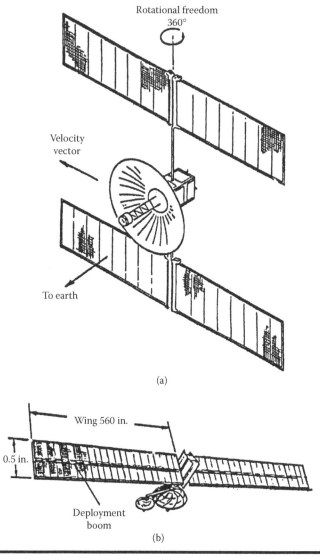

Figure 2.3 Solar array design configuration for (a) a 12-kW DirectTV broad-cast and (b) a 5-kW television satellite. (From Balombi, J.R., Wright, D.L. et al., "Advanced Communications Technology Satellite (ACTS)," *Proceedings of the IEEE*, © 1990 IEEE. With permission.)

high-power tracking lasers, and precision electro-optical sensors. On the basis of these power requirements, the spaced-based solar panels and batteries must be capable of meeting such power consumption requirements.

Satellite-based solar panels must be designed to meet the basic power consumption requirements as well as the extra power needed to compensate for any electrical damage to the solar cells that may be caused by the Van Allen radiation belts over the life of the satellite. This extra electrical power requirement can vary between 10 and 20% of the basic power level, depending on the expected life and the orbiting height of the satellite. Performance requirements for critical components associated with battery operation will be described under Section 2.4.

2.3.1 Solar-Array Performance Requirements

The solar-array performance requirements are strictly dependent on the following issues:

- Satellite orbit
- Spacecraft stabilization control and attitude control mechanism [3]
- Electrical power requirements over the operational life of the spacecraft or satellite
- Mission requirements (communications or surveillance-reconnaissance)
- Space available for installation of solar arrays on the launch vehicle
- Additional power needed for specific design features, such as a dual-spin approach or a three-axis stabilization technique

2.3.2 Electrical Power Requirements from the Solar Arrays during Dark Periods

LEO satellites could experience several dark periods during each 12-hour orbit depending on the orbital duration. Satellite designers claim that communications satellites launched in the equatorial orbit will experience the fewest dark periods. The designers further believe that because of the unique orbiting characteristics, a synchronous earth orbiting satellite will rarely experience a dark period, and therefore the solar panels will receive the unrestricted solar energy to generate the electrical energy needed for the onboard sensors and tracking devices. Power consumption requirements for stabilization control and attitude control are relatively high, and therefore the solar-array designer must ensure that such power levels are available on continuous basis. The deployment of sun-oriented solar panels provides nearly uniform power output throughout each orbit, which offers optimum electrical energy with high reliability. The temperature of the sun-oriented solar panels when illuminated can vary from 60° to 80°C, depending on the atmospheric temperature at the satellite's operating altitude.

2.3.3 Solar Panel Orientation Requirements to Achieve Optimum Power from the Sun

Solar panel orientation sometimes is necessary to achieve optimum electrical power output from the panels. In the case of a surveillance-reconnaissance spacecraft, the orientation of solar panels will be most attractive because more electrical energy is needed to power several microwave, high-resolution infrared cameras, high-power lasers for precision target tracking, attitude control mechanisms, and stabilization control sensors [3]. For equatorial orbits, a one-degree-of-freedom array orientation will maintain normal solar energy impingement within a seasonal variation of plus or minus 23.5 degrees. Two degrees of freedom may be necessary for other orbits in which the spacecraft-sun line may be at any angle with respect to the solar-array installation plane.

2.3.4 Solar-Array Configurations Best Suited for Spacecraft or Communications Satellite

The electrical power output capability of the solar array is strictly dependent on the number of solar cells used, cell conversion efficiency, array orientation with respect to the solar energy impingement direction, panel mounting scheme, and the maximum allowable size of the array. The body-mounted arrays shown in Figure 2.3 could provide several hundreds of watts, whereas the fixed-paddle array configuration is best suited for applications in which tens of kilowatts of electrical power levels are needed.

2.3.5 Direct Energy Transfer System

The direct energy transfer (DET) system plays vital roles when the satellite is operating in a dark period (absence of sun light) or during the sun-illuminated portion of the flight. Critical components of this system and its block diagram are clearly identified in Figure 2.4.

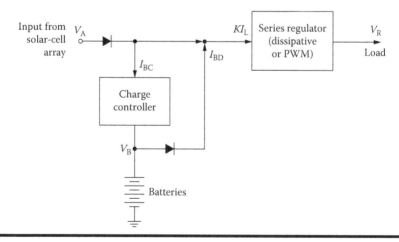

Figure 2.4 Critical electrical components of the basic spacecraft power system.

This particular DET system [1] eliminates the series insertion loss between the solar cell array and the load. The mode-select switch plays a key role in the system operation. This circuit ensures proper flow of the load current and a charge current in the system regardless of solar illumination condition. The shunt regulator is always on and bypasses a small amount of the current to the ground whenever the bus voltage V_R maintains an assigned magnitude. The shunt current generates a small threshold voltage across one of the resistive shunt elements. This DET system plays a performance stabilization role during the dark period and sun-illuminated portion of the flight.

During the sun-illuminated portion of the satellite flight, and if the array output level exceeds the electrical load requirement, the shunt regulator will draw additional current and cause the small threshold voltage (ΔV) to exceeds its predetermined threshold value. When the excess voltage is felt by the mode-select circuit, the circuit will then turn the charge regulator on. The excess voltage will allow the excess current to the battery.

If the communication satellite experiences a dark period, the threshold voltage (ΔV) will tend to drop below the designed threshold voltage level, thereby permitting the mode-select circuits to turn on the boost regulator and the charge regulator at the same time. This will allow for the minimum battery discharge essential to satisfy the dark period load. The shunt regulator function can be achieved in several ways, namely as a dissipative full shunt circuit or as a pulse-width modulation (PWM) shunt regulator or as a partial shunt circuit operating continuously or sequentially.

The node current equations that describe the DET system during the dark portion of the orbit and the illuminated portion of the orbit can be written, respectively, as follows:

$$e_{BR} (V_R/V_{BRO}) I_{BD} = [I_L + I_{SH}] \tag{2.1}$$

$$I_A = [I_{BC} + I_L + I_{SD}] \tag{2.2}$$

where V_R is the regulated load voltage, I_A is the input current from the solar-cell array, I_{BD} is the battery discharge current, I_{SH} is the shunt current, e_{BR} is boost regulator efficiency (shown in Figure 2.5), V_{BRO} is boost regulator output voltage, I_L is the load current, I_{BC} is the battery charge current, and I_{BD} and I_{SD} are the shunt regulator current. Typical boost regulator efficiency is about 90%. This efficiency can be enhanced to 95% if optimum values of circuit parameters are used.

The DET power system designers believe that for optimum system reliability, the minimum solar-array size should consist of 110 strings in parallel and 63 cells in series per string, which will involve a minimum of 6,930 solar cells. These solar cells will limit battery charge current to approximately 5.52 amperes.

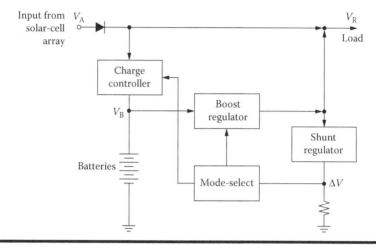

Figure 2.5 **Block diagram of a direct energy transfer power system aboard a spacecraft. (From Pessin, L., and Rusta, D., "A comparison of solar-cell and battery-type power systems for spacecrafts,"** *IEEE Transactions on Aerospace and Electronic Systems,* **© 1967 IEEE. With permission.)**

2.4 Cost-Effective Design Criterion for Battery-Type Power Systems for Spacecraft

The electrical power system for an orbiting spacecraft or a communication satellite consists of an energy-conversion device, such as a solar-cell array energy storage device like a battery, and a PC component, such as a DET subsystem. The energy-conversion components account for a large portion of the total spacecraft or satellite cost and weight. Therefore, reduction of its size and weight becomes an important design consideration for the energy-conversion component or the solar-cell array. Various system design configurations must be evaluated in terms of electronic circuit complexity, cost, performance, reliability, size, and weight. The overall system performance comparison must place maximum emphasis on the overall power system reliability. The addition of redundant elements to a system mat offset the cost or weight savings achieved in its energy-conversion components compared with those of other system configurations.

2.4.1 Method of Comparison for Optimum Selection of Power System for a Spacecraft

To keep the example of comparison simple, this particular method considers three distinct power system configurations, each composed of a solar-cell array as an energy-conversion device and a Ni-Cd battery as an energy storage device. The DET system is common to both design configurations. The overall system performance comparison is made in terms of weight and cost as a function of mission length or spacecraft operating life in years.

2.4.1.1 Step-by-Step Approach for Power System Performance

Each system must be designed to a common mission objective without regard for reliability design requirements. If a reliability goal is not satisfied, system reliability is increased to a specific goal, which could be accomplished by adding redundant components to the system. A redundancy optimization technique can be employed to minimize the resulting increase in system cost and weight. Finally, the systems are compared only in terms cost and weight, each as a function of mission duration or length.

The spacecraft power system to be considered is composed of a solar-cell array and a battery, in combination with a voltage regulator and a battery charger converter. Figure 2.5 illustrates a basic spacecraft power system block diagram. It is very important for the solar-cell array to supply sufficient electrical power during the illuminated portion of the orbit to meet the electrical load requirements aboard the spacecraft and to recharge the battery. The relation between the solar array and the battery is based on the energy balance equation, which can be expressed in terms of battery current. The energy balance equation can be written as follows:

$$\left[\int_o^T I_{BC} \, dt - e_B \int_o^T I_{BC} \, dt \right] \geq 0 \tag{2.3}$$

where I_{BC} is the battery charge current, I_{BD} is the battery discharge current, e_B is the battery-discharge factor with a typical value greater than unity, t is any time during the orbit, and T is the spacecraft or satellite orbit period ranging between 80 to 95 min. depending on the orbital axes. The relation as defined by Equation 2.3 must hold over any orbit. The battery-discharge factor is a function of battery temperature, depth of discharge (DOD), and the number of repetitive charge-discharge cycles.

The basic node current equations to be satisfied at any time t during an orbit, assuming negligible component losses during the dark portion and illuminated portion of the flight, can be written, respectively, as follows:

$$I_{BD} = [K I_L] \tag{2.4}$$

$$I_{ILL} = I_A = [I_{BC} + I_{BD}] = [I_{BC} + K I_L] \tag{2.5}$$

where K is a dissipative series regular (DSR) constant, which is equal to one.

For a PWM series regulator, the constant is defined as follows:

$$K_{PWM} = [V_R / e_{PR} \, V_A] \tag{2.6}$$

where parameter e_{PR} denotes the efficiency of the PWM regulator, V_A is the solar-cell array or unregulated voltage during the dark period of the flight, and V_R is the regulated voltage at the load as illustrated in Figure 2.4.

To solve Equations 2.4 and 2.5 with the load profile defined by the regulated load voltage (V_R) and the PWM regulator efficiency (e_{PR}) and subject to the condition imposed by Equation 2.1, it is necessary to use the current-voltage (I-V) characteristics of the spacecraft power system components. The current-voltage characteristics of the solar-cell array, the onboard battery, and the battery charge controller (CC) have been clearly shown in Figure 2.5. It is not easy to find simple solution of the power system equations due to the nonlinearities of the I-V characteristics, because both the current and voltage parameters are a function of time. In addition, the I-V characteristics of both the solar-cell array and the battery are a function of the operating temperature. Furthermore, the battery's I-V characteristic is dependent on the battery state of charge (SOC). The variation of the battery voltage with the SOC is evident from the curves shown in Figure 2.6. Typical battery voltage (V_B) versus the battery SOC characteristics as a function of battery discharge current and battery charge current can be seen in Figure 2.6. Battery overcharge as a function of solar-cell strings is illustrated in Figure 2.7. Variations in battery output voltage must be kept to minimum for optimum battery performance. Variations of battery cell voltage as a function of cell SOC and parameter I_{BC} are shown in Figure 2.8.

2.4.1.2 Modeling Requirements to Determine I-V Characteristics

It is necessary to use appropriate computer models composed of conventional circuit elements to determine the I-V characteristics of the solar-cell array and the battery deployed by a spacecraft or a satellite. At times, the approximate values of the modeling parameters or the assumptions made do not correlate with the measured component characteristics over the entire operating temperature range. For meaningful modeling, an energy balance computer program that contains subroutines for the storage and interrogation of the function of one, two, or three variables must be developed. This will permit the power system design engineer to define the component characteristics from the measured data as well as from the data obtained using the theoretical models.

The battery voltage can be determined as a function of battery current, SOC, and the battery operating temperature. Identification of the battery SOC is the key to the integration required by the energy balance as shown by Equation 2.3. Further discussion of the features of the energy balance program is beyond the scope of this chapter and of the book. The variations in voltages and currents as a function of temperature will affect the accuracy of the modeling parameters. In addition, the effects of temperature on cell capacity and current density must be taken into account for a specific electrolyte density. Typical temperature effects on a battery cell capacity (%) are evident from the data presented in Table 2.2.

It can be concluded from the data in Table 2.2 that if the current density of the battery is changed, either in the charge or discharge process or in both processes, the

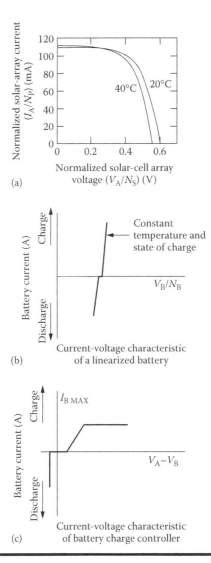

Figure 2.6   I-V characteristics: (a) solar array, (b) battery, and (c) battery charge controller.

battery capacity is affected. Rising current and falling temperature will decrease the battery capacity. With lower current levels, however, the battery capacity depends less on the electrical load, thereby increasing the capacity.

Battery designers believe that after the battery is discharged at low temperatures during formation, a large battery capacity can be achieved at higher temperatures. This trend indicates that the maximum battery capacity is achieved only when the battery undergoes low current discharges. Furthermore, the room temperature

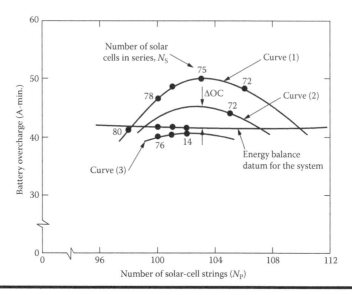

Figure 2.7 Parametric variations in solar-cell array. (From Pessin, L., and Rusta, D., "A comparison of solar-cell and battery-type power systems for spacecrafts," *IEEE Transactions on Aerospace and Electronic Systems*, © 1967 IEEE. With permission.)

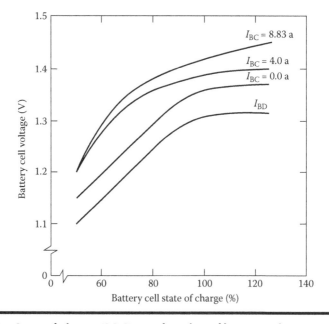

Figure 2.8 State-of-charge (SOC) as a function of battery voltage. (From Pessin, L., and Rusta, D., "A comparison of solar-cell and battery-type power systems for spacecrafts," *IEEE Transactions on Aerospace and Electronic Systems*, © 1967 IEEE. With permission.)

Table 2.2   Effects of Temperature on Battery Capacity

Temperature (°F)	Battery Capacity (%)
−10.5 (frozen)	9
23	57
34	68
50	82
60	87
68	94
77	100

capacity can be maintained at low temperatures only if the battery is discharged with smaller electrical loads. In brief, the variation of battery capacity and discharge rate are strictly dependent on battery temperature and electrical load condition.

2.4.1.3 Impact on Battery Electrical Parameters from Onboard Charging and Discharging

Stable open circuit voltage (OCV; V_{OC}) characteristic is considered advantageous with respect to on-boarding and discharging management, particularly in the case of batteries deployed by a satellite or a spacecraft. This permits easy and accurate estimation of the remaining battery capacity. Table 2.3 shows the actual OCV and internal resistance per battery cell as a function of battery SOC for a lithium-ion battery. These values will vary from one battery type to another as a function of SOC.

Values of both battery cell parameters remain practically constant over the SOC range from 80% to 100%. This means that the battery state of the charge must be maintained greater than 80% if optimum values of OCV and internal resistance are desired at a specified battery temperature.

2.5 Spacecraft Power System Reliability

The computation of spacecraft power system reliability is strictly based on the reliability data of the components involved. The reliability theory indicates that component reliability is defined as the probability that a specific component will perform as per its performance specifications during an interval of time under specific operating conditions. The determination of component reliability involves computing the component failure rate from the part failure rate data, applying this

Table 2.3   Open Circuit Voltage and Internal Resistance Characteristics as a Function of State-of-Charge for a Lithium-Ion Cell

SOC (%)	Open circuit voltage for the cell (V)	Internal resistance (Ω)
20	3.2	0.0027
30	3.5	0.0022
40	3.6	0.0020
50	3.8	0.0019
60	3.9	0.0018
70	4.0	0.0018
80	4.05	0.0018
90	4.10	0.0016
100	4.10	0.0015

rate to a mathematical model of failure, and conducting a failure-mode analysis. For a power system evaluation, it is sufficient to compute component reliabilities using the following exponential failure law:

$$R\,(t) = [e^{-\lambda t}] \tag{2.7}$$

where λ is the combined failure rate for N parts or elements involved over the operating period t. The combined failure rate is defined as follows:

$$\lambda = [\lambda_1 + \lambda_2 + \text{-------} \lambda_n] \tag{2.8}$$

2.5.1 Failure Rates for Various System Components

This section develops appropriate equations for failure rates for various spacecraft or satellite power system. The failure rate for a spacecraft power system presents a complex problem. The power system must consider failure rates for solar cells, for energy storage batteries, and for PC modules.

The failure rate expression for a solar-cell panel can be written as follows:

$$\lambda_{SP} = [\lambda_1 + \lambda_2 + \text{-------} \lambda_n] \tag{2.9}$$

where λ_{SP} is the failure rate for the solar panel. This parameter includes all of the components involved in the fabrication of the panel. The failure rate equation for the battery can be written as follows:

$$\lambda_B = [\lambda_1 + \lambda_2 + \text{---------} \eta_R] \tag{2.10}$$

where λ_B is the failure rate of the battery composed of all the R components involved.

The failure rate expression for the PC module can be written as follows:

$$\lambda_{PC} = [\lambda_1 + \lambda_2 + \text{------------ } n_S] \tag{2.11}$$

where λ_{PC} is the failure rate of the PC module composed of all the S components, devices, and circuits involved.

2.5.2 Failure Rate Estimation

Preliminary evaluation of a spacecraft power system reveals that the failure rate for the solar-cell arrays is moderate compared with the PC module. Theoretically, if the printing circuit is used, there is no failure mechanism in the solar cell. The cell output decreases as it ages, however. Solar power system engineers and designers predict a minimum operational life of 20 years for a solar cell and 10 years for an inverter. The failure rate for a spacecraft battery is strictly dependent on the type of electrodes deployed and the aging effects [4]. The operational life of the battery can exceed 15 years as long as the battery is charged and discharged according to guidelines issued by the supplier and the battery is not subjected to serious physical damage. There is higher probability for failure mechanisms in the basic spacecraft power system as shown in Figure 2.4. The block diagram of the spacecraft power system uses two critical system elements, namely the charged controller and the voltage regulator. This regulator can be a DSR or a PWM regulator. Particularly, the linearized battery charge characteristic and the temperature-dependent SOC of the battery characteristic is strictly a function of critical circuit parameters used in the PWM regulator. The failure rates of these critical parameters can seriously affect the failure rate of the spacecraft power.

2.5.3 Reliability Improvement of the Spacecraft Power System Using CC and PWM Regulator Techniques

To ensure power system reliability, it is important to estimate the OCV and the remaining battery capacity with high accuracy. The system reliability generally requires onboard battery charging and discharging. Furthermore, actual measurement of OCV is strictly related to the battery SOC. In brief, the OCV and the internal resistance of the battery are directly determined by the battery performance when the ampere-hour efficiency of the battery is assumed to be 100% of the charge-discharge process. If accurate reliability data and component failure analysis results on spacecraft power system components are desired, then these statements must be seriously considered. Reliability improvements for the spacecraft power system using redundant units of CC and dissipative PWM regulators as a function of mission duration are summarized in Tables 2.4 and 2.5.

Table 2.4 Reliability Improvement due to Redundant Unit Allocation for the Dissipative Regulator in the Spacecraft Power System as a Function of Mission Duration

Mission Duration (mo.)	Number of Redundant Units (CC, DSR)	System Reliability (%)
6	(0, 0)	78.5
	(1, 0)	88.9
	(1, 1)	97.4
12	(0, 0)	61.7
	(1, 0)	76.9
	(1, 1)	90.8
	(2, 1)	95.3
18	(0, 0)	48.4
	(1, 0)	65.2
	(1, 1)	82.1
	(2, 1)	89.4
	(2, 2)	94.2
	(3, 2)	96.8
24 (estimated)	(0, 0)	42.2
	(1, 0)	55.6
	(1, 1)	71.4
	(2, 1)	82.7
	(2, 2)	84.5
	(3, 2)	94.2

The data summarized in Tables 2.4 and 2.5 indicate that spacecraft power system reliability improves with an increase in redundant units, regardless of the type of regulator integrated in the power system. Furthermore, it is evident from the reliability data that the number of redundant units (2, 2) for the dissipative regulator offers higher reliability compared with the PWM regulator.

If a spacecraft or a satellite experiences a dark period, the mode-select circuits will turn on the boost regulator and the charge regulator, allowing for the minimum battery discharge that is needed to satisfy the power requirements during the spacecraft's dark period. The boost regulator and the mode-select circuits play

Table 2.5   Reliability Improvement for the Spacecraft Power System due to Redundant Unit Allocation for the Pulse-Width Modulation Regulation as a Function of Mission Duration

Mission Duration (mo.)	Number of Redundant Units (CC, PWM-R)	System Reliability (%)
6	(0, 0)	68.2
	(0, 1)	82.8
	(1, 1)	93.6
	(1, 2)	97.5
12	(0, 0)	46.4
	(0, 1)	64.2
	(1, 1)	79.7
	(1, 2)	88.4
	(2, 2)	92.8
	(2, 3)	96.5
18	(0, 0)	31.4
	(0, 1)	47.6
	(1, 1)	64.3
	(1, 2)	76.1
	(2, 2)	82.8
	(2, 3)	89.3
	(2, 4)	92.4
	(3, 4)	95.1
	(3, 5)	96.8
24 (estimated)	(0, 0)	17.4
	(0, 1)	41.6
	(1, 1)	46.2
	(1, 2)	61.2
	(2, 2)	72.5

continued

Table 2.5 (Continued)   Reliability Improvement for the Spacecraft Power System due to Redundant Unit Allocation for the Pulse-Width Modulation Regulation as a Function of Mission Duration

Mission Duration (mo.)	Number of Redundant Units (CC, PWM-R)	System Reliability (%)
	(2, 3)	82.1
	(2, 4)	84.4
	(3, 4)	87.8
	(3, 5)	94.7

a critical role in ensuring that all the spacecraft electrical systems, stabilization mechanisms, and monitoring sensors are getting the needed electrical power during the dark period.

2.5.4 Reliability Improvement of the Spacecraft Power System Using DET System, CC, and Battery Booster Techniques

When all three redundant units such as DET, CC, and battery booster (BB) are integrated in the spacecraft power, the data presented in Table 2.6 indicate that the reliability improvement is not impressive. The DET redundant option, however, offers the lowest system cost and system weight as a function of mission duration. The major reliability improvement is due to the shunt regulator circuit used in the DET system as illustrated in Figure 2.5.

2.5.5 Weight and Cost Penalties Associated with Redundant Systems

Weight and cost penalties could result from the deployment of redundant system elements in the spacecraft power system. The increase in both the total system cost factor and the total system weight is strictly dependent on the mission duration or length (months) and redundant system elements such as CC, BB, shunt boost regulator, PWM series regulator, DSR, and DET subsystem. The energy storage and PC components used in the redundant system or subsystems are deployed by mission-critical spacecraft or covert military communications geostationary satellites when the stated system reliability is of critical importance.

Brief studies performed by the author reveal that the conversion device accounts for a large portion of the total spacecraft or satellite cost and weight. A reduction in weight and size of redundant elements or components deployed by a spacecraft or

Table 2.6   Reliability Improvement of the Spacecraft Power System Using All Redundant Units Such as Direct Energy Transfer, Charge Controller, and Battery Booster

Mission Duration (mo.)	Number of Redundant Units (CC, BB, DET)	System Reliability (%)
6	(0, 0, 0)	75.2
	(0, 1, 0)	82.6
	(1, 1, 0)	88.5
	(1, 1, 1)	97.5
12	(0, 0, 0)	56.3
	(0, 1, 0)	67.2
	(1, 1, 0)	76.5
	(1, 1, 1)	91.2
	(1, 2, 1)	93.7
	(1, 2, 2)	96.8
18	(0, 0, 0)	42.2
	(0, 1, 0)	53.8
	(1, 1, 0)	64.4
	(1, 1, 1)	64.6
	(1, 2, 1)	87.2
	(1, 2, 2)	92.3
	(2, 2, 2)	95.3
24 (estimated)	(0, 0, 0)	21.8
	(0, 1, 0)	42.2
	(1, 1, 0)	52.3
	(1, 1, 1)	74.5
	(1, 2, 1)	78.6
	(1, 2, 2)	87.2
	(2, 2, 2)	96.5

satellite is critically important to mission length or duration. Multifunction covert military communications satellites offer key services, including secured broadband communications lines, and covert as well as uninterrupted communications between the government, military authorities, and important armed forces officials deployed around the world. Such broadband communications satellites are equipped with onboard power systems capable of providing 3.5 kW of electricity and estimated design lives exceeding 15 years. Onboard batteries play critical roles in meeting power requirements of electronic warfare systems, covert and uninterrupted communications, and irregular warfare support to ground forces.

2.5.5.1 Total System Weight and Cost as a Function of Mission Length

Rough estimates [1] for the increase in total system cost and weight as a function of mission duration are summarized in Tables 2.7 and 2.8, respectively. Based on the data presented in these tables, it can be concluded that the constant reliability of the spacecraft power system can be estimated by means of linear interpolation among the reliability figures that correspond to systems containing whole redundant components.

2.5.5.2 Reliability Degradation with the Increase in Mission Duration

Thus far, the emphasis has been placed on the increase in cost and weight as a function of redundant system options. Reliability degradation can occur in spacecraft

Table 2.7 Total System Cost Factor as a Function of Mission Duration and Redundant Options

Mission Duration (mo.)	DET Option	DSR Option	PWM-SR Option
6	1,238	1,314	1,282
12	1,284	1,346	1,349
18	1,340	1,387	1,430
24 (estimated)	1,408	1,420	1,498

Note: These projected values are based on the data collected in the late 1960s for a spacecraft power system. Furthermore, these values show merely a trend in cost factor as a function of various redundant options and mission duration and must not be taken as guaranteed values for other spacecraft or communication or reconnaissance satellite. Electrical load requirements are a function of electrical and electronic sensors and devices operating onboard the spacecraft. Additional electrical power is required for attitude and stabilization control mechanisms, which is significantly greater than the electrical power required for onboard electronic sensors and electrical devices.

Table 2.8   System Weight Projections as a Function of Redundant Options and Mission Duration

Mission Duration (mo.)	DET Option	DSR Option	PWM-SR Option
6	106	116	118
12	112	122	127
18	116	126	139
24 (estimated)	118	130	156

Note: The data presented in this table were collected in the late 1960s for a spacecraft power system. The tabulated values merely demonstrate a trend in increase in weight as a function of redundant options exercised and mission duration and must not be taken as guaranteed values for other spacecraft or communication satellite or tracking satellite. Power required for stabilization and attitude control mechanisms is significantly greater than power required for onboard electronic sensors and electrical devices.

power system as a function of mission duration, in spite of the various redundant system options that have been exercised. A reduction in power system reliability is evident from the data summarized in Table 2.9. The power system reliability will be lowest at the end of the spacecraft operating life.

Even though the reliability of the spacecraft power system decreases with the increase in mission duration, the power system will not experience catastrophic failure unless the spacecraft crashes or sustains serious structural damage. In the case of (CC, DSR) redundant system option (2, 1), the reliability improves to 95.3% from 91.8% for redundant system option (1, 1) regardless of mission length or duration. The power system reliability, however, will continue to degrade with the

Table 2.9   Degradation of Reliability in the Spacecraft Power System due to Redundant System Components and Longer Mission Durations

Mission Duration (mo.)	Redundant System Components			
	(CC, DSR)	Reliability (%)	(CC, BB, DET)	Reliability (%)
6	(1, 1)	97.4	(1, 1, 1)	97.5
12	(1, 1)	91.8	(1, 1, 1)	91.2
	(2, 1)	95.318	(1, 1)	82.1
	(1, 1, 1)	74.5	(2, 1)	89.4
24	(1, 1)	71.4	(1, 1, 1)	64.6

Source: Pessin, L., and Rusta, D., "A comparison of solar-cell and battery-type power systems for spacecrafts," *IEEE Transactions on Aerospace and Electronic Systems*, © 1967 IEEE. With permission.

increase in mission duration. Option (2, 1) means that two CCs have been integrated into the spacecraft power system, which improves reliability. The addition of extra CCs will increase both the power system cost and weight.

2.5.5.3 Increase in Weight and Cost due to Redundant Systems

Preliminary studies performed by the author indicate that implementation of redundant components in the spacecraft or satellite power systems does improve the system reliability, but it does so at the expense of higher weight and component costs as a function of mission length or duration. An estimated increase in total power system weight and cost as a function of mission duration is evident from the published data summarized in Tables 2.10 and 2.11, respectively.

Table 2.10 Estimated Increases in Total System Weight (lbs.) as a Function of Mission Length and Type of Redundant Systems Deployed

Mission Duration (mo.)	Redundant System Employed		
	DET System	*DSR System*	*PWM-SR System*
6	106	116	118
12	111	120	128
18	116	126	139
24	118	130	156
Increase in weight:	Maximum	Minimum	Moderate

Note: Based on these data it can be stated that the DSR redundant system offers minimum increase in weight. Table 2.10 summarizes the estimated increase in total system cost factor as a function of mission length and redundant systems deployed.

Table 2.11 Estimated Increases in Total System Cost Factor as a Function of Mission Duration

Mission Duration (mo.)	Redundant System Employed		
	DET System	*DSR System*	*PWM-SR System*
6	1,238	1,314	1,282
12	1,284	1,347	1,318
18	1,342	1,387	1,430
24	1,400	1,420	1,504
Increase in cost factor:	162	106	210

On the basis of the tabulated data it can be stated that the increase in cost factor is minimum with the DSR redundant system. These conclusions are valid for this spacecraft power system, and these conclusions indicate a trend in the increase in weight and cost factor as a function of mission duration using various redundant systems. In summary, performance data available from the power systems deployed by various spacecraft and satellites do indicate that redundant systems tend to yield higher reliability as a function of mission duration or length. Furthermore, the power system reliability does decrease with the increase in mission length regardless of redundant system deployed.

2.6 Ideal Batteries for Aerospace and Communications Satellites

This section describes the various battery types and configurations best suited for aerospace and satellite applications. Recent published literature indicates that sealed Ni-H$_2$ and sealed nickel-metal-hydride (Ni-MH) batteries are widely used in aerospace and satellite applications with a major emphasis on repeated charging and discharging cycles. Charging and discharging cycles are critically important in the case of satellite applications because of high reliability requirements, particularly over longer mission durations. Recent published aerospace reports indicate that aging effects in battery cells have been noticed sooner than the gradual loss of ampere-hour capacity, a standard performance criterion for space-based batteries. Extensive research and development activities with a focus on battery aging have been undertaken by Naval Weapons Support Center, Crane, Indiana, and other aerospace battery suppliers of Ni-Cd, Ni-H$_2$, and Ni-MH batteries best suited for aerospace and satellite applications. Some private companies and government research laboratories have done a great deal of life-cycle testing for Ni-H$_2$ and Ni-MH batteries to qualify them for deployment in GEO and LEO communication satellites. Critical performance parameters such as end-of-charge voltage (EOCV), end-of-discharge voltage (EODV), cell pressure, and cell ampere-hour capacity have been obtained on Ni-H$_2$ cells and batteries. Considerable battery performance data have been obtained as a function of temperature and DOD because aging effects are vital in predicting the performance of the space-based batteries. Battery performance decreases as a result of aging effects based on environmental factors, severe space conditions, and continuous use of battery over extended durations. Adverse effects of space radiation and nuclear radiation are more pronounced and have been not discussed in this chapter.

2.6.1 Typical Power Requirements for Space-Based Batteries

Power requirements for space-based or satellite-based batteries are strictly dependent on the type of satellite, such as commercial communications satellites, military

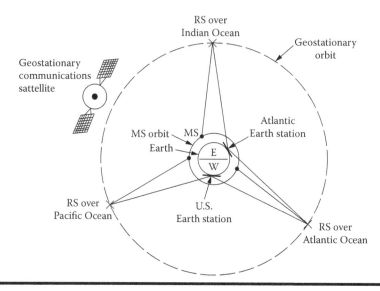

Figure 2.9 Orbital geometry with three relay satellites. (MS: manual satellite; RS: relay satellite.)

covert communications satellites, or Advanced Tracking and Data Relay Satellites (ATDRS; shown in Figure 2.9); the number of transponders used in the communication satellite; onboard antenna gain and uplink and downlink frequency; and stabilization configuration deployed (spin-type or body-mount). Battery power requirements are low for relay satellites, are minimum for space-to-space communications satellites, are moderate for commercial communication satellites, and are maximum for advanced commercial communication satellites and military communications and tracking satellites.

Battery power requirements to some extent depend on the mass of the satellite, which includes the weight of microwave transmitters, receivers, antennas, signal-processing equipment, electronic sensors, onboard electrical appliances, solar panels and associated components, and the stabilization system. Two distinct types of stabilization system design configurations and associated components for satellite control are shown in Figure 2.6. Table 2.12 summarizes the battery power requirement and other critical parameters of commercial and military communications satellites.

The battery power requirements are very high because the communication satellites are complex and are equipped with multiple voice and high data channels in addition to the large variety of electronic and electrical sensors and devices aboard the satellites [5]. Battery power requirements for commercial communication satellites launched before 1980 are moderate, as noted in Table 2.13.

These battery power requirements are moderate because the voice and video channels and the data requirements generally were very low from 1960 to 1980.

Table 2.12   Battery Power Requirements and Other Critical Parameters of Certain Commercial and Military Communications Satellites Launched after 1990 (Estimated Values)

Communications System	Battery Power (W)	Mass of Satellite	Band of Operation
Equatorial-based satellite	2,650	3,150	C and Ku
Thin-route satellite	2,782	3,229	Ka (30/20 GHz)
Skyphone system	2,997	2,862	Ka (30/20 GHz)
ACTS system	2,800	1,875	K and Ka (30/20 GHz)
MILSATCOM system	2,894	2,800	S, C, K, and K

Consequently, the satellite mass and battery power were moderate compared with communications satellites launched later on.

Most early launched communication satellites had an inclination range from 35 to 45 degrees relative to the Equator and orbiting heights generally between 250 to 350 miles, with orbit periods ranging from 94 to 97 minutes. The battery power requirements for a geostationary communications satellite are much higher compared with satellites traveling in other orbits. In general, communication satellites operating in a microwave frequency spectrum will require moderate to high power, depending on the electrical and electronic monitoring sensors and devices operating onboard the satellite and the number of channels dedicated to video, voice, and data transmissions. Battery power requirements will be very high for deep space tracking and surveillance satellites depending on their launch orbits and orbiting heights.

Table 2.13   Battery Power and Frequency Requirements for Satellites Launched before 1980

System Parameters	Frequency Band			
	S	C	X	Ku
RF frequency (GHz)	1.8–2.3	4–6	7–8	12–15
RF transmitter power (W)	150	125	100	50
Battery power output (W)	60	50	40	20

2.6.2 Aging Effect Critical in Space-Based Batteries

Battery designers have suggested the deployment of an impedance spectroscopic technique [4] to determine aging effects regardless of operating and environmental factors. It is critically important to estimate the space-based battery life before the launch of a communications satellite. The battery operational life must be greater than the satellite life to maintain the communications with the prescribed sources. Furthermore, the battery must be continuously charged using solar energy while the satellite is orbiting or parked in the prescribed orbit.

The impedance spectroscopy is an alternating current measurement technique in which the ratio of voltage and current associated with a two-terminal network is measured over a range of frequencies. Normally one is interested in steady-state applications, wherein the transient behavior of the system under investigation has decayed. This particular technique offers a nondestructive method, in which little energy is dissipated by the battery under test, leaving the battery capacity virtually unaffected. This technique is widely applied to electrochemical systems or batteries. Once the EOCV, which gradually increases, and the EODV, which decreases with the number of aging cycles, are known, one can determine the aging effects for the battery cells. In other words, once the EOCV and EODV parameters are known, is easy to predict the remaining age of the battery. These voltages must be measured before a particular battery is assigned to a communications satellite before its launch.

Large-scale accelerated life testing of space-based batteries, including Ni-H$_2$, Ni-Cd, and Ni-HM batteries, is necessary to achieve reliable data on aging effects. Simulations, both real-time and accelerated life-cycle testing data must be achieved on GEO and LEO communications systems. These data will identify early detection of aging effects in the electrochemical cells deployed in the space-based batteries. In brief, this type of information will form the basis for more robust and efficient batteries with high cycle life. Test results on specific batteries as a function of various pertinent parameters are summarized in Table 2.14.

2.7 Performance Capabilities and Battery Power Requirements for the Latest Commercial and Military Satellite Systems

Capabilities and battery power requirements for the latest commercial and military communications satellites will be briefly described in this section, with a particular emphasis on battery power requirements. As mentioned earlier, battery power requirements for military satellites are relatively higher compared with those for commercial satellites. This is strictly due to the deployment of powerful microwave, optical, and infrared sensors for precision tracking, surveillance, and detection of

Table 2.14   Summary of the Significant Features of Various Battery Test Populations

Cell Type	Capacity (Ah)	No. of Cells (Measured)	Storage (yrs.)	Storage Temperature (°C)	No. of Aging (Cycles)	DOD (%)
Ni-H₂	30	1	7	−5–35	10,800	30
	30	9	7	5	1,500	35
	50	18	0	−3–5	62,000	15
	65*	4	0	5	600	15–75
	75†	4	0	0–5	43,000	15
Ni-MH	4†	8	0	22	3,800	37
	8†	2	0	22	3,800	37

Source: Adapted from Smith, R.L., and Bray, A., *Impedance Spectroscopy as a Technique for Monitoring Aging Effects Ni-H₂ and Ni-MH Batteries,* Texas Research Institute Inc., Austin, TX, 1992.

* Accelerated life test data taken in real-time low-earth orbiting simulations by the Eagle-Picher test facility in 1991 in Joplin, Missouri.
† Accelerated life test data taken in real-time geostationary earth orbiting at the same test facility [4].

space-based targets that are of significant interest to space mission and data transfer requirements.

2.7.1 Commercial Communication Satellite Systems

Several communications satellites were launched by the United States, the former Soviet Union, Japan, and European countries between 1970 and 2000. It is impossible to describe the capabilities and battery power requirements for so many satellites within this chapter. Therefore, only select communications satellites will be described with an emphasis on their battery power requirements.

The United States launched an advanced communications satellite known as the INTESAT-IV system in the 1990s, which represented a fourth-generation commercial communications satellite, incorporating the latest radiofrequency and digital components with minimum weight, size, and power consumption. This system offers specified communications service requirements and can be used in support of future manned space flight missions. Frequency operating bands and secondary battery power requirements for this particular communications satellite are summarized in Table 2.15.

These power estimates assume a traveling wave tube amplifier (TWTA) efficiency of 30% for S-, C-, and X-band units and 25% for the Ku-band unit. To

Table 2.15   Operating Bands and Battery Power Requirements for the INTESAT-IV System

	Frequency Band			
System Parameters	S	C	X	Ku
RF power (W)	150	125	100	50
Battery power (W)	450	375	300	400
Solar array output (W)	2,700	2,250	1,800	900

charge the secondary batteries to their full output capacities, the solar arrays must be designed to the power ratings specified in row three of Table 2.15.

2.7.1.1 Performance Capabilities of the Commercial Communications Satellite System

This particular satellite communications system can permit relaying communications between an earth satellite (ES) system and the manned space (MS) system. The INTESAT-IV was launched shortly after 2000. The system offers several voice channels, secured communication lines, and high data transfer capability. Its upgraded version was launched around 2004 with significant improvements in voice, video, and data channels. The improved communications satellite deploys high-efficiency and compact continuous wave-TWTAs with significant reduction in phase noise. The total battery power requirements are still less than 525 W. The secondary batteries are recharged by the solar array with power ratings ranging from 5 kW for television satellites to 12 kW for television broadcasts and data transmissions, as illustrated in Figure 2.10 [6]. These satellites were launched between 1965 and 1980. Communications satellites launched thereafter deployed solar arrays with ratings ranging from 10 kW to 25 kW, depending on the number of voice, video, and data channels; the launch orbit and orbiting height; and mission duration. The solar panel or array power output level can be optimized using one-dimensional tracking technology and concentrating devices, if additional cost and complexity are acceptable [6].

In general, the rechargeable batteries must be capable of providing direct current (DC) power levels of 50 to 100 W for space-to-space communications links, 100 to 350 W for ground-to-space communications links, and 350 to 500 W for space-to-ground communications links. These power levels may increase for military satellites because of the deployment of various electrical, electronic, and electro-optical devices and sensors deployed by the satellite. Total power requirement could further increase depending on the reliability requirements using redundant systems and mission duration. Reduction in DC power consumption can be minimized using alternate voice and data channels.

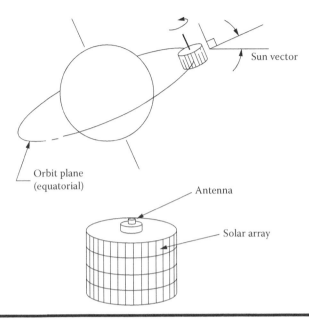

Figure 2.10 Schematic representation of orbital characteristics and solar array location for a spinning satellite. (Adapted from Smith, R.L., and Bray, A., *Impedance Spectroscopy as a Technique for Monitoring Aging Effects Ni-H$_2$ and NiMH Batteries*, Texas Research Institute Inc., Austin, TX, 1992.)

2.8 Military Satellites for Communications, Surveillance, Reconnaissance, and Target Tracking

Several communications satellites have been launched by Department of Defense (DoD) and NASA over the past four decades. DoD has launched numerous surveillance, reconnaissance, and target tracking satellites since 1965 or so. Specific details on performance and functional capabilities on DoD-launched military satellites sometimes are not readily available. In general, DoD communications satellites have several voice channels and high data transmission capabilities. The military satellite to meet surveillance, reconnaissance, and target tracking requirements is launched in GEO (see Figure 2.10) and at optimum orbital height. The objective of launching a satellite in geosynchronous orbits (GSOs) is to achieve optimum overall performance in terms of tracking accuracy, longevity, and reliability. The efficiency, reliability, and longevity of the solar system is of paramount importance for an MS system. Such spacecraft must carry powerful and efficient batteries because various electronic, electro-optical, mm-wave, and microwave systems are operating to maintain critical functional and performance capabilities during the mission.

2.8.1 Military Communications Satellites and Their Capabilities

Published literatures indicate that about 30 simple spinning-type satellites were launched by the United States, which were placed in near synchronous orbits for specific reasons. Defense Satellite Communications Systems (DSCSs), such as DSCS I, DSCS II, DSCS III, and DSCS IV, were deployed to provide communications and telecommunications services to U.S. armed forces. Continuous improvements are made in the satellite design configurations by incorporating state-of-the-art antenna and sensor technologies. This leads to significant improvements in performance, weight, reliability, and power consumption. MILSATCOM (military satellite communications) is the communications satellite system capable of meeting the military communications and telecommunications needs.

2.8.1.1 DSCS-III Communication Satellite System

The DSCS-III satellite offers six covert communications channels and is equipped with spatial user-jammer discrimination capability, which sometimes is known as antenna nulling capability, to achieve high discrimination performance. According to the published literature, two channels in this satellite system deploy high-efficiency 40 W TWTAs and the remaining four channels each use only 10 W TWTAs. This means that the total RF output of the TWTAs is 120 W. Assuming a DC-to-RF efficiency of 40% each, the DC input power required for the TWTAs will be close to 300 W. The batteries must supply this amount of DC power. In addition, additional DC power is required to operate the antenna stabilization mechanisms and the various electronics, electro-optic, and microwave sensors onboard the satellite. The solar panels must be capable of charging the batteries to meet all the DC power consumption.

In the case of multicarrier operation, back-off power is required, which tends to lower the DC-to-RF efficiency for the TWTA units to around 25%. Under such operation, the secondary batteries must be able to supply DC power in excess of 1,200 W for the TWTAs alone. Low-power TWTAs are deployed in DSCS-III systems because of high-gain antennas and the integration of advanced technologies in the receivers.

2.8.1.2 Power Generation, Conditioning, and Storage Requirements

The stated performance requirements that include the reliability for the DSCS-III satellite are clearly defined in this section for power generation, PC, and storage devices. Specific details for these requirements are explicitly defined under each category.

2.8.1.2.1 Power Generation

The solar-array output power is reduced after each year of operation in space regardless of satellite category. To explain the solar-array power degradation, the output power of the solar array for a communications satellite with operational life of seven years is assumed to be 650 W in the beginning of the first year of operation. This power output is reduced to 610 W after one year of operation, to 570 W after two years of operation, to 548 W after three years of operation, to 525 W after four years of operation, to 510 W after five years of operation, to 480 W after six years of operation, and finally to 450 W after seven years of operation.

Studies performed by the author on solar cells [6] reveal that the degradation of the silicon solar cells is caused by the defects in the semiconductor structures resulting from electron and proton bombardment in the space environments. The studies further reveal that the major reduction in power output occurs in a GEO from protons during solar flares. It is estimated that the normal increase of solar-array output power is about 9% during equinox at either of the two times each year, namely on March 21 and September 23. Eclipses occur during this time when the sun crosses the Equator (day and night) and, consequently, more power is required for battery recharging, which comes from the solar arrays. The battery recharging power may be up to 25% of the equinox sunlight power requirement, and therefore, the 9% increase in natural power may not be sufficient. Under these circumstances, additional solar cell area is necessary to meet the recharging power for the battery. In brief, the equinox power condition becomes the most critical solar-array design. Space scientists believe that the degradation of the solar-array output can be matched to the degradation of the payload capacity, which can limit the channels to be deployed.

2.8.1.2.2 PC Function

PC function plays a key role in all spacecrafts regardless of the spacecraft category. The power available from the solar arrays depends on the normal array area for the incident solar radiation and the extent of irradiation degradation. Furthermore, the solar cell temperature is strictly dependent on the array temperature, which could vary from –180° to +60°C during the eclipse period, which can exceed the normal operating voltage up to 2.5 times the normal value. Therefore, regulation of the array voltage is necessary to avoid overvoltage conditions. PC can be achieved through the proper use of voltage regulation and unregulated bus configuration to keep the voltage variation within 1 to 2%.

2.8.1.2.3 Energy Storage Devices

Brief studies conducted by the author indicate that the geostationary satellite experiences 90 eclipses during one year on the station with a maximum eclipse duration of 72 minutes per day. This leads to a relatively low number of charge-discharge cycles of the energy storage system (i.e., battery). This allows for the application of a high DOD. This leads to 50 to 70% of the battery capacity being used during each discharge cycle compared with 10 to 20% for low-orbit satellites, which have thousands of cycles per year. Under these operating conditions, energy storage devices or batteries must have high reliability and long lifetime compared with the solar cell devices. Ni-Cd batteries have demonstrated high reliability and longevity for space applications. Because of improved overall battery performance and use of advanced electrode materials, the Ni-Cd batteries are widely used for satellite applications. The battery power requirements are at least 3.8 kW.

2.8.2 MILSATCOM System

The MILSATCOM system offers covert, reliable, and interconnecting services to military users covering wide geographic regions. This satellite meets high data transmission rates involving beyond line-of-sight (LOS) mobile services to military aircraft, helicopters, ships, and man pods. Both the DSCS IV and MILSATCOM offer reliable and covert communications functions, serving strategic and tactical communications requirements. The system provides a high degree of antijam, nuclear scintillation protection, and physical survivability under harsh space environments. Its battery-bank power requirements are greater than 3.75 kW, and solar panels must be designed to deliver output power exceeding 12.5 kW for a seven-year satellite life and exceeding 15 kW for a 10-year satellite life.

2.8.3 European Communications Satellite System

The European communications satellite system (EUROPSAT) offers communication and telecommunications services for the European countries and North Atlantic Treaty Organization (NATO) forces. The system provides covert communications services in particular to NATO armed forces. This system was launched in 1994 and provides telecommunications and data transmission services over the entire European continent. TWTAs with output power ratings of 580 W and 620 W constantly charge the onboard batteries by the large solar panels. The satellite operating life is about seven years. Intersatellite relay can be accommodated at mm-wave frequencies (36–38 gigahertz [GHz]) with minimum cost and complexity. In every communications satellite, once the satellite is on the designated station, its attitude must be held fixed to keep the antenna beams directed as desired. As stated, attitude control and stabilization of every satellite or spacecraft are of paramount importance despite the adverse effects of gravity gradient, earth's magnetic field, solar radiation, and uncompensated motion generated by disturbing forces acting on the satellite. To eliminate the effects of uncompensated motions, some sort of stabilization mechanism is required, which is accomplished through spinning the satellite in the orbit at a rate of 30 to 100 RPM. Spin stabilization means that a given solar panel, if effectively illuminated by the sun only $1/\pi$ or 32.8%

of the time, causes the battery power to be only $1/\pi$ of the value it would have been if the solar cells were not spinning. This particular issue must be taken into account during the battery design selection. Larger solar panels may require dual-spin configuration. Even higher solar panel output might require three-axis stabilization technique to achieve higher power output from the solar panels to charge the onboard batteries. This satellite communication system uses Ni-Cd batteries to meet DC power requirements of 2.5 kW, which would require a solar panel power rating exceeding 20 kW.

2.9 Batteries Best Suited to Power Satellite Communications Satellites

This section identifies rechargeable or secondary battery requirements best suited for communications and surveillance and reconnaissance satellites. The battery power requirements are strictly dependent on several factors, including launch orbits such as LEO, elliptical, or GSO; orbital height; the type of stabilization technique used (i.e., mono-spin, dual-spin, or three-axis configuration); satellite operational life; attitude control system; and the overall DC power requirements needed to power the electronic and electrical subsystems, the electro-optical and microwave sensors, and the attitude and stabilization control mechanisms.

2.9.1 Rechargeable Batteries Most Ideal for Communications Satellites

Satellite that deploy Ni-Cd batteries have the power-specific mass at 70% discharge, which is about 12 watts/kilogram (W/kg) when calculated for the maximum duration eclipse. The mass and inefficiency of charge and discharge cycles will reduce the power-specific mass to below 10 W/kg for the Ni-Cd batteries. Published literatures claim that Ni-H$_2$ and silver-hydrogen (Ag-H$_2$) batteries could offer a mass saving, ranging from 30% to 60%, respectively, compared with Ni-Cd batteries. Saving in power-specific mass is of paramount importance, because it offers significant reduction in launch cause, improves the satellite longevity, and enhances the overall reliability of the satellite. Selection of battery banks for a space program is the most critical factor, and therefore, all pertinent issues must be carefully examined including cost, reliability, and operational life.

2.9.1.1 Performance Capabilities of Ni-Cd Rechargeable Batteries for Space Applications

The spacecraft or satellite power system is composed of a solar-cell array and a rechargeable battery in conjunction with a voltage regulator and a battery charge current. As far as the rechargeable storage batteries are concerned, Ni-Cd batteries have been used in several space-based systems over the past two decades. Battery designers indicate that Ni-Cd batteries can be safely operated in the temperature range from −5° to +10°C. The storage temperature is critical because a majority of the battery life in orbit is in the storage mode. The output of the solar panels is used only to recharge the batteries when their output power drops below the safe value. Battery stress must be avoided particularly for long missions. Furthermore, during the eclipse season, about 23 h are available for recharging the batteries regardless of the type of batteries. Performance requirements of such batteries will be described with particular emphasis on charge and discharge rates and energy capacity at lower ambient temperature. The safe upper limit of cell voltage of 1.55 volts (V) is necessary to avoid overcharging the Ni-Cd cell voltage. SOC and DOD characteristics of Ni-Cd batteries are most attractive for space applications. Ni-Cd batteries can be discharged to 100% DOD capability. This is a unique performance parameter of this battery. Typical charge method requires constant current, followed by tickle charge. Fast charge is preferred to limit the self-discharge and can be completed within 1 h. Slow-charge method can be used, but it requires 14 to 16 h. Slow charge is not recommended when rapid and reliable battery performance is the principal requirement. The storage requirements for this battery are fairly simple. The battery can be stored at 40% SOC. This particular battery can be stored for five years or more at room temperature or below with no compromise in battery performance. Charge, discharge, and storage conditions for Ni-Cd rechargeable batteries are impressive. Its nominal voltage is about 1.2 V. Principal advantages of this battery include its high-efficiency charge, its moderate charge cycle rate, and its ability to be recycled easily with minimum cost and complexity. Earlier versions of this battery indicated the adverse effects, namely the memory and toxic effects. But the latest Ni-Cd battery designs have minimized such effects. In the past decade, several space-based programs have deployed such batteries because of impressive portability, improved reliability, and high-efficiency features of this battery. Communications satellite designers believe that the battery consumption could be minimized using alternate voice and data channels. High-data transmission satellites generally require more power from the batteries. In the case of MS systems, a minimum use of lighting, heating, air conditioning, and other high-power-consuming appliances, if possible, will significantly reduce battery power consumption. Electronic and electrical sensors that monitor the critical mission parameters of the MS system must continuously receive the required power from battery banks [7].

2.9.1.2 Performance Parameters of Ni-H$_2$ Batteries

Space program managers have seriously considered Ni-H$_2$ batteries for communication satellites. These batteries were originally developed for aerospace applications and have been in continuous development since the early 1970s for satellite applications. This cell utilizes hydrogen as a negative electrode, and the positive electrode is nickel-oxide-hydro-oxide (Ni-O-OH). The hydrogen electrode consists of a thin-film of platinum (Pt) black catalyst supported on a nickel foil substrate, backed by a gas diffusion membrane. The preferred Ni electrode consists of a porous sintered nickel power substrate, supported by a nickel screen and electrochemically impregnate with nickel hydro-oxide ($Ni(OH)_2$). The separator is a thin and porous zirconium-oxide (ZrO_2) ceramic cloth, which supports a concentrated potassium-oxide (KOH) solution.

Ni-H_2 cells are manufactured in various design configurations. The individual pressure vessel (IPV) cell contains an electrode stack in a cylindrical pressure vessel contained within a common pressure vessel (CPV) cell. Two stacks in a series in a vessel yield 2.5 V. In a single-pressure vessel (SPV) battery, a number of cells (typically more than 20) are connected in a series and are placed in a single vessel.

This type of cell is widely used for GEO satellites and LEO satellites. These cells are manufactured by Eagle-Pitcher Technology Corp. in the United States. In both cases, primary power is supplied by solar panels. When orbit brings the satellite into the earth's shadow, which represents an eclipse, the battery starts to deliver electrical energy to various electrical components and electronic and electro-optic sensors. The solar panels will subsequently recharge the secondary batteries during the sunlit periods.

A 16 ampere-hour (Ah) Ni-H_2 battery composed of 11 CPV cells has been used in LEO satellites as well as in the Hubble Space Telescope. These batteries are operating continuously with no compromise in electrical performance and reliability. The Ni-H_2 batteries have been used in several planetary missions, especially to Mars, and have demonstrated reliable performance under harsh space environments. The battery's deployment in terrestrial applications is limited, however, because of the high initial cost of such batteries as well as additional drawbacks discussed in the next paragraph.

Low volumetric energy level, high thermal dissipation at high rates, and safety hazards are major drawbacks of this battery. In addition, to meet longevity requirements, the Ni-H_2 battery has to work in a tight temperature range, preferably from –10° to +15°C. Self-discharge for this particular battery is rather fast even at +10°C. Battery capacity loss could be as high as 10% after three days. This means that this battery is suited for very short missions.

2.9.1.3 Performance Capabilities of Ag-Zn Batteries

Since the early 1990s, few battery suppliers have been actively engaged in the research and development activities for Ag-Zn rechargeable batteries. These batteries have been highly recommended for space applications. Ag-Zn batteries are produced with prismatic form factors incorporating flat electrodes wrapped with multiple layers of separator. This particular battery is highly preferred because of its high specific power. Conventional high-rate cells typically yield specific power between 1.5 and 1.8 kW/kg, which could be increased to 3.7 to 4.3 kW/kg by using thin electrodes and thin separators. This specific power can be further improved to 5.5 kW/kg using bipolar electrodes. This high specific power capability is especially exploited for space applications, where high specific power, small size, and low weight are of critical importance. The Atlas-V launch vehicle used for the Mars Reconnaissance Orbiter mission in 2005 deployed 28 V, 150 Ah Ag-Zn battery banks. These batteries have been used by the astronauts of several missions in their extravehicular activities. Secondary Ag-Zn batteries can be used in portable applications, such as medical equipment, television cameras, and remote communication systems, by virtue of their high energy and specific power.

2.9.1.4 Space Applications of Lithium-Ion Batteries

Since 2000, significant research and development efforts have been directed toward improving the performance improvement of lithium-ion (LI) batteries. The latest production tests on these batteries seem to predict their suitability for electric car and satellite applications. These batteries are best suited for heavy-duty use. High nominal voltage (3.7 V), the absence of memory effect, and low-self discharge are the principal advantages of LI rechargeable batteries. Major drawbacks include their high cost, some safety problems, and battery performance reduction at high temperature and control requirements for charge-discharge limits. Despite these drawbacks, LI batteries are highly recommended for portable electronic and electrical components, military and space applications, several consumer devices, power tools, and electrical vehicles and hybrid electric vehicles.

2.10 Conclusion

This chapter identified various rechargeable batteries best suited for aerospace and satellite applications. Performance capabilities and major drawbacks are briefly described for each rechargeable battery. Battery power requirements are summarized for communication satellites and surveillance-reconnaissance-tracking satellites orbiting in GEO and LEO. Battery performance requirements for short and long mission durations are identified, pointing out the reduction in battery power output as the mission duration increases. Performance requirements for onboard rechargeable batteries during the dark period known as satellite eclipse experienced by the satellites are identified. Solar-array power output requirements to charge the rechargeable batteries whenever required are discussed in detail. Whenever the battery-rated power falls below the designated level as a function of operational life, the solar arrays must be designed to compensate for the loss in battery capacity. Estimated loss in battery capacity as a function of operational life is specified. Preliminary computations indicate that a 650 W battery drops to 93.5% after one year, 92% after two years, 80.6% after five years, 55.2% after eight years, and ultimately to 35% after 10 years. As discussed, the solar arrays must be overdesigned to compensate for the drop in battery capacity as a function of time. The power system aboard a satellite or a spacecraft consists of a solar array, rechargeable battery, and the associated electrical components, such as a voltage regulator and battery charge converter. Potential power generation, PC, and storage scheme performance requirements are identified. Performance requirements of these components are specified, with an emphasis on reliability and longevity. Effects of temperature on the battery capacity are described. Modeling data indicate that the battery capacity is about 57% at 23°C, 68% at 34°C, 82% at 50°C, and 100°C. Effects on battery performance due to charge-discharge variations are identified. Variations in OCV and internal battery resistance as a function of SOC for lithium-ion cells are specified. Spacecraft power system reliability is discussed in great detail, with an emphasis on power system component failure rates. Reliability improvement of spacecraft power system as a function of mission duration is identified using CC and PWM techniques. Power requirements for satellite attitude control and stabilization mechanism are briefly mentioned. Estimated weight and cost penalties resulting from the deployment of redundant systems are summarized in detail as a function of mission duration and type of redundant systems deployed. These redundant systems considered include the DET, DSR, and PWM-SR. Critical roles of boost and charge regulators are highlighted to achieve high charging with enhanced reliability. Electrical power requirements from solar arrays to recharge the batteries in various orbital orbits are summarized. Important electrical performance parameters of $Ni-H_2$, Ni-MH hybrid, and polymer-based LI rechargeable batteries are specified. Rough power estimates for rechargeable batteries for possible deployment in military and commercial communications satellites are provided for microwave downlink and uplink operating in various frequency bands.

References

1. Leo Pessin and Douglas Rusta, "A comparison of solar-cell and battery-type power systems for spacecrafts," *IEEE Transactions on Aerospace and Electronic Systems* Vol-AES-3, no. 6 (November 1967), p. 889.
2. A. R. Jha, *Solar Cell Technology and Applications,* Boca Raton, FL: CRC Press (2010), p. 200.
3. R. Vondra, K. Thomassen et al., "A pulsed electric thruster for satellite control," *Proceedings of the IEEE* 59, no. 2 (February 1971), p. 271.
4. R. L. Smith and Alan Bray, *Impedance Spectroscopy as a Technique for Monitoring Aging Effects Ni-H₂ and NiMH Batteries,* Austin: Texas Research Institute Inc. (1992), p. 156.
5. J. R. Balombi, D. L. Wright et al., "Advanced Communications Technology Satellite (ACTS)," *Proceedings of the IEEE* 78, no. 7 (July 1990), p. 1174.
6. A. R. Jha, *Solar Cell Technology and Applications*, Boca Raton, FL: CRC Press (2010), p. 181.
7. S. H. Durrani and David Pike, "Space communications for manned spacecraft," *Proceedings of the IEEE*, 59, no. 2 (February 1971), p. 129.

Chapter 3

Fuel Cell Technology

3.1 Introduction

Demand for a long-duration, portable power source has grown rapidly in the past few years. Furthermore, uninterrupted power requirements for high-power electrical components and electronic devices used in battlefield environments cannot be met by conventional batteries for long military missions. In addition, conventional high-power lithium-ion batteries suffer from weight, size, reliability, discharge rates, disposal issues, and recharge capacity. In light of these issues, manufacturers of portable electrical power sources are increasingly leaning toward fuel cells to replace the lithium-ion batteries. Fuel cells generate electrical power by an electrochemical conversion technique, which can be replenished with minimum time and effort. The latest studies performed by various fuel cell scientists indicate that direct methanol fuel cells (DMFCs) are best suited for portable, high-power applications, because the DMFC technology offers the most promising, practical solutions and unique benefits, such as compact form factor, improved reliability, and significant reduction in weight and size. Furthermore, the methanol fuel is widely available without any restriction. A fuel cell is a system that combines an oxidation reaction and a reduction reaction to produce electricity.

Comprehensive research and development activities were undertaken by European scientists in 1990 using high temperature and semisolid electrolytes [1]. The Bacon HYDROX fuel cell was designed to operate at medium temperatures and high pressures. Electrochemical energy converters were designed to operate at ambient temperature and pressure. German scientists developed the double-skeleton catalyst (DSK) fuel cells using a liquid carbonaceous fuel such as methanol and catalytically different electrodes, while the Swiss scientists designed the monoskeleton catalyst (MSK) fuel cells using cheap fuel (hydrocarbons) and electrochemically active metal electrodes.

Certain terms are generally used in dealing with fuel cells. The most common terms used are as follows:

- *Anode*: This is the negative or fuel electrode that gives up electrons to the external circuit. The hydrogen is oxidized in the process.
- *Cathode*: This is the positive or oxidizing electrode that accepts electrons from the external circuit and the oxygen is reduced in the process.
- *Membrane electrode assembly (MEA)*: This is a laminated sandwich of two porous electrodes separated by an ion-conducting polymer electrolyte. The catalyst is a part of MEA assembly.
- *Proton exchange membrane (PEM)*: A polymer film is used to block the passage of gases and electrons while allowing the hydrogen ions known as protons to pass.
- *Reformer*: A small chemical reactor carried on board some fuel cell vehicles used to extract hydrogen from alcohol or hydrogen fuel.

3.1.1 Classifications of Fuel Cells

In the 1960s and 1970s, three distinct fuels cells were designed, developed, and evaluated mostly by scientists and engineers in the United States and European countries. But their performance parameters, such as current densities, terminal voltages, and continuous operating hours, were marginal. Early developed fuel cells can be briefly described as follows.

3.1.1.1 Aqueous Fuel Cell Using Specific Electrolyte

The first fuel cell of this type was developed by German scientists and professors working at the Braunschweig Institute of Technology. The key feature of this fuel cell is the DSK electrode. The device consists of two electrodes: one to provide structural support and the other to provide high electrical conductivity. The catalytically active skeleton is embedded into the supporting skeleton. The fuel electrode is made from hot-pressing nickel (Ni) under controlled conditions. Optimum cell performance of fuel electrode is possible using pure hydrogen. Current densities as high as 400 mA/cm^2 have been achieved at ambient temperature (65°C) and low pressure (5 psi). A four-cell device demonstrated ultra-high reliability over a continuous operation exceeding one full year. Both fuel and oxidant electrodes can be made using catalyzed porous nickel. The aqueous fuel cell uses alkaline electrolyte, hydrogen as fuel, and oxygen or air as the oxidant.

3.1.1.2 Fuel Cells Using Semisolid Electrolyte

Fuel cells using semisolid electrolyte consist of porous magnesium oxide (MgO) with a mixture of sodium (Na), potassium (K), and lithium carbonate ($LiCO_3$).

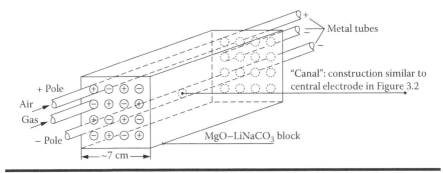

Figure 3.1 **High-power fuel cell consisting of MgO–LiNaCO₃ semisolid electrolyte and containing metal tubes as electrodes for air and fuel gas.**

The cell is essentially an MgO disk impregnated with a mixture of slats. Specific details on various elements of this fuel cell are shown in Figure 3.1.

3.1.1.3 Fuel Cells Using Molten Electrolyte

The fuel cell configuration with tubular cells uses fine-grain solid MgO electrolyte and molten electrolyte. The electrode surfaces are coated with metallic powders, such as silver (Ag) for the air or oxygen cathode and iron (Fe), Ni, or a zinc oxide/silver (ZnO/Ag) mixture for the fuel anode. The structural details of a fuel cell with tubular configuration are illustrated in Figure 3.2. This particular fuel cell

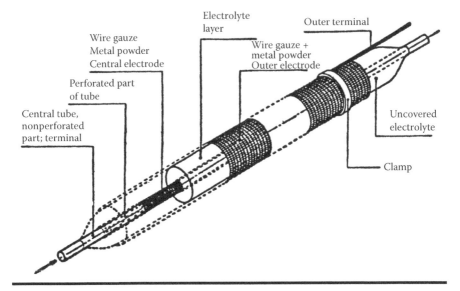

Figure 3.2 **Critical elements of a high-capacity fuel cell using an electrolyte paste consisting of fine-grain solid MgO and molten electrolyte for optimum performance.**

Table 3.1   Electrical Performance of a Fuel Cell with Tubular Configuration and Using Various Fuels

Current Density (mA/cm²)	Polarization Voltage (V) with Fuel Used		
	Hydrogen (H₂) (65%/35%)	Methanol (CH₃) (13%/87%)	Methane (CH₄) (65%/35%)
0.2	0.90	0.56	0.20
0.4	0.82	0.36	0.07
0.6	0.63	0.25	0.05
0.8	0.52	0.19	0.02
1.0	0.35	0.11	0.01
1.2	0.25	0.05	—
1.4	0.20	0.00	—

is capable of offering a current density of 100 mA/cm² at a polarization voltage of 0.7 volts (V). Even at high current densities, hydrogen exhibits very low polarization. In most low-temperature fuel cells, carbon dioxide and water produced in a high-temperature fuel cell are evolved in the fuel rather than in the electrolyte. This causes a dilution of fuel, which makes it difficult to achieve both the high current density, high fuel utilization without obtaining strong polarization. The electrical performance of a fuel cell with tubular configuration operating at 700°C, using semisolid electrolyte, flaked Ni anode, Ag cathode, and various fuels, is summarized in Table 3.1.

Upper percentage values are associated with the fuels used, and the lower percentage values indicate the carbon-dioxide gas produced from the electrochemical reaction. For example, in the case of hydrogen fuel, 65% is the hydrogen, whereas 35% is the carbon dioxide. In another example of methane fuel, the percentages of fuel to carbon dioxide are identical, but the polarization voltage is extremely low.

3.1.2 Classifications of Fuel Cells Based on Electrolytes

Just before and after the World War II, design engineers and material scientists extensively focused their efforts on fuel cells to improve the overall performance of the devices. In doing so, they concentrated their design efforts on the development of fuel cells using various types of fuels, electrode configurations, and electrolytes. Using all these disciplines, the scientists have classified six distinct types of fuel cells, which are identified as follows:

■ Alkaline fuel cell
■ Phosphoric acid fuel cell

- Molten-carbonate fuel cell
- Solid-oxide fuel cell (SOFC)
- Solid-polymer fuel cell (SPFC)
- DMFC

High-temperature fuel cells using hydrogen fuel offer optimum performance irrespective of current density in terms of moderately higher efficiency, reliability, and operational costs. In addition, methane does not participate directly in the electrochemical reaction. Both methanol and methane yield poor performance when the current density exceeds 0.6 mA/cm^2.

3.2 Performance Capabilities of Fuel Cells Based on Electrolytes

Practical fuel cells designed and developed during the 1960s and 1970s focused on three types of fuel cell design configurations using various electrolytes that will be described in this chapter, with an emphasis on performance, reliability, and longevity. Three distinct fuel cell designs are available depending on the types of electrolytes used by the cells. Potential advantages and disadvantages of each type are briefly described with an emphasis on performance on efficiency, safety, and reliability. The three distinct fuel cell types are as follows:

- Those using semisolid molten electrolytes
- Those using solid electrolytes
- Those using aqueous electrolytes

According to published literature [2], various scientists initially designed, developed, and tested 10 aqueous electrolyte cell systems, six molten electrolyte cell systems, and only three solid electrolyte cell systems. As mentioned previously, the duel cell is an energy conversion device in which the chemical energy is isothermally converted into direct current (DC) electricity. In addition, it can convert the chemical energy into the electricity without involving the thermodynamic relation demonstrated by the Carnot cycle to limit the efficiency of heat engines.

3.2.1 High-Temperature Fuel Cells with Semisolid Molten Electrolyte

High-temperature fuel cells contain molten salt electrolytes, usually a mixture of salts such as Na, K, and LiCO$_3$. The electrolyte acts electrochemically as a liquid melt consisting of porous MgO with a mixture of Na, K, and LiCO$_3$. The MgO disc is impregnated with a molten salt. The electrode surfaces consist of metallic powder, such as Ag for the air or oxygen cathode and Fe, Ni, and ZnO/Ag mixtures for the fuel anode. The perforated sheets are firmly pressed on the powder to hold the metallic powder tightly against the electrolyte and also to conduct the current, but they do not serve as electrodes as illustrated in Figure 3.3. At high operating temperatures (500–750°C), substances are usually difficult to react electrochemically to yield high current density close to 100 mA/cm² and polarization voltage of 0.7 V, with noticeable polarization only on the carbon monoxide (CO) side but not on the air or oxygen side with an Ag electrode. As stated previously, hydrogen produces electricity in these fuel cells, but this is probably due to the formation of immediately preceding thermal decomposition or formation of CO in the presence of steam. The operating efficiencies of these cells range from 30 to 35%, which is about half of the efficiencies possible with primarily cells using such fuels as coal, natural gas, and petroleum [2]. On the basis of the technical literature published in the 1970s, scientists were hesitant to pursue extensive research because of many technological difficulties associated with high-temperature fuel cell technology. Instead, the fuel cell designers focused on low-temperature cells.

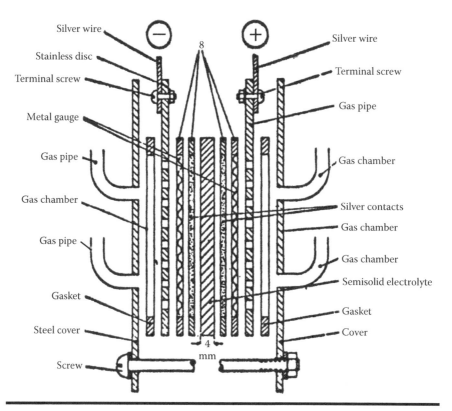

Figure 3.3   High-temperature fuel cell with semisolid electrolyte and its critical components.

Table 3.2 Polarization Voltage as a Function of Temperature and Current Density for a Fuel Cell Using Semisolid Electrolyte (V)

Current Density (mA/cm²)	40°C	80°C
50	0.04	0.02
100	0.08	0.04
150	0.14	0.08
200	0.17	0.10
250	0.22	0.14
300	0.30	0.18

When operating under mild temperature and pressure environments, DSK electrodes would require less weight and size without compromise in cell electrical performance and reliability.

Typical electrical performance of a fuel cell using semisolid electrolyte in a tubular configuration, flaked Ni anode, Ag cathode, and DSK electrodes is indicated from the data summarized in Table 3.2.

The data presented in Table 3.2 show how 100% Faraday efficiency, i.e., complete hydrogen utilization, could be reached with little sacrifice of polarization and limiting current density by optimizing the fine-pore layer coating. Finer pores on the electrolyte are required to prevent gas leakage. The Faraday efficiency is known as the gas consumption efficiency to scientists and engineers working in the electrochemical field.

3.3 Low-Temperature Fuel Cells Using Various Electrolytes

In this category, hydrogen-oxygen (H_2-O_2) fuel cells operating at ambient temperatures were given serious consideration. The H_2-O_2 fuel cell is known as the HYDROX cell. Scientists believe that among all possible electrochemical fuels, hydrogen is the most desirable fuel because it reacts very fast; its reaction by-product is water, which does not corrode the electrodes; and it is useful as a solvent for the electrolytes. In addition, the hydrogen molecule can be easily dissociated and ionized into protons by chemisorptions and subsequent desorption from a simple catalyst such as commercial Ni at temperatures more than 200–250°C. Over this moderate temperature range, this particular cell can generate current densities ranging from 1,200 mA/cm² at 0.7 V to 2,000 mA/cm² at 0.46 V. The HYDROX fuel cell is considered the most powerful modern fuel cell. Critical components of the HYDROX fuel cell are shown in Figure 3.4, and the device uses semisolid

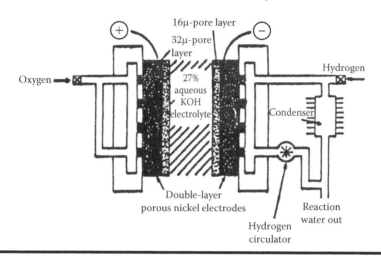

Figure 3.4 High-pressure H₂-O₂ fuel cell using double-layer Ni electrodes and operating at 200°C and 600 psi.

electrolyte. The operating temperature and pressure are 200°C and 600 psi, respectively. Despite its high current density and high power capacity, the device suffers from a relatively long preheating time, high operating pressure, and ultra-high purity requirements for hydrogen and oxygen, which can significantly increase the operating costs. Significant design improvements in the HYDROX fuel cell have been made by Pratt and Whitney Aircraft Company of the United States.

3.3.1 Performance of Low-Temperature and Low-Pressure Fuel Cells Using Aqueous Electrolyte

In this particular fuel cell, the electrons liberated during the oxidation of the fuel pass through an external circuit and generate electricity. The merit of this type of cell lies in the fact that the fuel oxidation is done electrochemically with minimum dissipation of energy through random heat loss. But the most distinct disadvantage is that the power output of this cell is materially degraded by use of any fuel other than pure hydrogen. In addition, the carbon oxide produced in the fuel oxidation will react adversely with the alkaline electrolyte, which will not only use up the electrolyte but also could clog electrodes pores with a carbonate salt, thereby reducing the reliability and efficiency of the fuel cell.

The key feature of this fuel cell is known as the DSK electrode, because it consists of two metallic skeletons—one is substrate and structural, and the other is active and highly dispersed skeleton. The catalytically active skeleton is embedded into the supporting skeleton that is electronically highly conductive. The fuel electrode is made from hot-pressing Ni. In the case of the oxygen electrode, Ag is used to make the active skeleton to form a substrate. The fuel electrode works efficiently on pure hydrogen, and it has demonstrated a current density as high as 400 A/cm^2 at ambient temperature and low pressure. However, a lower current density around 50 A/cm^2 yields high reliability. Assuming this current density, a cell voltage of 0.5 V, efficiency of 50%, and electrode area of 1 square foot (929 cm^2), the electrical power generated by this fuel cell can be determined.

Allis-Chalmers Energy Inc. has designed, developed, and tested several such fuel cells. The company used a fuel that was a mixture of gases. The oxidant was oxygen. The fuel was used to operate a heavy-duty tractor. Electrical power generated by this particular fuel cell = [50 × 0.5 × 0.5 × 929] = 11.5 kilowatts (kW). This sample calculation shows that the power output capability of a fuel cell is a function of electrode area, cell efficiency, electrolyte properties, and cell output voltage. Such a fuel cell with 1 square foot area, efficiency of 60%, polarization voltage of 0.7 V, and current density of 50 A/cm^2 could generate a power output close to 19.5 kW, which could be used by a heavy-duty tractor for various industrial applications.

The company has developed a four-cell power module using a low-temperature (65°C), low-pressure (5 psi) cell that uses hydrogen or ammonia as fuel and oxygen or air as oxidant [3]. This power module has demonstrated a 4,500-h nonstop operation with no failure. The company has developed and marketed 20-cell power modules complete with controls, condensers, and circulation systems. The commercial power module demonstrated a power output of 1 kW for continuous operation. Both the fuel and oxidant electrodes are made from catalyzed porous Ni, and each electrode has a thickness of 0.028 in. The alkaline electrolyte is held in asbestos or similar porous matting with a thickness of 0.030 in. The matting or the asbestos must withstand a differential pressure exceeding 1,000 psi. The fuel cell design requires low-electrode-holder resistance and precision machining, which will distribute fuel or oxidant uniformly to the electrodes for optimum performance.

3.3.2 Output Power Capability of Aqueous Fuel Cells

With hydrogen as fuel and with each cell at a polarization voltage of 0.78 V and with a current density of 130 A/ft.^{2}, this particular fuel cell has a watt per pound ratio of about 22 and a volumetric power density of 1.5 kW/ft.^{3} If the power module has 20 cells and each cell has an output voltage of 0.76 V, a power density of 130 A/ft.^{2}, and a module efficiency of 65%, the total power output will be around 1.32 kW.

With ammonia at 77°C and a caustic electrolyte, a current density of 160 A/ft.^{2} at 0.3 V can be achieved, and an output power per volume ratio can be achieved initially. During the life tests, however, the current density drops to less than 50 A/ft.^{2} at 0.3 V after continuous operation of 700 h.

This fuel cell can be operated with methanol and other alcohols as fuels along with acid and alkaline electrolytes and air or oxygen as the oxidant. Prototype cell designs have demonstrated the cell's continuous power exceeding 1 kW. Such devices have demonstrated current densities better than 80 A/ft.^{3} at 0.3 V and a power output per volume ratio of 0.8 kW/ft.[3], if the electrodes used have lifetimes exceeding 250 h or higher. The cell structure is identical to that of the H_2-O_2 cell.

3.4 Fuel Cells Using a Combination of Fuels

Fuel cell scientists in the United States, the Russian Federation, and Europe have conducted comprehensive research studies and laboratory experiments using a combination of two types of fuels best suited to improving electrical performance and device reliability without increasing manufacturing costs. Fuel cell designers and scientists conducted experiments on pressure-sensitive cell designs [4] consisting of a liquid and a gas. The scientists simultaneously started working on fuel cells using two different types of fuels. Specific details on the types of fuels, electrolytes, and oxidants used to fabricate fuel cells are described in the following sections.

3.4.1 Liquid-Gas Fuel Cell Design

Some scientists have designed fuel cells simultaneously using both liquid and gas (LG). This particular device uses a metallic envelope that is filled with air or oxygen and is dipped into an electrolyte fuel mixture. One side of the envelope is porous Ni and the other side is solid foil. The Ni acts as the air or oxygen electrode, and the catalyst-coated outside surface operates as a fuel electrode. The net result is that an array of such cells can produce a bipolar battery. Preliminary calculations indicate that a module composed of 80 cells could produce output power exceeding 750 watts (W). Higher module power can be obtained by deploying an increased number of cells.

3.4.2 Performance of Liquid-Liquid Fuel Cell Design

Scientists have designed liquid-liquid (LL) fuel cell devices in kit form. This LL fuel cell has a bipolar construction and uses alcohol or ammonia for fuel and hydrogen peroxide or other oxidizing liquids for oxidants. Either the acid or alkali can be used as electrolyte. Then fuel electrode is platinized Ni. The oxidant electrode is made out of Ag for higher efficiency and minimum loss. This device has demonstrated a current density in excess of 50 A/ft.[2] at the cell output voltage of 0.3 V and a cell output power per weight ratio of 3 W per pound. Modules with power ratings of 0.5 kW, 1.0 kW, and 1.5 kW or higher have been designed and developed.

Both the LG and LL cells can be designed for higher current densities and higher polarization voltages. These cells are best suited for battery applications. These cells have demonstrated operating hours in excess of 1,200 h without a change in electrode performance or properties. Batteries composed of up to 80 single cells and producing an output power level of 200 W have been operating for more than 1,000 h with no compromise in performance or reliability.

3.5 Fuel Cell Designs for Multiple Applications

In rare cases, the fuel cell is required to meet specific critical performance parameters such as reliability, power output capability, and longevity. In such cases, the cell designers will try to meet those critical performance specification parameters, while barely meeting other performance requirements. Sometimes a particular application requires that the fuel cell must meet those critical performance requirements under harsh operating environments.

3.5.1 Fuel Cells for Electric Storage Battery Applications

Material presented in this section will be useful in the design and development of fuel cells best suited for electric storage battery applications. Fuel cells will be most useful in applications in which reliability and continuous power output are of critical importance. The Electric Storage Battery Company in Philadelphia was the pioneer in developing the alkaline, low-temperature, H_2-O_2 cells for this particular application. The company used microporous electrodes with high electrochemical activity, excellent mechanical strength, and high electrical conductivity. The electrodes used were not pyrophoric and thus could be exposed to atmosphere without experiencing any damage. These cells have demonstrated continuous operation for more than 25,000 h at a current density of 2.5 mA/cm^2, for more than 6,000 h at a current density of 70 mA/cm^2, and for more than 4,000 h at a current density of 200 mA/cm^2. These continuous operating hours demonstrate clear evidence of high reliability and minimum maintenance requirement. As mentioned, both electrodes are made of porous Ni. The hydrogen electrode contains a palladium-silver catalyst, and the oxygen electrode contains a silver-nickel catalyst. A concentric H_2-O_2 fuel cell has been designed, developed, and tested over thousands of hours. This particular fuel cell has demonstrated a power density in excess of 1 kW/ft.^{2} with no compromise in reliability and electrical performance.

3.5.2 DSK-Based Fuel Cells Using Hydrogen-Based DSK Electrodes and Operating under Harsh Conditions

A fuel cell known as the DSK system can operate using solid electrolyte, such as highly brittle Ni alloy. This Ni alloy is prepared by melting 50% aluminum with 50% Ni by weight. The residual grains found in the metallic powder consist of Ni with a large internal surface. The metallic powder can be pressed sintered to obtain electrodes of suitable form and dimensions. Such electrodes provide sufficient mechanical stability, excellent electrical conductivity, and remarkable thermal conductivity. To solve the susceptibility to poisoning, a macroskeleton for mechanical support and electrical and thermal conductivity must be introduced. The pores of the macroskeleton contain the catalyst grains, which are bonded through a diffusion process to retain catalyst activity. The alloy used here is composed of equal

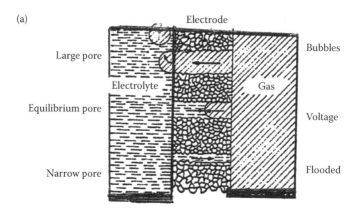

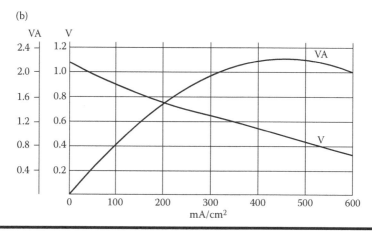

Figure 3.5 Electrical performance of hydrogen/oxygen fuel cell showing (a) behavior of different pore sizes and (b) characteristics of the fuel cell as a function of current density. Note that the "V" curve shows the terminal voltage, whereas the "VA" curve shows the power output.

parts by weight aluminum grains of purity better than 99.8%, and the anode Ni is fused in a carbon crucible under a protective layer of calcium chloride ($CaCl_2$) at a temperature greater than 1,300°C. Fuel cells using hydrogen-based DSK electrodes (Figure 3.5a) with fine porous coating yield optimum performance in terms of current density and polarization voltage as illustrated by Figure 3.5b.

3.5.2.1 Performance of DSK-Based Fuel Cells with Monolayer DSK Electrodes

The performance of DSK-based HYDROX fuel cell with monolayer DSK electrodes is a function of operating temperature. Polarization voltage, power density (W/cm^2), and current density (mA/cm^2) characteristics of oxygen and hydrogen electrodes are strictly dependent on operating temperatures. Experimental test data obtained by various designers indicate that the oxygen electrode has a finite polarization voltage even when the current is zero. The test data further indicate that the polarization can be reduced to 100 mV at 85°C. This cell attains a reversible potential of 1.23 V. The conversion efficiency of this cell from chemical energy into electrical energy can be measured by the product of the voltage ratio times the charge ratio. The total efficiency of 92% can be achieved even at the current density of zero. Both the voltage and efficiency decrease with increasing current density. Maximum output power from the cell is possible if the discharge voltage is maintained at 50% of the cell electromotive force, which in this particular cell design happens to be 0.65 V. At this terminal voltage, the current density will be about 250 mA/cm² at an operating temperature of 68°C with a gas-tight electrode, even at the maximum current density of 500 mA/cm², with a 1-mm-thick electrolyte layer of potassium hydro-oxide. The calculated power density is about 154 mW/cm². Preliminary calculations using appropriate values of the parameters involved indicate that the power to weight ratio for this particular cell comes to about 77 mW/g without any accessories, assuming the weight of each electrode and a 1-mm potassium-hydroxide (KOH) layer of 2 g/cm².

It is important to point out some of the critical design aspects of a fuel cell. Fuel can be a mixture of gases, and the oxidant can be oxygen. Generally, both fuel and oxidant electrodes are made of catalyzed porous Ni with a typical thickness of ~0.028 in. [1] If the cell uses alkaline electrolyte, it should be held in asbestos or a similar porous matting with a thickness of 0.030 in., which can withstand a differential pressure of 100 psi or more. Cell stacking and adequate structural support to the module must be given serious consideration, if highly reliable, continuous operation over extended durations is the principal design requirement. H_2-O_2 fuel cells using DSK electrodes and alkaline electrolyte offer the most efficient and reliable electrochemical conversion with minimum cost and complexity. Performance characteristics of an H_2-O_2 DSK cell with alkaline electrolyte at ambient temperatures are summarized in Table 3.3.

Table 3.3   Performance Characteristics of a Hydrogen-Oxygen Cell with Alkaline Electrolyte at an Ambient Temperature of 70°C

Current Density (mA/cm²)	Terminal Voltage (V)	Power Output (W)
0	1.10	0
100	0.91	0.82
200	0.78	1.57
300	0.64	1.94
400	0.55	2.21
500	0.42	2.19
600	0.35	2.00

These computed results indicate that as the current density increases, the terminal voltage decreases, but the power output level increases. The drop in terminal voltage at higher current densities is due to resistive loss in the electrical circuits.

3.6 Ion-Exchange Membrane Fuel Cells

The ion-exchange membrane (IEM) fuel cell was designed and developed in 1962 by General Electric (GE) Company. The GE scientists made significant progress in optimizing the current density parameter of the cell. Use of a highly improved electrolyte demonstrated a significant increase in the current density from 17.5 W/ft.^{2} in 1960 to 35 W/cm^2 in 1962, both at 0.7 V. The new membrane material was responsible for achieving a power density per unit volume better than 2 kW/ft.[3] Despite significant improvement in other performance parameters, the steady-state output voltages of these H$_2$-O$_2$ fuel cells never exceeded 0.93 V. The GE scientists found out that optimization of an oxygen electrode will improve both cell efficiency and open-circuit voltage.

3.6.1 Performance Specifications for IEM Fuel Cells and Batteries for Space Applications

Polarization data collected by various scientists on propane, propylene, and cyclopropane indicate that cyclopropane produces the highest current density of these hydrocarbons. Studies perfumed by the scientists reveal that around 90°C or so, with a platinum catalyst and three-molar sulfuric acid electrolyte, a current density of 38 mA/cm^2 can be achieved, which will be reduced to about 35 mA/cm^2 because of resistive loss in the cell. The latest research data indicate that, at present, current densities of 90–100 mA/cm^2 have been achieved at 0.3–0.4 V using propane.

High-capacity IEM batteries were developed for deployment at the electrical power source in a NASA project called Gemini. NASA scientists stated that a 30-min, 600-mile, high-orbital rocket flight in 1960 demonstrated the initial feasibility of using the IEM fuel cells in space. Two such batteries, each having 35 cells with a 50 W output power at 28 V, were tested in an orbiting satellite. These fuel cells operated intermittently for 30 days under full-load conditions, which could be interpreted as the equivalent of seven days of nonstop operation in the satellite.

A portable 200 W power source consisting of IEM fuel cells operating on air and hydrogen demonstrated its capability to power a mobile Army radio receiver, a transmitter set, and a field surveillance radar unit. A power module consisting of fuel cells and using diesel oil or methanol as fuel to generate hydrogen and using liquid oxygen as an oxidant has demonstrated its application in submarine propulsion system. This particular power system demonstrated a volumetric power density in excess of 2 kW/ft.[3] and that it would meet hull space requirements.

A high-capacity power system using fuel cells was developed by Thompson Ramo Wooldridge (TRW) Inc., Cleveland, Ohio. This power system consisted of two electrodes each pressed against an IEM and separated from each other by the membranes and intervening electrolyte. The power system was designed strictly using the dual-membrane cell design configuration for possible application in manned and unmanned space vehicles. Its performance parameters included a volumetric power density of 2 kW/ft.[3], low-pressure operation, quick start-up capability at 90% of full power, reliability under space conditions, and power source efficiency of 55 to 70% under severe space environment.

3.6.2 Fuel Cells Using Low-Cost, Porous Silicon Substrate Materials

Studies performed by the author on moderate-capacity cells indicate that silicon-based fuel cells offer several advantages, such as simple design, minimum cost, and small size. Such fuel cells are best suited for domestic applications, where power consumption ranges from 3 to 5 kW approximately. The substrate is porous, is highly structured, offers faster electrode reaction, and can use liquid or gaseous electrolyte. This type of fuel cell offers unique design configuration because of the characteristics of the well-controlled geometry of the porous silicon structures. The porous silicon offers significant benefits because of the extremely small deviations in the porous size, ranging between 10 and 50 mm, and in its distribution. Fuel cell designers and scientists consider this particular fuel cell design concept to be a perfectly engineered device that is most attractive for domestic applications. The device fabrication process is fully matured because it is dependent on well-defined silicon processing techniques and methods that are widely used in high-volume manufacturing. In brief, silicon-processing methods and quality control techniques currently deployed for manufacturing of silicon devices, such as transistors and diodes, can be used to manufacture fuel cells with minimum cost and complexity.

Fuel cell designers believe that no other fuel cell technology is available to demonstrate the feasibility of high-volume manufacturing with minimum cost. This design concept offers the opportunity to deliver a versatile and highly scalable fuel cell for applications for which cost, reliability, weight, size, and longevity are the critical requirements.

3.6.2.1 Hydrogen-Oxygen Power Fuel Cell Using Porous Silicon Structure

Fuel cell scientists are currently working on the development of an H_2-O_2 fuel cell using a porous silicon structure and liquid and using acidic as an electrolyte. The gas diffusion (GD) interface is inside the pores of the silicon substrate that extends throughout the pore. This particular fuel cell design concept offers a self-contained power-generating source and an electrolyser unit. This device deploys silicon electrodes modified for gaseous reactants that create the GD interface inside the pores of the silicon structure that extends throughout the pore. This type of fuel cell is best suited for applications for which electrical power requirements are low to moderate, ranging from 1 to 5 kW. Such fuel cells are most ideal for motor scooters and domestic electrical power modules. A retired NASA scientist recently revealed a prototype fuel cell design involving hundreds of such cells in a stack format and using porous silicon electrodes and methanol fuel. The dimensions of the porous silicon disc are approximately 4.5 in. × 4.5 in., and the overall dimensions are close to that of a shoe box. The scientist expects a current density of about 180 mW/cm^2. Assuming this current density and the porous silicon electrode of 4.5 in. × 4.5 in., the powder generated by a single element will be about 23.5 W. Assuming a stack consisting of 100 porous silicon disc elements, the electrical power generated by the 100-element fuel cell will be close to (100 × 24 = 2,400 W) or 2.4 kW. If the stack contains 200 elements, then the power output will be about 4.8 kW, which is sufficient to meet one household's electricity consumption requirements. Computed electrical power generated a single silicon porous element with electrode dimensions of 4.5 in. × 4.5 in.; total electrical power generated as a function of the number of disc elements is summarized in Table 3.4.

These calculations offer complete information on the physical dimensions of the fuel cell package capable of generating a specific amount of electrical energy. The power generated by various elements does not take into account various losses or cell efficiency. If these two factors are taken into account, the net electrical power generated will be about 85% of the values shown in Table 3.4.

As far as the fabrication cost of this fuel cell and total weight are concerned, no reliable estimates are available to date because no manufacturer has been contacted

Table 3.4 Electrical Power Generated per Disc Element and Total Power Generated by Number of Discs in the Fuel Cell Assembly

Disc Element Size (assumed, inch)	Element Area (sq. in.)	Element Area (sq. cm)	Power Generated by Each Element (W)	Power Generated by Elements	
				100 (kW)	200 (kW)
2 × 2	4	25.81	4.65	465	930
3 × 3	9	58.05	10.44	1,044	2,088
4 × 4	16	103.22	18.58	1,858	3,716
4.5 × 4.5	20.25	130.64	23.52	2,352	7,432
5 × 5	25	161.25	29.03	2,903	5,806

thus far. However, the author's best engineering judgment indicates that the weight of this particular fuel cell package will not exceed approximately 20 pounds. The fuel cell described above operates like the DMFC. The DMFC power sources are best suited for portable power segments, because DMFC technology offers several advantages, such as compact form factor, high reliability, affordability, and minimum weight and size.

3.6.2.2 Fuel Cell Reactions and Thermodynamic Efficiencies

Fuel cell reactions and thermodynamic efficiencies must be given serious consideration regardless of the cell type. These two issues have an impact on the electric energy that can be obtained from the cell and the amount of heat delivered by the chemical reaction. In a classical thermodynamic theory, the reaction enthalpy (ΔH_r) is the heat delivered by the chemical reaction and the reaction entropy (ΔS_r) indicates the measure of the change in the order of the system during the reaction. The parameter (ΔH_r) is negative when the heat is given out. Therefore, the reaction entropy term (ΔS_r) becomes less than zero or becomes a negative term.

For a clear understanding, these parameters consider a simple H_2-O_2 fuel cell. A part of the chemical energy is transferred to heat in the case of the electrochemical energy conversion process, which is in addition to the heat generated via the potential drop in the resistances inside the electrochemical cell as shown in Figure 3.6. The electrolyte resistance also contributes to potential drop.

In the case of H_2-O_2 fuel cells, the thermodynamic efficiency or the ideal efficiency of the energy conversion is always related to the reaction enthalpy of the chemical process involved. The ideal thermal efficiency of the cell can be expressed as follows:

$$E_{th} = E^0 / E_h^0 \qquad (3.1)$$

where E^0 is the thermodynamic cell voltage and E_h^0 is the thermal cell voltage as shown in Figure 3.6.

For this particular fuel cell reaction, the chemical reaction equation can be written as follows:

$$[H_2 + (1/2)\, O_2] = H_2O \qquad (3.2)$$

This clearly shows that the by-product of this chemical reaction is water.

In an H_2-O_2 fuel cell, typical cell voltage (E^{00}) under standard conditions or under normal temperature and pressure conditions is typically 1.23 V, and the thermodynamic cell voltage (E_h^0) is about 1.48 V. For almost all fuel cell reactions, one can expect the reaction entropy (ΔS_r) to be less than zero. This means the amount of heat generated in the surrounding areas will be a product of temperature (T) and reaction entropy (ΔS_r). With the conversion of the chemical energy into the

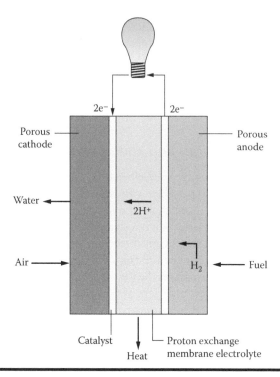

Figure 3.6 **Elements of a compact fuel using the PEM design architecture and porous anode and cathode electrodes.**

electrical energy, additional energy is gained via absorption of the heat from the surroundings, thereby making the ideal cell efficiency theoretically greater than 100%. Thermodynamic data on potential fuel cell reaction under standard conditions are summarized in Table 3.5.

Data presented in Table 3.5 indicate that formic acid yields the highest cell efficiency, because the oxidation of the free energy change is larger than the reaction enthalpy. In other words, with the conversion of the chemical energy into the electrical energy, additional energy is gained via the absorption of the heat energy from the surroundings. When the reaction entropy (ΔS_r) is less than zero or negative, the cell efficiency is less than 100% regardless of the fuel used. That is why hydrogen offers the lowest cell efficiency. Please refer to the 51st edition of the *Handbook of Chemistry and Physics* for indexes, symbols, and abbreviations appearing in Table 3.5.

3.6.2.3 DMFC Devices Using a PEM Structure

DMFC devices that deploy a PEM structure as shown in Figure 3.6 use air as a reactant [5]. DMFC devices have improved significantly since their initial development

Table 3.5   Important Thermodynamic Data on Various Fuel Cell Reactions under Standard Conditions

Fuel Cell Type	Valence	Reaction Equation	E^{00} (V)	Ideal Cell Efficiency (%)
Hydrogen	2	$H_2O + (1/2)\ O_2 = [H_2O]$	1.229	83.0
Carbon oxide	2	$CO + (1/2) = [CO_{2]}]$	1.066	90.8
Formic acid	2	$HCOOF + (1/2)\ O_2 = [CO_2 + H_2O_1]$	1.480	105.6
Formaldehyde	4	$CH_2\ O_g + O_2 = [CO_2 + H_2O_1]$	1.351	93.2
Methanol	6	$CH_3OH + (3/2)\ O_2 = [CO_2 + 2H_2O_1]$	1.214	96.7
Methane	8	$CH_4 + 2O_2 = [CO_2 + 2\ H_2O_1]$	1.060	91.8
Ammonia	3	$NH_3 + (3/4)O_2 = [(1/2)N_2 + (3/2)H_2O_1]$	1.172	88.5
Hydrazine	4	$N_2H_4 + O_2 = [N_2 + H_2O_1]$	1.558	96.8
Zinc	2	$Zn + (1/2)\ O_2 = [Zn\ O]$	1.657	91.3

in the 1960s, but they continue to have some technical and commercial limitations, including operational issues, water, water management issues, poor reliability, low power density, higher procurement costs, and lower conversion efficiencies. Most of the drawbacks have been eliminated or minimized by using an improved design that involves a porous silicon structure and a liquid electrolyte that allows the fuel cell to operate at higher efficiency in non-air-breathing environments. Power output capacity of a PEM-based fuel cell can be significantly improved using a stacking technique as illustrated in Figure 3.7. Operation and reaction schemes of an MEM-based fuel cell (PEMFC) operating on $C_xH_yO_z$ fuel are quite evident from Figure 3.8. Hydrogen utilization factor has an impact on both the current density and terminal voltage as illustrated in Figure 3.9

3.6.2.3.1 Design and Operational Aspects of PEM-Based DMFC Devices

All fuel cells have two electrodes: one negative electrode and one positive electrode. In the case of DMFC, the electrical power is generated by the methanol fuel at the negative electrode, which is known as the anode, with the oxygen from the air at the positive electrode, which is called the cathode. Electrical energy is produced by the chemical reaction that occurs when the fuel, catalyst, and the electrolyte

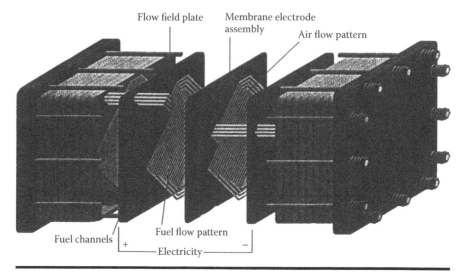

Figure 3.7 **Specific architectural details for a fuel cell stack composed of several hundred individual cells stacked in a series to generate high-power density in excess of 1 kW/L. Higher power levels in excess of 5 kW are possible with this technique.**

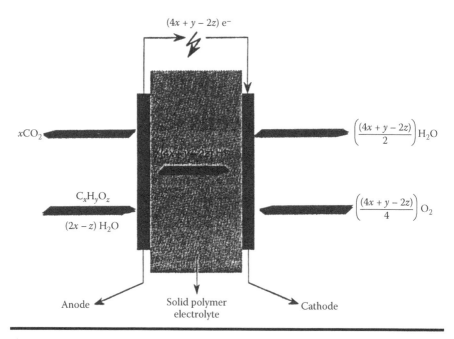

Figure 3.8 **Concentration and reaction scheme of a PEM-based fuel cell using porous cathode and anode electrodes and hydrogen as a function of fuel along with a catalyst that aids the breakdown of hydrogen atoms into protons and electrons. (x: atoms of C; y: atoms of H_2; z: atoms of O.)**

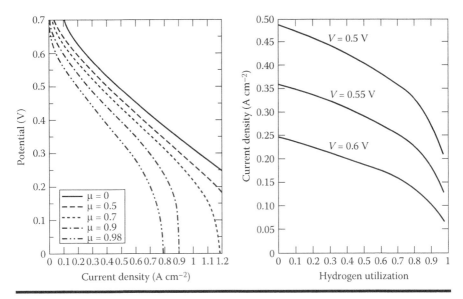

Figure 3.9 Cell potential and current density as a function of hydrogen utilization factor.

come together at a common point called the three-phase interface. Current DMFC designs create this three-phase interface at the surface of a polymer material called the PEM. But the membrane limits the reaction zone to a facial or two-dimensional area, which restricts the electrical power output of the cell, because power generated is strictly dependent on the active area. Full cell designers have observed typical power densities between 65 and 85 mW/cm². The PEM-based devices have some inherent technical and commercial limitations [3], which can be briefly summarized as follows.

3.6.2.3.1.1 Electrochemical Efficiency Depends on the Utilization Rate of Methanol Fuel — Methanol (CH_3OH) crossover occurs where the CH_3OH passes from the anode through the PEM medium and reacts with the catalyst at the cathode. This reduces the voltage of the cell and wastes some amount of fuel. Although the use of diluted CH_3OH minimizes the crossover, it requires carrying additional water, which reduces the gravimetric energy density of the device.

3.6.2.3.1.2 Causes for Reduced Conversion Efficiency and Power Density — Fuel cell scientists believe that PEM-based DMFC devices using CH_3OH as fuel typically operate at efficiencies anywhere between 20 and 30% because of limitations in the chemical reaction with the CH_3OH. In particular, CH_3OH crossover not only reduces the voltage and power density but also degrades the conversion efficiency of the device.

3.6.2.3.1.3 Impact of Surrounding Environments on PEM-Based DMFC Device Performance — Control of moisture content is of critical importance, if achieving the maximum electrical power from this device is the principal design requirement. According to DMFC design scientists, if the DMFC devices generate excessive moisture in the proximity of the device, it can degrade the device's performance. Even European fuel cell scientists have expressed serious concerns about the water vaporization for the membrane-based cells. Fuel cell scientists have observed various problems resulting from surrounding environments. For example, carbon-dioxide contamination problems have been observed for alkaline cells using thin electrodes. The presence of heat in the proximity of the cell structure affects the cell's electrical performance and reliability.

3.6.2.3.1.4 Factors That Affect Structural Integrity of PEM-Based DMFC Devices — DMFC cell designers feel that performance of such devices could degrade because of the factors outlined in this paragraph. Because these devices use atmospheric air as a reactant, any toxicity and contamination present in the air will significantly degrade the cathode. Presence of high humidity can degrade PEM performance. Even low humidity can produce cracks in PEM. In brief, operating such cells under high temperatures, high humidity levels, and excessive shock and vibration levels could significantly degrade the PEM-based device's electrical performance, reliability, and longevity.

Despite these shortcomings, this type of fuel cell offers some advantages. It can use the cheapest fuel and could use a gaseous fuel, liquid fuel, or a combination of the two under moderate-to-low-temperature environments [6]. This fuel cell offers the following advantages:

■ It uses CH_3OH fuel, which is abundant.
■ DMFC operates below 150°C.
■ Zero product of NO_X.
■ CH_3OH is stable in contact with the acidic membrane and is easy to manage.
■ Changes in electrical power generated can be accomplished simply by an alternation in the supply of CH_3OH feed, which is the major advantage of this device.

The anode chemical reaction can be defined as follows:

$$CH_3OH + H_2O = [CO_2 + 6H^+ + 6e^-] \qquad (3.3)$$

The cathode chemical reaction can be defined as follows:

$$(3/2)O_2 + 6H^+ + 6e^- = [3H_2O] \qquad (3.4)$$

The sustaining cells have yielded current densities greater than 150 A/ft.² at 0.7 V open-circuit voltage and have demonstrated operating hours exceeding 3,000 without cell failure or performance deterioration. Maximum cell efficiency that can be achieved with this cell with a carbon anode is about 35%

3.6.2.4 Silicon-Based DMFC Fuel Cells

Preliminary studies conducted by the author indicate that silicon-based DMFC fuel cells (Figure 3.10) seem to be free from most of the problems associated with PEM-based DMFC devices. The studies further indicate that silicon-based DMFC devices are capable of providing higher power densities with liquid electrolytes and badly wanted portability, particularly in battlefield environments. One can achieve a larger reaction zone with a liquid electrolyte, which will yield higher power densities over PEM-based devices. These fuel cells could also use suitable oxidants, and the devices can be manufactured with lower costs and higher reliability. Typical power output levels available from PEM-based and silicon-based DMFC devices as a function of fuel cell surface area are summarized in Table 3.6.

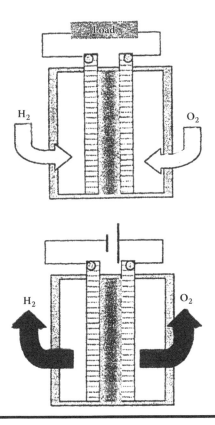

Figure 3.10 **Architectural details of a silicon-based H_2-O_2 fuel cell operating at ambient temperatures.**

Table 3.6 Output Power Generated by PEM-Based and Silicon-Based DMFC Devices as a Function of Surface Area Provided by the Electrodes (W)

Electrode Surface Area (sq. in./sq. cm)	PEM-Based DMFC	Silicon-Based DMFC
2.5 in. × 2.5 in. (6.25/40.32)	2.62/3.43	7.46/8.23
3.0 in. × 3.0 in. (9.00/58.06)	3.77/4.94	10.74/11.90
3.5 in. × 3.5 in. (12.25/79.03)	5.14/6.72	14.62/16.20
4.0 in. × 4.0 in. (16.00/103.22)	6.71/8.773	18.58/21.16
5.0 in. × 5.0 in. (25/161.2)	10.41/13.70	29.01/32.48
10.0 in. × 10.0 in. (100/645)	41.92/54.82	116.10/132.34
25.0 in. × 25.0 in. (625/4,031)	262.64/342.68	725.58/826.35

These calculations assume a power density of 65 to 85 mW/cm² for PEM-based devices and about 180 to 205 mW/cm² for silicon-based DMFC devices. These calculations reveal that a silicon-based DMFC device can meet the power requirements for several domestic and industrial applications.

It is obvious from the calculations above that the silicon-based DMFC devices can be manufactured with lager electrodes to achieve much higher volumetric energy density. The fuel cell can be stacked to meet specific power requirements. Furthermore, dimensions of the surface areas can be altered to accommodate the fuel cell in a given space. In other words, one dimension can be slightly smaller or larger than the other to accommodate the device in a given available space.

3.6.2.4.1 Unique Design Features and Potential Benefits of Silicon-Based DMFC

Following are the unique design features and potential benefits of silicon-based DMFC:

1. *This DMFC provides a closed-system operation*: Fuel cell designers believe that the porous silicon electrodes can be easily assembled into cells and stacks with minimum separation. The CH₃OH fuel and the oxidant react at the catalyst locations in the porous silicon structure to generate electrical energy. After completion of the electrochemical reaction, leftover or residual fuel can be removed from the cells using a continuous flow of liquid through the electrodes.

2. *This DMFC permits efficient non-air-breathing operation*: Because the fuel does not use air as an oxidant, it can be used under special operating environments, such as underwater or in a smoked-filled building, with no compromise in reliability or longevity.

3. *This DMFC utilizes all the CH_3OH fuel cell*: The DMFC design uses a recirculation operating process that continues to run until all available CH_3OH fuel in the replaceable fuel container or cartridge has been consumed. If further CH_3OH supply is needed, another cartridge with CH_3OH can be inserted to retain a constant electrical energy source.

4. *This DMFC acts as a reservoir to collect excess water*: The fuel cartridge is the source of oxidant needed by the cell and also serves as a water reservoir to collect the excess water produced during the electrochemical process. Because of this, these fuel cells do not expose the stack to ambient air and also do not vent hot water vapors into the surrounding areas close to the fuel cells.

5. *This DMFC can be customized for any application with minimum cost and complexity*: Critical components such as electrodes, micro-pumps (MPs), heat exchanges, and fuel cartridges associated with DMFC assembly can be sized and modularized for any application. This flexibility provides optimum device performance, maximum design flexibility, and minimum design cost. To illustrate the design flexibility, the fuel cartridge can yield a continuous operation of 4, 8, 12, or 16 h simply by changing the cartridges. This allows the electronic, electrical, and mechanical components suppliers to provide a wide selection or mix of products best suited for a particular application. This type of customized approach lets the fuel cell customers select the size of fuel cell best suited for their application.

6. *This DMFC deploys a mature, no-risk technology*: On the basis of statements made regarding the merits of silicon-based DMFC, silicon-based fuel cells can be manufactured using the existing mature and cost-effective infrastructure of the semiconductor processing industry. In addition, quality control techniques and reliability criterion tests widely used in the semiconductor industry can be applied with no additional costs. In other words, the manufacturing costs for these devices will be extremely low. In addition, manufacturing of some fuel cell components such as heat exchangers, MPs, electrodes, fuel cartridges, and printed circuit boards can be outsourced, if necessary, to reduce further manufacturing costs and procurement delays. Because of several advantages in terms of cost, performance, and reliability, silicon-based DMFC is a key technology in the future for portable electronic, electrical, and mechanical devices for a variety of applications ranging from automobiles to portable electrical power sources to unmanned air vehicles (UAVs) to communication satellites.

7. *Component requirements for the silicon-based DMFC and their design aspects*: Critical components of the silicon-based DMFC include an MP with fluidic manifold, anode and cathode electrodes, power management battery, scrubber device, fluidic sensor, a monitoring device capable of continuously monitoring the temperature and pressure, active elements for the stacks, and a cartridge containing the CH_3OH fuel.

Among all these components, the MP is the most critical element of the silicon-based DMFC, and its fabrication uses the nanotechnology principle. This MP deploys either a electrostatic or piezoelectric actuation mechanism. This pump requires few milliwatts of electrical power, comes with or without valves, and has no moving parts, thereby offering ultra-high reliability. This kind of pump is best suited for fuel delivery applications in cases in which constant and uniform fuel flow is the principal requirement. Most of the other components are readily available in the market with immediate delivery. Cartridges with different fuel capacities can be procured in free markets.

3.7 Potential Applications of Fuel Cells

Fuel cells have several domestic and industrial applications that range from motor scooters to electrically driven bikes to minicars and vans. Studies performed by the author on electrical requirements for UAVs indicate that this device is best suited for applications involving intelligence gathering, surveillance, and reconnaissance missions over a period of time ranging from approximately 5 to 10 h. These fuel cells are most suitable in cases in which continuous supply of electrical power over a specified duration, high reliability, lower procurement cost, and minimum weight and size are the principal design requirements.

3.7.1 Fuel Cells for Military and Space Applications

There are numerous military and space applications for these fuel cells. Technical articles published occasionally in *Military and Aerospace Electronics* outline the applications of fuel cells for battlefield sensors and weapons. Some vehicles, such as UAVs, unmanned ground vehicles (UGVs), and unmanned air combat vehicles (UACVs), can play important roles in battlefield environments [7]. These vehicles are not manned by soldiers, yet they perform the functions or missions as desired by the field commanders. Portable power sources using fuel cells are most attractive to power these vehicles as long as the electricity requirement does not exceed 5 kW or so. Fuel cells with power capacity exceeding 5 kW or so tend to become bulky. Portability is the most critical requirement for most battlefield sensors and weapons, UAVs, airborne systems, and underwater sensors. Therefore, fuel cells with the desirable power capacity, minimum weight and size, high reliability, and improved efficiency will be most ideal for these applications.

3.7.1.1 Fuel Cells for Battlefield Applications

Fuel cells can be used for a variety of military missions in future battlefield applications, including robot-based applications; UGV- and UAV-based reconnaissance, surveillance, and target acquisition (RSTA); logistics and cargo transportation; remote-controlled security inspection; obstacle breaching; medical applications involving removal of wounded soldiers; route clearance; and destruction of mines and improvised explosive devices (IEDs). Although military robotic applications were considered to be the futuristic stuff of science fiction, Army robotics is here now and will stay in the future. Even in the 21st century, the robots are effectively detecting and destroying IEDs in Iraq and Afghanistan.

The latest technical articles appearing in *Military and Aerospace Electronics* reveal that future battlefield conflicts will heavily involve armed UAVs incorporating missile attack capability along with RSTA missions [5], UACVs capable of undertaking combat missions deep in enemy territories in addition to RSTA activities, and UGVs for a variety of battlefield missions. The latest article published in *Military and Aerospace Electronics* suggests the development of state-of-the-art electro-optic and microwave sensors to help UAVs detect and attack hostile submarines and surface warships, as well as attack ground targets and participate in electronic warfare operations against IED devices [8].

The power capacity of a fuel cell is strictly dependent on the mission duration, payload capacity, number of sensors deployed, and types of battlefield tasks to be performed by a specific UAV. Preliminary studies performed by the author on the power output capacities of fuel cells for these battlefield applications indicate that fuel cells with power outputs between 3 and 5 kW are adequate for UAVs with basic sensors to perform the RSTA and counterinsurgency tasks. Provision for additional power close to 2 kW may be required to power side-looking radars. To undertake various missions with durations less than 5 h by a UACV, fuel cells with power output between 4 and 6 kW are essential. Finally, fuel cells with power capacities between 5 and 7 kW may be required for the UGVs to perform assigned tasks over mission durations not exceeding 4 h. The longer the mission duration, the higher the payload, and the greater the numbers of sensors involved, the higher the power-capacity requirement for the fuel cell. Power capacity, payload requirement, and mission duration will determine the type of fuel cell and its performance requirements.

3.7.1.2 Deployment of Fuel Cells in UAVs Acting as Electronic Drones Capable of Providing Surveillance, Reconnaissance, Intelligence Gathering, and Missile Attack Capabilities

On the basis of successful attacks by UAVs on hostile targets in Iraq and Afghanistan, it can be stated that the electronic drones fully equipped with electro-optical systems, side-looking radars, and other radiofrequency/infrared radiation sensors most likely will be deployed in future military conflicts regardless of locations. Precision target location and missile attack by these pilotless electronic drones will be the preferred way to fight future military conflicts. These drones can accomplish the military mission objectives with minimum cost and complexity and with no loss of drone operator.

3.7.1.2.1 Fuel Cell Requirements for Electronic Attack Drones

Fuel cells with unique design concepts can be developed with an emphasis on reliability, weight, size, and power capability. Fuel cells with power capacities exceeding 5 kW, minimum weight, compact size, and operational lives exceeding 10,000 h are possible. The weight and size will be compatible with the payload capacity of the drone. Fuel cells with closed-cycle operation are best suited for this particular pilotless vehicle.

3.7.1.3 Why Fuel Cells for Counterinsurgency Applications?

Since World War II, several armed conflicts have been experienced around the world. During these conflicts, the occupying power or country has implanted armed mines and IEDs. During the Soviet occupation of Afghanistan, the *mujahideens* (terrorists) implanted IEDs throughout the country, which have crippled local people and injured U.S. personnel. In Iraq, Pakistan, Somalia, and other Islamic countries, IEDs are implanted by the terrorists, and these devices have caused serious injuries to U.S. personnel when they step on the pressure plates associated with IEDs [8]. Military planners and defense experts believe that the insurgent fighters use IEDs as "weapons of strategic influence." Military experts further believe that the IEDs and the casualties they cause cannot be eliminated. Because these devices are implanted on hidden spots on roadsides, even armed personnel carriers equipped with powerful lasers may miss the chance to neutralize them. Field commanders generally deploy soldiers equipped with the latest detection sensors to neutralize the IEDs, but these soldiers have to carry heavy packages consisting of detection sensors, batteries, ammunitions, and other essential items on their backs. Monthly global acts carried out by terrorists (excluding those committed in Iraq and Afghanistan) are summarized in Table 3.7

It appears from these data that terrorist activities are at their peak during the months of January, October, November, and December. To avoid physical injuries, property damage, and battlefield death, it is absolutely necessary to undertake counter-IED activities regardless of month or year. Battlefield commanders are looking for efficient and reliable counter-IED equipment with minimum cost, size, weight, and power consumption. Whether the counter-IED equipment is carried by a foot soldier or a heavily armored carrier, the specified design requirements are essential for a cost-effective and reliable operation in the battlefield.

Battlefield commanders are looking for efficient and reliable counter-IED equipment with minimum cost, size, weight, and power consumption. Regardless of whether the counter-IED equipment is carried by a foot soldier or a heavily

Table 3.7   Monthly Global Terrorist Acts, Excluding Iraq and Afghanistan

Month	2006	2007	2008
January	485	445	505
February	367	315	390
March	425	412	388
April	268	385	460
May	265	435	438
June	318	450	462
July	466	268	336
August	344	415	424
September	322	423	435
October	435	462	482
November	448	554	565
December	335	538	562

armored carrier, the design requirements as stated above are essential for a cost-effective and reliable operation in the battlefield.

3.7.1.3.1 Specific Requirements for Counter-IED Equipment

Deployment of compact and efficient fuel cells will not only significantly reduce the equipment weight and size but also increase the mission duration. Furthermore, deployment of fuel cells in military ground robots will significantly increase their mobility and ability to perform demanding missions over long durations. Army scientists expect massive deployment of small unmanned ground vehicles (SUGVs) in hostile regions to execute surveillance and reconnaissance missions. Preliminary studies suggest that fuel cells with output capacities between 5 and 15 kW will be most desirable depending on the tasks involved. Electrical power requirements for these SUGVs can be satisfied by reliable and low-cost fuel cells. Besides the U.S. Air Force and Army, the Navy is also considering the deployment of unmanned underwater vehicles (UUVs) to perform special missions. Here again, fuel cells can play a critical role in providing a clean and noise-free energy source.

Experts in this particular area indicate that it is better to detect and neutralize the IED devices before they become active operationally. This way, bodily injuries and deaths can be avoided in the battlefield, thereby significantly reducing personal suffering experienced by the military personnel working in these hostile areas as well as decreasing insurance and medical costs. It is essential to detect and neutralize these devices so that the medevac helicopter can safely land to remove the injured soldiers for medical treatment. The United States and other countries are pouring billions of dollars into counter-IED activities [8].

According to defense experts, the deep-buried IEDs and explosively formed penetrators (EFPs) are the most lethal forms of IEDs. These devices can be triggered using remote-control techniques. These experts further believe that the deep-buried IEDs often contain hundreds of kilograms of explosives capable of destroying heavily armored vehicles, including the mine-resistant ambush-protected (MRAP) vehicles, and killing all onboard. Counter-IED experts mention that the kinetic energy of the fragments is so high that it can penetrate most kinds of armor. Five distinct functions, including prediction, detection, prevention, neutralization, and mitigation, must be addressed in the design and development of foolproof counter-IED equipment. In addition, selection of a reliable high-power battery or fuel cell is of critical importance. This power source could be designed either using a bank of batteries or a stack of miniaturized fuel cells. A bank of batteries poses weight and size problems in addition to logistic problems in charging the batteries in battlefield environment. Under these circumstances, the choice of fuel cells offers cost-effective and reliable operation besides significant reduction in weight and size of the power source. In addition, higher electrical energy is available from fuel cells compared with rechargeable batteries.

Studies performed by the author indicate that only a few design configurations of fuel cells are best suited for UAV applications. Compact size, light weight, enhanced conversion efficiency, low-cost fuel, high structural integrity, and ultra-high reliability are the most demanding requirements for the selected fuel cell for successful accomplishment of the tactical missions under battlefield environments. The fuel used by the cell must be free from toxic gases, high-temperature fumes or vapors, and hydrocarbon contents.

3.7.1.3.2 Low-Cost Fuels for UAVs

Low-cost fuels include hydrogen, methane, gasoline, and kerosene. Hydrogen can be generated at the lowest cost and can be stored to satisfy a 4 to 5 h UAV mission. Hydrogen fuel cells offer the highest conversion efficiency, because hydrogen electrodes yield the most efficient electrochemical process. Hydrogen has the lowest density among the most gaseous media as shown in Table 3.8.

The density of air is based strictly on the percentage of nitrogen and oxygen contents in the air at a particular height above ground. These contents vary as a function of altitude. The oxygen content in air decreases as the height increases above the earth. On the basis of these statements, the performance of an oxygen-based fuel cell is strictly dependent on the percentage of oxygen content in the air. Because UAVs typically operate at various altitudes and under battlefield

Table 3.8   Density of Various Gases Widely Used in the Design of Fuel Cells Most Ideal for UAV Applications

Gaseous Medium	Density (g/cm³)
Hydrogen	0.0837
Nitrogen	1.1652
Oxygen	1.3318
Air (~79% N_2 and 21% O_2)	1.1712

environments ranging from 2,000 to 5,000 ft., fuel cell designers have an option to design an appropriate configuration without involving oxygen.

As mentioned previously, a UAV drone is equipped with electro-optical sensors, infrared cameras, RF/mm-wave systems, and a host of other sensors to accomplish mission objectives during a specified duration that could last from 1 to 4 h. A PEM-based fuel cell (shown in Figure 3.5) with 3 to 5 kW capacity is adequate for a UAV provided it does not deploy a side-looking radar aboard the vehicle. If a side-looking radar is essential for a specified mission, then a fuel cell with a capacity of 7.5 kW is required. The PEM-based fuel cell can maintain the same output power level over 4 to 6 h.

3.7.1.3.2.1 Impact of Cell Parameters on the Performance of PEM-Based Cells — The PEM-based fuel cell with moderate power capacity is most the compact and efficient. Critical elements of this cell are depicted in Figure 3.5. The PEM fuel cell consists of two porous electrodes separated by a polymer membrane. The electrodes can be separated by a porous silicon membrane, but the cost will be slightly higher. The membrane allows hydrogen ions (H^+) to pass through the membrane, but it blocks the flow of both electrons and gases. In other words, the fuel electrode is known as anode, and the oxidant electrode is known as cathode. As illustrated in Figure 3.4, the hydrogen fuel flows along the surface of the anode, and the oxygen or air flows along the surface of the cathode. The catalyst helps in the breakdown of the hydrogen atoms into protons and electrons. The electro-chemical process in the cell produces the electricity directly by oxidizing the hydrogen. The chemical process occurs around 80°C, and the cell uses a thin plastic sheet acting as their electrolyte. The by-products of this cell are heat and water and, thus, both are harmless. This device is safe to manufacture and operates at moderate pressure, which increases the power density, simplifies the construction, and decreases the manufacturing cost.

The open-circuit voltage of a single cell is approximately 1 V. Cells can be connected in a series to form a cell stack that produces higher voltage as illustrated in Figure 3.6. It is possible to build stacks consisting of several cells to meet a desired power density and power output. The current rating of a fuel cell is a function of the size of the active surface area of the cell. This means that the voltage and the current density of a fuel cell can be varied by deploying the fuel cells in stacks using series or parallel configurations to obtain the desired power characteristics of the fuel cell. Current cell designs using silicon membranes offer current densities close to 200 mA/cm^2.

Such fuel cells were used in the 1960s and 1970s to power Gemini and Apollo spacecraft. These cells are still widely deployed aboard the space shuttles, which clearly indicates their remarkable mechanical integrity and reliable electrical performance over extended periods. Because of compact size, improved reliability, and cost-effective performance under severe operating environments, these fuel cells are best suited for military and space applications involving UAVC, UUG, UUV, UAV, MRAP, and anti-IED operations.

3.8 Fuel Cells for Aircraft Applications

Several aircraft companies such as Boeing, Cessna, Sky Spark, and others are planning to deploy fuel cell–based electric propulsion systems to operate their all-electric aircraft for short flights [9]. The all-electric technology offers small, light, practically noise-free, and maneuverable aircraft. This all-electric aircraft is considered a good military stealth vehicle and can be used for surveillance, reconnaissance, and intelligence-gathering missions in the battlefield. Hybrid technology involving lithium-polymer batteries and hydrogen-based fuel cells has been considered by various aircraft designers for all-electric vehicles.

3.8.1 Performance Capabilities and Limitations of All-Electric Aircraft or Vehicles

It is absolutely necessary to deploy the hybrid technology in the design of an all-electric vehicle to ensure continuous propulsion capability and reliable flight performance. Some aircraft designers have selected 20 kW hydrogen-based fuel cells with PEM configurations and 20 kW lithium-polymer batteries that guarantee alternative or supplementary power capability during takeoff and initial clime operations [9]. The hydrogen-based PEM fuel cells are capable of delivering a current level more than 100 amperes at an output voltage ranging from 200 to 240 V [9]. This type of fuel cell converts hydrogen directly into electricity with the highest conversion efficiency and coverts the heat without combustion. Furthermore, such fuel cells are emission free and quieter than hydrocarbon fuel–powered engines. Air and water vapors are emitted by these cells at environmental temperatures that do not pose environmental problems. Sky Park is a fixed-wing all-electric aircraft that uses hydrogen fuel cells capable of providing electrical power exceeding 65 kW, which is sufficient to power this aircraft, with a lithium-polymer battery to provide additional power, if needed under emergency conditions. Such all-electric aircraft [9] powered by fuel cells can travel at 100 mph with a flight time ranging from 1 to 3 h. These performance capabilities were demonstrated in March 2008 by Boeing Research and Technology organization in Europe. As mentioned earlier, the fuel cells provide all the power for the cruise phase of the flight. During takeoff and climb, more power is required, which can be supplemented by light-weight lithium-ion batteries. The key to future deployment of all-electric aircraft is the battery technology. Battery development must focus on increased storage density and enhanced life expectancy, while reducing the recharge time. The possibility of high-energy storage density (per unit volume and per unit mass) battery technology will definitely justify their use in small all-electric aircraft.

Studies performed by all-electric aircraft design engineers indicate that a solar system consisting of 12,000 solar cells can power four electric motors, each rated for 7.5 kW. There are four nacelles, each with a set of lithium-polymer batteries, a 10 horsepower (HP) electric motor, and a two-blade propeller. This all-electric aircraft has demonstrated a successful 26 h flight, including a 9 h night flight on July 7–8, 2010. Scientists believe that an aircraft using regenerative soaring can potentially remain aloft indefinitely at higher altitudes in cases in which the solar energy available from the vertical atmospheric motion can exceed solar power by a factor of 10 or more.

3.8.2 Fuel Cells for Electric Vehicles and Hybrid Electric Vehicles

Not every fuel cell is best suited for electric vehicles (EVs) and hybrid electric vehicles (HEVs). Technical articles published in *IEEE Spectrum* in June 2001 indicate that metal-based fuel cells would be most suitable as backup and emergency sources with high reliability and longevity. Studies performed by the metal-based fuel cells propose that these fuel cells are best suited for powering the cars, trucks, and electric scooters operating in crowded cities as well as for reducing pollution. These fuel cells would particularly be found most cost-effective in third world countries where scooters and motorbikes are powered by low-cost, two-stroke engines that run on cheap fuels. Noise-free, pollution-free, and low-cost operations are only possible with metal-based fuels (Figure 3.11).

3.9 Fuel Cells for Commercial, Military, and Space Applications

Deployment of fuel cell technology has been accelerated in commercial, military, and space applications. Fuel cells were widely used in space applications as early as the 1960s and demonstrated remarkable reliability and longevity. Fuel cells have been widely used by buses, automobiles, and scooters since 1990s. Fuel cells have

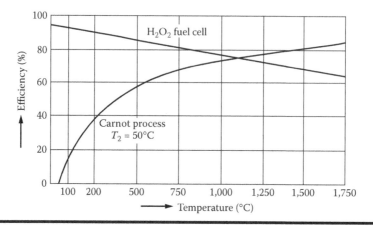

Figure 3.11 Comparison between the thermodynamic efficiency of a heat engine (Carnot cycle efficiency) and the ideal efficiency of an H₂-O₂ fuel cell.

demonstrated their applications in small experimental aircraft and UAVs. In the case of military applications, significant test and evaluation efforts must be undertaken to demonstrate high reliability, constant electrical power availability, and foolproof longevity.

3.9.1 Fuel Cells for Automobiles, Buses, and Scooters

The big three automobile manufacturers, namely General Motors, Ford Motor Company, and Chrysler Corporation, were the first companies to develop fuel cell technology as a viable automotive power source in their products. Thereafter, other automobile companies, namely Toyota, Nissan, Mazda, and Volvo, began aggressively designing cars featuring the hybrid technology using fuel cells along with lithium-ion batteries and lithium-polymer batteries. The goal of these automobile companies is to develop affordable fuel-efficient, low-emission vehicles. Confidence in these vehicles started building when a Canadian company called Ballard Power Systems Inc. demonstrated a reliable stack design configuration capable of generating a power density close to 1,000 watts per liter (W/L) of hydrogen fuel. This means that 30 L of hydrogen is needed to convert into electricity at a rate of 30 kW, which is adequate for most cars and small buses. The hydrogen-based PEM fuel cells were developed to replace the existing internal combustion engines (ICEs), and currently these PEMs are widely deployed by buses and automobiles. The hydrogen-based PEM fuel cell consists of a fuel electrode (anode) and an oxidant electrode (cathode), which are separated by a porous polymer electrolyte called PEM. The catalyst is integrated between electrodes, and the entire laminated structure is known as the membrane electrode assembly. The PEM yields low emissions, high efficiency, fast start-up capability, and rapid response to transients.

The 200 kW bus engines are fueled by hydrogen stored in cylinders located on the top of buses. The PEM fuel cells use hydrogen as a fuel. Several types of fuels are available for use in fuel cell devices, such as hydrogen, CH_3OH, natural gas, and gasoline, and the conversion efficiency of the cell is dependent on the type and purity of the fuel. Preliminary studies undertaken by the author indicate that public transportation buses can travel at a maximum range of 400 km without further supply of hydrogen and automobiles can travel more than 200 km without further supply of hydrogen. PEM-based fuel cells with a 30 kW capacity are sufficient for cars, and fuel cells with a capacity around 1 kW are sufficient for scooters. Fuel cells with a 20 to 25 kW capacity are sufficient for small passenger cars. Developing countries, such as Vietnam, China, and Korea, and some African countries, have shown great interest in using fuel cells for cars and scooters, because the operating cost using fuel cells will be much lower compared with gasoline-driven cars or scooters, provided the cost of manufacturing the cells is reduced by 25 to 40% and the weight and size of the cells are reduced proportionately compared with conventional automobiles. Fuel cells can be categorized by the operating temperature and by the type of electrolyte. Not all fuel cells are appropriate for automotive applications. The performance characteristics summarized in Table 3.9 clearly indicate

Table 3.9 Performance Comparison for Various Fuel Cells

Parameter	Proton Exchange Membrane (PEM)	Phosphoric Acid	Alkaline	Molten Carbonate	Solid Oxide
Electrolyte	Polymer	H_3PO_4	KOH/H_2O	Molten salt	Ceramic
Operating Temperature (°C)	80	190	80–200	650	1,000
Fuel	Hydrogen reformate	Hydrogen reformate	Hydrogen	Hydrogen reformate	Hydrogen reformate
Reforming	External	External	N/A	External, internal	External, internal
Oxidant	Oxygen, air	Oxygen, air	Oxygen	CO_2, O_2, air	Oxygen, air
Ffficiency (%)	45–60	40–50	40–50	>60	>60
Application	Small utility, cars	Small utility	Aerospace	Large utility	Large utility

that the hydrogen-based PEM fuel cells are best suited for automotive applications, where efficiency, cost, longevity, and reliability are the principal design requirements [10]. The PEM fuel cell directly converts the hydrogen into electrical energy with the highest conversion efficiencies. In addition, a reformer, which is a small chemical reactor, is carried on board some fuel cell vehicles to extract hydrogen from the alcohol of other hydrocarbon-based fuels such as gasoline.

The data presented in Table 3.9 clearly identify the type of fuel, electrolyte, conversion efficiency, power-generating range, operating temperature, and application of a specific fuel cell. Electric vehicle designers feel that for the PEM fuel cells to become truly competitive with the ICEs, the stack cost containing a hundred cells must be reduced to about \$20 per kW. Because the typical fuel cell capacity requirement for an automobile is at least 20 kW, the fuel cost must not exceed \$400 to be most cost-effective. This clearly indicates that current fuel cell design and development efforts must address not only performance, size, and weight but also procurement cost. Our studies on fuel cells reveal that fuel cells operating at low or moderate temperatures yield low production cost, high conversion efficiency, improved voltage performance, and high reliability over extended durations. In the case of fuel cells using hydrogen and oxygen as fuel, thermodynamic efficiencies greater than 80% are possible only when the operating temperature does not exceed 850°C (as shown in Figure 3.11). In the case of a galvanic cell, the reaction entropy (DS) plays an important role. Figure 3.11 illustrates the heat and electric energies generated in such a cell assuming a zero electrolyte resistance (R_E). Figure 3.12 illustrates the thermodynamic cell voltage (E^0), thermal cell voltage (E_H^0), and current levels as a function of various parameters.

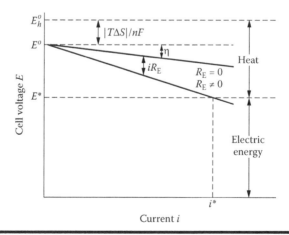

Figure 3.12 **Fuel cell voltages and current levels as a function of various parameters involved.** (ΔS: reaction entropy; T: temperature; R_E: electrolyte resistance; E_h^0: thermal cell voltage; E^0: thermodynamic cell voltage; E: cell voltage under load; i: load current; η: overvoltage at the electrodes; n: number of cells; F: configuration factor.)

Table 3.10 Voltage Performance of a Hydrogen-Oxygen Fuel Cell as a Function of Operating Hours and Using Various Electrolytes

Electrolyte	Cell Resistance	Current Density	Cell Voltage (V) after			
			0 h	1,250 h	2,500 h	5,000 h
NaOH	High	25	0.85	0.83	0.81	0.76
KOH	Low	50	0.95	0.91	0.83	0.80
KOH	Low	100	0.86	0.82	0.80	0.77

3.9.1.1 Low-Cost, High-Efficiency, Low-Temperature H_2-O_2 Fuel Cells

It is evident from the data summarized in Table 3.9 that an H_2-O_2 fuel cell with PEM design configuration offers a device with low cost (currently about $100 per kW), low-temperature operation (80°C), high electrochemical efficiency (>55%), and improved reliability over extended durations (>16,000 h). The electrical performance of an H_2-O_2 fuel cell developed by Union Carbide in 1960 with porous carbon electrodes, and as a function of various electrolytes and continuous operating hours, is shown in Table 3.10.

The test data presented in Table 3.10 were collected by the Union Carbide scientists in 1960 using KOH and sodium hydroxide (NaHO) electrolytes. It is believed that the voltage data significantly improved when tests were conducted earlier than 1970. The Union Carbide scientists believe that carbon itself is a good catalyst and offers good results. The scientists further believe that current density with hundreds of amperes per square foot is possible using a small amount of catalysts on the anodes only. Union Carbide developed another H_2-O_2 fuel cell with 1 ft.^{2} electrodes each with a thickness of 0.25 in.. Its voltage performance as a function of current density is shown in Table 3.11. A fuel cell was designed by a European scientist to demonstrate the catalytic dehydrogenation of the liquid fuel dissolved in the electrolyte. In this fuel cell oxygen or air was used as a cathode, whereas the dehydrogenation acted as an anode as illustrated in Figure 3.13. Four critical steps of dehydrogenation and oxidation of the ethylene-glycol (CH_2OH–CH_2OH) fuel are illustrated in Figure 3.14 for the benefit of readers who are anxious to pursue design and development activities using this fuel.

The tabulated data in Table 3.11 indicate that near-constant cell voltage is possible with this particular fuel cell design configuration. Higher current densities can be obtained from metallic fuel cells with larger electrode surfaces [11]. Furthermore, the voltage drop at increasing current densities is small and, therefore, these cells can operate continuously over several thousands of hours. This particular fuel cell design is most ideal for applications for which reliable performance, quasi-constant cell voltage, high conversion efficiency, and operational safety over extended durations are the principal requirements.

Table 3.11   Cell Voltage as a Function of Current Density of a Hydrogen-Oxygen Fuel Cell with a 1-sq. ft. Electrode Made from Carbon

Current Density (A/ft.²)	Fuel Cell Voltage (V)
0	1.10
50	1.00
100	0.98
150	0.97
200	0.95
250	0.94
300	0.92

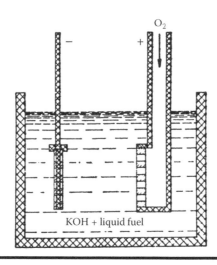

Figure 3.13   Fuel cell operation based on the catalytic dehydrogenation of liquid fuel dissolved in the electrolyte showing the oxygen or air cathode at right and dehydrogenating electrode at left.

3.9.1.2 Design Aspects and Performance Parameters of a Low-Cost, Moderate-Temperature Fuel Cell

DMFC offers a low-cost, portable, reliable fuel cell operating at moderate temperatures not exceeding 100°C and under one atmospheric pressure. It uses dissolved fuels, such as CH_3OH, which are fairly inexpensive and available without any restriction. A dissolved fuel cell contains a mixture of the usual alkaline aqueous electrolyte and a soluble cheap fuel. The electrodes must show different specific catalyst activities to produce voltage. The cell must contain a highly

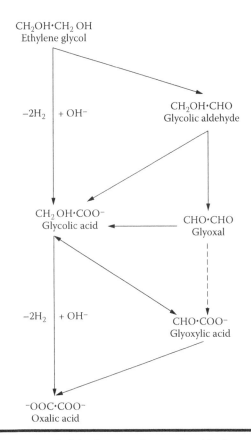

Figure 3.14 Four steps of dehydrogenation and oxidation of ethulene glycol fuel and production of glycolic oldehyde, glyoxol, and glyoxylic acid as final products.

active dehydrogenation catalyst. Platinum or palladium must be used to remove hydrogen atoms from the hydrocarbons and to act as a hydrogen electrode. Obviously, such an anode needs no complicated and expensive system of pores. A plain sheet with a cheap thin active layer will suffice. Therefore, there are no more geometrical limitations by three-phase boundaries, because both the catalytic dehydrogenation and electrochemical reaction will take place in a two-phase boundary. In addition, hydrogen may be absorbed immediately in the atomic state without intermediate molecular recombination, thereby saving dissociation energy as well as enhancing conversion efficiency. For these reasons, the dissolved fuel cell can achieve limiting current densities exceeding 1,000 mA/cm^2 under mild temperatures (65–100°C) and unity atmospheric conditions. This cell uses a soluble cheap fuel, which can be either a formic acid, CH_3OH, or ethylene glycol. In the case of complete combustion, water and carbon dioxide are the final

Table 3.12 Typical Dimensions Suggested for Various DMFC Components

Cell Component	Dimension (μm)
Flow channel thickness	3,000–4,000 (0.3–0.4 cm)
Diffusion layer thickness	100–350
Catalyst thickness	5–25
Proton conducting membrane	20–200

chemical reaction products. Both the liquid fuel and water contribute hydrogen. In the case of ethylene glycol, four moles of absorbed hydrogen fuel are produced. Moreover, because the liquid fuel is inflammable, it is highly attractive for space and underwater system applications.

The potential advantages of DMFC devices can be summarized as follows:

■ DMFC operates below 100°C.
■ Zero product of NO_x.
■ Minimum cost.
■ Compact packaging.
■ The devices offer safety, reliability, longevity, and portability.
■ Methane is stable and can be easily stored and transported with no risk.
■ Combustion products include carbon dioxide and water, which are relatively harmless.
■ A cartridge filled with methane fuel can be inserted to replace the empty cartridge, if continuous operation is desired.

Dimensional requirements for critical components for DMFC devices using methane as fuel are strictly dependent on the electrical power-generating capacity of the cell. DMFC cell designers recommend the typical dimensions for various components as shown in Table 3.12.

Note the recommended dimensions are valid for the DMFC devices operating at ambient temperature and pressure and with moderate power capacity only. Maximum power output and optimum cell conversion efficiency occur when all methane passing through the membrane is consumed on the cathode side in a direct catalytic burning with carbon dioxide and water as output products.

3.9.1.2.1 Applications of Dissolved Fuel Cells

These fuel cells are compact, relatively cheap, more reliable, and safe to operate. Such fuel cells are best suited for powering small vehicles such as scooters to operate in urban traffic environments, because the power-to-weight (P/W) ratio is large and the CH_3OH is rather cheap. The P/W ratio is better than

16.50 mills/kWh compared with 3.80 mills/kWh for gasoline. Small cars could travel the same distance with half of the fuel consumption and without excessive noise and odor.

The high-temperature fuel cells using natural gas are most ideal for large power stations, whereas the low-temperature fuel cells using hydrogen or alcohol are best suited for small domestic power plants, powering satellites, and propulsion of EVs or HEVs.

3.9.1.3 Design Requirements for Cost-Effective Fuel Cells

Fuel cells using hydrogen as fuel and carbon electrodes with porous Ni offer the most cost-effective design. Currently, this fuel cell design can be achieved with a cost as low as $100 per kW. Hydrogen is the cheapest fuel, and it must use a chemical converter for smooth and efficient operation. In the case of some cheap fuels, chemical converters are used to convert the hydrocarbons into hydrogen. This conversion process tends to increase the conversion efficiency of the fuel. In some hydrogen-based fuel cells with PEM configurations, water vaporization is a problem for the membrane. As stated earlier, the PEM cell consists of a fuel electrode (anode) and an oxidant electrode (cathode). The two electrodes are made of porous carbon material and are separated by an ion-conducting polymer electrolyte. The polymer electrolyte is also called the PEM. Integrated between each electrode and the electrolyte is a catalyst. Preliminary cost estimates indicate that the production cost of a PEM-based fuel cell is the lowest compared with other fuel cells with the same power capacity. Cost-effective fuel cell offers lower cost, high reliability, high longevity, and remarkable portability.

3.9.2 Ideal Fuel Cells for the Average Homeowner

Currently, all of the fuel design configurations will not meet the most appropriate requirements of a homeowner, such as low cost, compact size, and portability. Studies performed by the author indicate that for homeowners, the cost of the fuel cell is a matter of serious concern. Dr. Sridhar, an ex-NASA scientist, has proposed a fuel cell design configuration best suited for homeowners in terms of cost, size, and weight. This scientist calls this particular cell design a "bloom box." This fuel cell can fit into a shoe box with dimensions of 13 in. × 8 in. × 7 in. This design uses cheap natural gas as a fuel and converts the gas into electricity with drastically lower carbon emissions. The scientist claims that this fuel cell design configuration offers electrical power capacity close to 3.5 kW, which is sufficient to meet the power requirements for an average household with no central air-conditioning system. The author is seriously interested in design, development, test, and evaluation of this fuel cell design configuration in the immediate future. The author plans to use and evaluate other cheap fuels using this design with an emphasis on cost, conversion efficiency, safety, reliability, and carbon emission contents. Chemical converters are used in the case of certain cheap fuels to convert the hydrocarbons into hydrogen.

3.9.2.1 Design Requirements for Fuel Cells for Homeowners

Requirements for the most cost-effective fuel cell must be defined for each application. Such requirements will not only be different from one application to another, but also the design requirements will be more stringent for a fuel cell designed for space or satellite application. In this particular application, the availability of certain electrical power levels is needed to accomplish the specified space missions, and fuel cell longevity and power plant safety are of paramount importance. In the case of nonspace applications, the requirements will be relatively less stringent. For a fuel cell to power the lights and domestic appliances in a household, maximum emphasis must be placed on cost, performance, reliability, safety, portability, and longevity. In general, the following design requirements are given serious consideration for the portable, cost-effective design of a fuel cell:

- Use of the cheapest fuels, such as natural gas, kerosene, petroleum, and hydrogen, should be given preference.
- Use of a chemical converter is essential to extract the hydrogen for alcohol or hydrocarbon fuels, such as gasoline or kerosene oil.
- Use of carbon electrodes with porous Ni types is recommended to keep the device cost well below $100 per kW of output power. Use of other electrode materials should be considered if the cost of carbon electrodes is too high.
- Water vaporization problem must be avoided, if membrane-based cells are used.
- Optimum thickness of electrodes based on performance trade-off studies for the electrodes must be selected.
- Higher temperature operations must be avoided to eliminate the heat flow and extract problems, because it will increase the cell procurement cost.
- Computational fluid dynamic (CFD) software is recommended to solve the hydrodynamic flows of the gases in the case of high-capacity fuel cells, if natural gas is used as a fuel.

3.9.2.2 Compact Fuel Cells for Cars, Scooters, and Motor Bikes

Most citizens of third world countries cannot afford the high cost of gasoline for their automobiles and scooters. Fuel cells using cheap fuels offer the most economical solutions for those driving cars, scooters, or motor bikes. According to published reports, some citizens of Asia, Africa, and Latin America have shown great interest in the deployment of fuel cell technology. Fuel cells will one day help improve the air quality by powering electric cars, trucks, and scooters in crowded cities around the world.

Material scientists believe that an immediate application of metal-based fuel cells is as an emergency electricity source that will replace the generators driven by the internal combustion engines that produce the maximum amount of greenhouse gases. A zinc-air fuel cell has demonstrated to be reliably capable of powering computers, lights, printers, radios, fax machines, and other low-power appliances during a power outage while generating neither noise nor pollutants. Aluminum-air fuel cells have demonstrated a cost-effective operation compared with conventional lead-acid batteries, which occupy more space and cost more. Fuel cell designers claim that the metal-based fuel cells yield much higher volumetric energy densities, thereby requiring much less space to provide the same amount of backup electric power. The cell designers further claim that low-power (less than 1 kW) metal-based fuel cells will be most ideal for powering cell phones, iPods, and laptop computers, whereas the medium-sized devices (less than 5 kW) and large-sized cells (less than 1 MW) will be most ideal for stationary power system applications.

Preliminary calculations indicate that a hydrogen-based fuel cell offers a specific energy of 42 kWh/kg, whereas a gasoline-based fuel cell offers a specific energy of 14 kWh/kg compared with a specific energy of 0.042 kWh/kg from an acid-based battery. Material scientists claim that aluminum-based fuel cells can store a specific energy density of 4 kWh/kg compared with 1 kWh/kg for zinc-based devices. Corrosion is a major problem with aluminum-based devices and requires preventive maintenance in the case of aluminum-based fuel cells. In case of zinc-air batteries, zinc can be left in contact with a corrosive electrolyte. This means that the "spent" fuel (zinc oxide), along with some of the liquid electrolyte solution, must be removed, and a zinc pallet must be inserted in to the fuel cell along with the replacement liquid electrolyte. This presents a maintenance problem in the case of zinc-air fuel cells. In summary, maintenance problems in both metal-based fuel cells must be addressed using a low-cost maintenance approach. Once the maintenance problems are minimized or eliminated, the use of metal-based fuel cells must be deferred until an appropriate solution to the maintenance problems outlined above are solved.

The author has given some thought on how to solve the maintenance problems in these devices. Research studies on gelled liquid electrolytes seem to offer reasonable and low cost solution to the above mentioned problem [11]. It makes sense to use gelled liquid electrolytes in both metal-based fuel cells to eliminate the maintenance problems. A Taiwanese manufacturing facility is actively engaged in manufacturing zinc-air fuel cells using gelled liquid electrolytes for scooters. This manufacturer has a goal of developing an aluminum-based fuel cell involving a three-layer laminate of aluminum, membrane electrolyte, and air cathode that can be cheaply produced in large quantities. If this design approach is successful, metal-based fuel cells will find a place in electric cars, trucks, and scooters.

3.9.2.3 Fuel Cells for Portable Electric Power Systems

Portable electric power systems are most attractive for applications where an electric power source can be deployed from one place to another regardless of the distance or location. Portable applications include lighting and music needed for wedding parties, laptops, and remote locations where commercial power lines are not available. Power capacity for such portable power sources could vary from hundreds of watts to tens of kilowatts. Silicon-based fuel cells with moderate power capacity are best suited for this particular application. Silicon-based fuel cells use liquid electrolyte to achieve fast electrode reaction and a highly structured, porous silicon substrate to obtain improved cell performance in terms of conversion efficiency and fast electrochemical response. A porous silicon structure offers several advantages because of the extremely small deviations in the porous size and distribution. The porous size typically varies from 5 to 15 mm. The cell can be fabricated using the well-defined silicon-processing technology that has been successfully deployed in the fabrication of microwave solid-state devices. This silicon-processing technology will allow the device fabrication with minimum cost and high yield. In summary, the cell can be manufactured in high volumes with the lowest cost, thereby leading to widespread commercial applications of fuel cells where affordability, portability, and longevity are principal design requirements.

3.9.2.3.1 Hybrid Versions of Silicon-Based Fuel Cells

The fuel cells described in Section 3.9.2.3 are silicon-based devices whose principal function is to generate clean electrical energy at a minimal cost. The hybrid system uses a porous silicon structure, a flow-through methanol anode, and the nitric acid cathode to achieve high electrochemical conversion efficiency. The silicon electrodes are modified for the gaseous reactants to create a gas-liquid interface within the pores of the silicon structure. This approach is most practical and cost-effective for hybrid-based energy sources where the hydrogen-oxygen fuel cells could be integrated with renewal energy sources such as solar cells or microwind turbines and electrolyzers to produce hydrogen at lower costs. This technology, which involves a liquid electrolyte and a porous electrode structure, yields the most efficient method in the development of a cost-effective reversal cell capable of operating both as a fuel cell and as an electrolyzer simultaneously.

3.9.2.3.2 Applications of Reversal Fuel Cells

Providing electrical power at remote and inaccessible regions is not easy if commercial power lines are not in those locations. Currently, the power at these locations in most cases is provided by high-power batteries or conventional diesel generators. Both options have several fundamental problems, such as crude oil cost, greenhouse gas emissions, and continuous power capacity in the case of batteries. If a gas engine is used at such locations, then a commercial gas line is needed at that location in addition to a gas pipe installation cost plus carbon dioxide emissions. Because of the previously mentioned problems, even the conventional fuel cell technology alone cannot meet the present and future power requirements at inaccessible locations.

3.10 Fuel Cells Capable of Operating in Ultra-High-Temperature Environments

Research scientists working at GE and Westinghouse Research Laboratories have done outstanding work on ultra-high-temperature cells. German and Russian scientists have performed limited experimental work on high-temperature fuel cells. No other known scientists were actively engaged in the design and development activities on such fuel cells in the 1900s. Types of materials and their availability at lower costs are of critical importance because of high-temperature operations close to 1,100°C. In addition, the purity of fuel and electrolyte is a serious concern in the design and development of such fuel cells.

3.10.1 Types of Materials Used in Ultra-High-Temperature Fuel Cells

GE scientists developed the cell using natural gas as fuel. The gas was enclosed in a heating jacket. The cell's operating temperature was 1,093°C (2,000°F). When this temperature is reached, the natural gas is fed directly into the cell. The natural gas decomposes into carbon and hydrogen at this operating temperature. This carbon deposits on the outside of a long cylindrical cup made of the solid electrolyte. The deposited carbon acts as a fuel electrode. The electrolyte is a solid gas-impregnated zirconia (ZrO_2), suitably doped with calcium oxide (CaO) to supply enough oxide ions to carry the cell current. The oxidant, air or oxygen, is bubbled through the molten Ag cathode, which is held inside the ZrO_2 cup. The by-products, CO and the hydrogen that was formed in the initial fuel decomposition, are then burned outside the cell to keep the cell at the operating temperature. The hydrogen is not directly involved in the electrochemical reaction.

The Westinghouse scientists have developed a solid electrolyte for their high-temperature cell. ZrO_2, doped either with 15-mol% CaO or 10-mol% yttrium oxide (Y_2O_2), is used as the electrolyte. The scientists used two types of fuels, either 7% hydrogen in nitrogen or pure hydrogen only. Air was used as an oxidant. Both air and fuel electrodes were made out of platinum, initially applied to the ZrO_2 as platinum paint. The ZrO_2 electrolyte contains typically 10-mol% Y_2O_2. A fuel cell using pure hydrogen as a fuel and using air as an oxidant can yield the current density constant at 50 mA/cm^2 and 1,000°C, according to material scientists.

3.10.2 Solid Electrolyte Most Ideal for Fuel Cells Operating at Higher Temperatures (600–1,000°C)

Two distinct electrolytes, namely molten electrolyte and solid electrolyte, have been investigated for high-power sources. A fuel cell using molten electrolyte, hydrogen fuel, and air as oxidant that produced a power density of 45 W/ft.^{2} at 0.7 V has demonstrated a remarkable reliability. A two-cell device demonstrated a power density of 58 W/ft.^{2} at 1.3 V and a continuous operation exceeding 4,000 h with no deterioration in cell performance.

GE Company and Westinghouse Corporation have done considerable work on such fuel cells. The fuel cell using solid ZrO_2 as the electrolyte and air as an oxidant has demonstrated a current density more than 150 A/ft.^{2} at 0.7 V and at an operating temperature of 900°C. Tests conducted by GE indicate that this particular device has a conversion efficiency better than 30% in the beginning, but in later tests (after doping the ZrO_2 electrolyte with Y_2O_2), the efficiency improved, exceeding 48%. The cell designers predict that the power density of this device could exceed 2.5 kW/ft.^{2} after optimizing the geometrical cell parameters. Test conducted by GE showed the continuous operating life initially better than 3,500 h with no performance degradation. The cell designer predicts the watts per pound to be better than 125. This estimate is strictly based on the volume and weight of the active portion of the cell, geometrical area of the two electrodes, and the distance between their inner surfaces. The design configuration of this fuel cell is best suited for applications in which large output power levels are required over short durations close to 1,000 h.

3.10.2.1 Molten Electrolytes Offer Improved Efficiencies in High-Temperature Operations

High-temperature fuel cells initially conceived using molten salt electrolytes were later modified at the University of Amsterdam and at the Technical Institute in the Netherlands [10]. The electrolyte used appears to be a solid electrolyte but acts electrochemically as a liquid melt electrolyte. The electrode surfaces usually consist of metal powders. This includes Ag for the air or oxygen cathode and Fe, Ni, and ZnO/Ag mixtures for the anode. At high temperatures (600–800°C) of these molten electrolytes, substances that usually are difficult to react electrochemically easily yield high current densities better than 100 mA/cm^2 at 0.7 V. Even at high temperatures, hydrogen exhibits only low polarization. Hydrocarbons also produce electricity in these cells, but this is strictly due to the formation of hydrogen during the thermal decomposition or formation of carbon dioxide. High-temperature fuel cells offer the possibility of using cheap fuels such as coal, natural gas, and petroleum to generate electricity in competition with the usual thermal methods but at least twice the conversion efficiency, ranging from 65 to 75% versus 30 to 40% for conventional thermal power plants. This means that high-temperature fuel cells are best suited to generate large amounts of electricity at much higher efficiency compared with thermal-based electrical power plants.

Fuel cells offer a technology for more efficient energy conversion from thermal energy to electrical energy. Regardless of operating temperatures, fuel cells offer the following benefits:

- Thermal energy conversion efficiency ranging from 60 to 85%
- Weight and volume reductions in the order of one-tenth to one one-hundredth
- Low cost to produce electricity
- Most cost-effective portable electrical power-generating source

3.10.2.2 Performance Capabilities of Porous Electrodes

Fuel cell designers have recognized the most important fact: Performance capability is significantly improved when porous electrodes are deployed. Designers believe strongly that porous electrodes are used to maximize the interfacial area of the catalyst per unit of the geometrical area of the electrode, leading to a significant increase in the power output of the device. In fuel cells, an electrolytic and dissolved gas acting as a catalyst are used to improve efficiency. The electrode must be designed to maximize the available catalyst area, while minimizing the resistance to mass transport in the electrolytic and gas phases and the electronic resistance in the solid phase. Clearly, this is a stringent set of performance requirements. The porous electrode theory provides the formation of a three-dimensional structure with continuous transport paths in multiple phases as well as provides a reliable mathematical framework to model the complex electrode structures in terms of the well-defined macroscopic variable involved.

Material scientists believe that porous electrode theory has been used to describe a variety of electrochemical devices, including fuel cells, rechargeable batteries, separable devices, and electrochemical capacitors. In many of these devices, except fuel cells, the electrode contains a single solid phase and a single fluid phase. In the case of fuel cells, the electrode contains more than one fluid phase, which not only introduces additional complications but also degrades the cell conversion efficiency. The classical gas diffusion electrode contains both an electrolytic phase and a gas phase in addition to the solid, electronically conducting phase. In summary, porous electrodes tend to increase both the thermodynamic efficiency and the electrochemical efficiency, leading to significant performance improvement of a fuel cell.

3.10.3 Electrode Kinetics and Their Impact on High-Power Fuel Cell Performance

Electrode kinetics plays a key role in designing the most efficient electrodes for high-capacity fuel cells operating at high temperatures. Scientists and engineers believe that the application of electrode kinetics to porous electrode structures is of paramount importance. Problems associated with engineering design, materials of construction, catalyst, temperature, and mass transport control can be solved if the theory of electrode kinetics is applied rigorously during the design phase of the electrodes. Electrochemical kinetics and mass transport within the cell determine the voltage-current characteristics and ultimately the power output of the cell [10]. The effectiveness of a given electrode structure and material and the reasons for short electrode life or limited power output are all related closely to the electrochemical kinetics on the electrodes. Electrode kinetics can explicitly explain why the cell voltage falls when the current is drawn for the cell. This loss of voltage is called polarization. In fact, there are three distinct reasons for the decrease in cell voltage. First, when the current flows, there will be potential drop within the fuel cell due to ohmic resistances within the electrodes, leads, and electrolyte. Second, the slowness of chemical and electrochemical reactions will give rise to a form of polarization known as activation polarization. Third, the mass transport effects are responsible for a rise in polarization.

Scientist and design engineers working on programs of empirical research must use the concepts of electrode kinetics, if meaningful evaluation and improvement of the performance of fuel cells are the principal design goals. The theory of electrode kinetics reveals that the electrons will stabilize with positive ions in the electrolyte under the equilibrium condition. Under such conditions, the distance between the electrode surface and the plane of closest approach is determined by the molecular interaction forces.

3.10.4 Polarization for Chemisorption-Desorption Rates

Determination of chemisorption-desorption rates under electrochemical reaction and under equilibrium condition is possible. These rates can be determined using the Temkin Isotherm theorem and high-speed computers. Polarization-current relation of a chemical reaction can be investigated using large concentrations of reactants and products. Furthermore, impurities in the electrode material and other critical elements in the overall system can yield two pronounced effects. First, the electrochemical reactions in general proceed via active surface sites on the electrode. Any impurity in the electrode material may chemisorb strongly on these sites and block off the surface. Second, the impurity current interference will produce irregular current directions, which is extremely difficult to predict. Even the existing models are in no position to determine accurately the chemisorption and desorption effects. More research work is needed to determine these rates as accurately as possible. These two effects should be given serious consideration, particularly in the design phase of high-power and high-temperature fuel cells.

3.11 Fuel Cell Requirements for Electric Power Plant Applications

Fuel cells with output capacity ranging from 1 to 50 MW were designed between 1960 and 1990. Recently, higher capacity fuel cells have been designed and developed that are compatible with commercial gridline requirements. Fuel cells using natural gas with an output capacity ranging from 40 to 60 kW have been designed for hospital applications. Five fuel cells each with a power output capacity of 250 kW have been developed and are connected to a commercial gridline to demonstrate the power capability for high-power generation exceeding 1 MW or higher. Distributed electric power-generating plants typically have power capacities ranging from 2 to 20 MW (minimum). Base-load electric power plants typically have capacities ranging from 100 to 500 MW while operating on natural gas. Electric power modules with low-power capacities have been developed to operate under minimum noise and a low thermal signature.

3.11.1 Performance Requirements of Fuel Cells for Power Plants

Power output capacity of fuel cells for power plants could range anywhere from 1 to 100 MW depending on their ability to meet customer electric power needs. Therefore, the design configuration for the electrodes, types of electrode, the type and amount of fuel needed, and the physical parameters of the cell will strictly depend on the capacity of the fuel cell needed to meet the power plant electrical load requirements. Designers of high-capacity fuel cells reveal that the larger the fuel storage tank or larger the electrodes the greater the energy available from the fuel cells will be. Furthermore, by using dry or metallic fuels, high reliability and safe operation are possible compared with liquid fuels used in the fuel cells where leakage could be a serious maintenance problem [11]. Metallic-based fuel cells using aluminum or zinc tend to yield higher electrical energy per unit mass. Aluminum offers an energy density of ~4 kWh/kg of mass, whereas zinc offers an energy density of ~1 kWh/kg. High capacity from fuel cells can be achieved by stacking the fuel cells and connecting them in appropriate configurations.

A company in Taiwan is developing the metallic fuel cells based on a three-layer laminate of aluminum, membrane electrolyte, and air cathode [11]. The company claims that such fuel cell configuration can be manufactured at the lowest cost. These fuel cells will be most ideal for electric vehicles and portable power sources. The output capacity of such devices can exceed 150 kW. Capacity as high as 250 kW or more is possible using the stacking technique. Metal fuel cells are best suited for powering electric cars, buses, and backup electric power generators. These metal fuel cells are most ideal for power plants with moderate capacities ranging from 100 to 500 kW. However, power capacity can be increased to more than 1 MW if the stacking technique—involving hundreds of such devices—is used.

3.12 Summary

The chronological developmental history of fuel cell technology is briefly summarized. The operating principle of fuel cells is explicitly defined, and the critical role played by each cell element is identified. Performance capabilities and limitations of fuel cells using aqueous, semisolid, molten, and acidic electrolytes are summarized by identifying the temperature and pressure conditions. Calculated values of current density (mA/cm^2) and polarization voltage (V) for fuel cells using hydrogen, methanol, and methane fuels are provided for the benefit of fuel cell designers. Fuel cells for residential applications with power capacities ranging from 3 to 8 kW can be developed using natural gas. Natural gas used as a fuel offers minimum operating cost. Polarization voltage and current density as a function of temperature for various fuel cells using semisolid electrolytes are specified. Performance parameters of various types of fuel cells, including alkaline fuel cells, phosphoric-acid fuel cells, molten carbonate fuel cells, SOFCs, and DMFCs are summarized with an emphasis on output power capacity, reliability, and electrochemical conversion efficiency. Typical current density and terminal voltage estimates for low- and high-temperature fuel cells must be provided by the fuel cell designers. Critical performance parameters of HYDROX fuel cells are summarized as a function of operating temperature and pressure. Power density and current density estimates for fuel cells using aqueous electrolyte are provided for the benefit of fuel cell designers. Electrode dimensional requirements for liquid-gas-based fuel cells are identified. Architectural design parameters for hydrogen-based double-skeleton catalyst fuel cells are summarized, with a particular emphasis on electrochemical conversion efficiency and current density as a function of operating duration. Test data on terminal voltage and output power of a low-temperature hydrogen-oxygen fuel cell are summarized at 70°C. Performance specifications for an IEM fuel cell are provided to prove the suitability of this device for space system applications in terms of reliability, safety, and continuous power capability over extended durations exceeding 5,000 h and without any degradation in cell performance. Performance parameters of fuel cells using low-cost, porous silicon substrate materials are discussed, with an emphasis on reliability and longevity.

Mathematical expressions for fuel cell reactions and associated thermodynamic efficiencies are derived for the benefit of readers. Byproducts of various electrochemical reactions are identified. PEM-based DMFC is described in great detail with a particular emphasis on conversion efficiency and power output. Adverse effects of surrounding environments on reliability, structural integrity, and longevity associated with PEM-based fuel cells are summarized. Design aspects and performance parameters of DMFC devices are identified. Typical dimensional parameters of a DMFC device are provided for the customers to set aside space for installation. Fuel cells most ideal for anticounterinsurgency, space, battlefield, and underwater vehicle applications are briefly described. Fuel cell types and design configurations for UAVs, UGVs, attack electronic drones, and underwater vehicles are indentified.

Fuel cell design configurations for battlefield-based UAVs are discussed, with a major emphasis on cost, size, weight, and structural integrity. Estimates of conversion efficiency and power output for various fuel cells are summarized. Estimated values of current density and terminal voltage for sodium hydro-oxide Na OH and potassium hydro-oxide electrolytes are provided.

Current density and terminal voltage of a hydrogen-oxygen fuel cell using carbon electrodes are summarized under ambient operating conditions. Performance capabilities and applications of dissolved fuel cells are specified with an emphasis on cost, safety, and reliability over long durations. Design requirements for most cost-effective fuel cells are defined. Design parameters of fuel cells best suited for scooters, electric vehicles, and motor bikes are identified with a major emphasis on cost, reliability, and structural integrity.

Design configurations and performance requirements for portable electric power systems are identified, with a principal emphasis on weight, size, and longevity. Advantages of hybrid-version, silicon-based fuel cells are briefly summarized. Design requirements for fuel cells capable operating under high-temperature and -pressure environments are also outlined. Both the electrical and mechanical reliability are of paramount importance under high-temperature and -pressure environments. Performance parameters and major benefits of porous electrodes are identified. Electrode kinetics and their impact on performance of high-power fuel cells are discussed in great detail.

References

1. Edward W. Justi, "Fuel cell research in Europe," *Proceedings of the IEEE* (May 1963), pp. 784–791.
2. C. Gordon Peattie, "A summary of practical fuel cell technology," *Proceedings of the IEEE* (May 1963), pp. 795–804.
3. Tom Gilcrist, "Fuel cells to the fore," *IEEE Spectrum* (November 1988), pp. 35–40.
4. Chief Editor, "Extreme pressure sensitivity," *Machine Design* (March 2010), p. 55.
5. Chief Editor, "Better power from DMFC for commercial, military applications," *Electronic Products* (March 2010), pp. 51–54.
6. K.V. Kordesch," Low temperature fuel cells," *Proceedings of the IEEE* (May 1963), pp. 806–819.
7. John Kelley, "Unusual vehicles leave boot camp to join the regular forces," *Military and Aerospace Electronics* (July 2009), pp. 17–18.
8. Glen Zorpette, "Countering the IEDs," *IEEE Spectrum* (September 2008), pp. 27–29.
9. Editor-in-Chief, "Electric power aircraft," *Power Electronics Technology* (October 2010), pp. 19–21.
10. L. G. Austin, "Electrode kinetics and fuel cells," *Proceedings of the IEEE* (May 1963), p. 820.
11. Editor, "Metal fuel cells," *IEEE Spectrum* (June 2001), pp. 55–59.

Chapter 4

Batteries for Electric and Hybrid Vehicles

4.1 Introduction

Electric vehicle (EV) and hybrid electric vehicle (HEV) technologies are on a roll and are currently receiving great attention. Furthermore, the high cost of gasoline, low mileage per gallon of gasoline, and health hazards from greenhouse gas effects associated with gasoline-based vehicles are forcing drivers to consider the use of EV and HEV vehicles. For these reasons, automobile manufacturers are eager to adopt EV technology, and many automobile manufacturers around the world have recently unveiled their latest models of electric cars with many options. EV technology offers environmental-free, noise-free transportation as well as complete independence from hostile gasoline suppliers. These EVs and HEVs require high-capacity, reliable rechargeable batteries with specific performance requirements in terms of depth of discharge (DOD), state of charge (SOC), open-circuit voltage (OCV), and rate of discharge (ROD). Because of high consumer interest in EVs and HEVs, several battery-manufacturing companies are looking at the latest designs of rechargeable batteries most suited for these vehicles.

Invention in EVs and HEVs was demonstrated as early as 1900. The Baker EVs, which demonstrated a top speed exceeding 22 miles per hour (mph), were popular in the early 1900s [1]. Jay Leno, the host of *The Tonight Show*, owns one such vehicle, which is still operational. A hybrid EV using batteries and chargers was developed and tested by Walter Baker in the first hybrid-based electric car (the Torpedo). This particular hybrid electric car demonstrated a dash speed of 75 mph in 1902. As far as battery capabilities are concerned, a nickel-iron (Ni-Fe) battery was developed by Thomas Edison with performance characteristics similar to a lithium-ion (Li-ion) battery in terms of energy density and production cost. The battery designer predicted that Li-ion batteries could last for decades, showing acceptable degradation in performance.

In 2010, every major automobile manufacturer had a glut of gas-based vehicles, but several types of HEVs and several improved versions of electric cars were advertised in leading newspapers. Even upstarts like Zero Air Pollution (ZAP) and Tesla Motors vehicles are flexing their muscles and showing their intentions to join the automobile market in the near future. Studies performed by the author on these cars seem to indicate that they have major advantages, such as a lower center of gravity, higher gas mileage, improved mechanical integrity, and optimum performance. Because the batteries in electric cars are generally stored under the floor, a lower center of gravity is possible in EVs and HEVs, which is a major advantage demonstrating improved safety, high reliability, and excellent road ability. Despite several advantages of EVs and HEVs, the availability of charging stations or locations could be the greatest challenge in mass-scale acceptance of EVs.

Despite their higher selling price and limited operating range compared with gas vehicles, EVs offer several advantages. The higher cost of EVs is strictly due to the recharging infrastructure and lower sales volume at appropriate time frames. The HEV technology concept allows the owner or the customer of the HEV or an alternate-fuel vehicle to sell back the electricity to utility power grids as demonstrated in Figure 4.1. Potential advantages of EVs are much greater than their disadvantages. EVs are best suited for city driving involving stop-and-go operation. These vehicles offer optimum torque capability and have the lowest center of gravity. The electric motors can be placed next to the wheels. Active stability control becomes much easier and more effective because the electric motors are much easier to control with minimum time, including the ability to provide driving power or breaking capability. The most important design feature of all-electric vehicles includes the complete elimination of belts, oil changes, and often the costly transmission. The all-electric vehicles offer noise-free and pollution-free operation. The car wiring becomes significantly simple, reliable, and least expensive. In the case of all-electric vehicles, heating and air conditioning do not depend on the engine, thereby allowing other critical hardware to be placed in different areas of the vehicles, possibly in unused space. This design aspect of all-electric vehicles opens the possibility for optimized distributed systems. The battery placement in all-electric cars is more flexible than gasoline engines and drivetrains. The internal temperature of all-electric vehicles is relatively much lower because of the absence of the internal combustion engine (ICE), which produces the maximum heat, thereby requiring no or little air conditioning while driving.

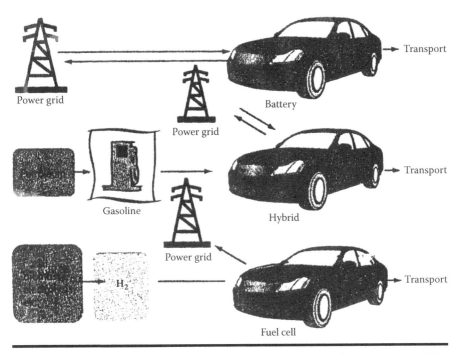

Figure 4.1 **Extended range of hybrid electric vehicle technology concept, which allows the use of electric, hybrid electric, and alternate-fuel vehicles to sell back the electricity to commercial grids.**

4.2 Chronological Development History of Early Electric Vehicles and Their Performance Parameters

Several contenders have shown great interest in pushing their EVs in the automobile market. Manufacturing companies such as ZAP and Baker Electric are the early pioneers in developing the electric cars. Current electric car manufacturers, namely Aptera and Tesla Motors, are the current players that have designed electric and hybrid electric cars that have demonstrated impressive gas mileage even under short-distance driving conditions. Leading automobile manufacturers, including General Motors (GM), Ford, Toyota, Nissan, Honda, and Tesla Motors, are actively engaged in the design and development of EVs and HEVs using advanced induction electric motor technologies and high-capacity, reliable battery banks.

4.2.1 Electric-Based Transportation Means

Low-cost and extremely compact electric-based transportation vehicles will be rolling out through 2012. The electric-based vehicles summarized in Table 4.1 will change the way we drive. These electric-based vehicles offer the lowest transportation

Table 4.1 Ultra-Compact Transport Vehicles for City and Short-Distance Driving

Vehicle Type	Specifications	Advantages	Drawbacks	Models	Price (US$)
PTVs	Battery powered, two-seater	Easy to navigate and easy to park	Limited space	Duo, Tango	$23,000–$25,000
Electric cars	Plug in and drive	$0.02 to $0.03/mile	Recharge after 50 to 100 miles	Nissan LEAF, Chevy Volt	$25,000–$30,000
Minicars	Fuel efficient, four-seater	Can go up to 50 miles between fill-ups	Noisy at high speeds	Ford Fiesta, Chevy Spark	$12,000–$16,000

cost, offer moderate procurement cost, and occupy minimum parking space. Such vehicles are best suited for short driving distances in the city. These electric-based vehicles do not offer a comfortable ride, which is normally possible in midsize or full-size automobiles.

4.3 Electric and Hybrid Electric Vehicles Developed Earlier by Various Companies and Their Performance Specifications

EVs and HEVs manufactured by various automobile companies will be briefly discussed in this section with an emphasis on battery types used, recharging time frame, types of electric motors deployed, miles per gallon, retail or estimated current price of the vehicle, and critical performance parameters of the vehicles.

4.3.1 ZAPTRUCK

Published articles indicate that ZAP has been active in turning out a range of EVs for road transportation with minimum cost and passenger comfort. ZAP has manufactured electric cars and trucks. Early versions of ZAPTRUCKs mostly use lead-acid batteries and were best suited for short hauls not exceeding 25 miles. Currently, ZAP uses lead-acid batteries as well as Li-ion batteries to achieve a travel range close to 50 miles. The top speed of these trucks using

Li-ion batteries is approximately 50 mph. The ZAPTRUCK uses a direct current (DC) motor. But, when the platform switches to an alternating current (AC) motor, the travel range is extended to 50 miles, and regenerative braking becomes possible with this package configuration. These trucks offer the most cost-effective performance for round-trips not exceeding 25 miles. Most stop-and-go travel is ideal for EVs of either configuration, but the worst option is always associated with gas vehicles. The ZAPTRUCKs deploy batteries with recharge times less than 1 h, regardless of battery type. This allows battery swap if necessary. Most customers feel that the gas vehicles will always be the best alternative for long-range travel. Studies performed by the author indicate that the EVs will be most appropriate options for customers whose daily travel does not exceed 35 miles.

4.3.2 ZAP ALIAS

ZAP has developed a three-wheeled vehicle known as ZAP ALIAS, which has a pair of electric motors to drive the front wheel. The three-seat, three-wheeled driving machine is slim and stable. ZAP ALIAS has a top speed exceeding 100 mph and a travel range of more than 100 miles. The car uses Li-ion battery packs located under the passenger cage, which can be charged from a 100-volt (V) electric source. The 216 V AC induction drive motors are located in the front section and are capable of driving the front wheels. The control electronics are fairly advanced. Battery management and battery-charging technology are impressive. The ALIAS retails at about $35,000.

4.3.3 Aptera Motors

Aptera Motors has developed both EVs and HEVs. The three-wheeled EV has a travel range of 100 miles, employs an AC induction motor, and runs on Li-ion battery packs. The battery packs can be charged from a 110 V electric source within less than 1 h.

The hybrid version of this three-wheeled electric vehicle has a 50-mile electric range (electric range means the car can operate without recharging the battery) and a 350-mile gasoline range. In brief, this particular car offers hybrid-based travel close to 400 miles. Its aerodynamic body design offers Formula One-inspired passenger cell, recessed windshield wipers, and a low-rolling resistance tires. The vehicle is equipped with a bank of batteries located below the passenger seat, an induction motor, and an Li-ion battery pack. The hybrid version offers cost-effective performance both in city driving and long-distance travel on highways. The older models of this car use an Li-ion battery, whereas new models (series 2) will deploy Li-ion-phosphate batteries to achieve overall better performance. The estimated retail price is more than $55,000.

4.3.4 Tesla Motors

The Roadster (electric car) is considered to be a high-performance sports car from the ground up. The company spent a lot of engineering time and effort to create its motor and battery pack system. The 375 V, three-phase, four-pole, air-cooled AC induction electric motor generates a peak output power in excess of 288 horsepower (HP). The induction motor has a speed of 14,000 revolutions per minute (rpm) and weighs about 70 pounds (lbs.). The minimum estimated price is $128,500 [1]. The stator of the motor has high-density winding to achieve improved performance in terms of low-resistance, high-peak torque, and a top speed of less than 3.5 seconds. High-peak torque delivers the kind of performance expected from an expensive EV.

This sports car has stylish interior and exterior designs and has a travel range close to 235 miles. The lithium-based battery pack employs 6,831 individual cells, which are needed to meet the electrical power input requirements of the three-phase induction motor. The battery pack can be fully charged in 3.5 h using a 240 V, 70-ampere (A) power source. The battery pack is designed to last for a minimum of seven years or 100,000 miles, whichever comes first. For optimum reliability performance, the battery pack is water cooled and microprocessor managed. This particular Tesla Roadster model is expensive and is also designed for two persons: hence, the company does not expect to sell a large number of such cars. Tesla Motors plans to introduce a seven-passenger, affordable electric car (Model S) to the market in 2012.

Tesla Motors Model S can accommodate seven people (five adults and two children) and is priced under $50,000. This particular model uses a three-phase, water-cooled, AC induction motor and has an operating range of more than 300 miles. The model has a top speed of 120 mph as well as some unique features, such as a 17-in. infotainment touch screen with 3G Internet connectivity.

Several battery-charging options are possible with this model. The quick charging time of 40 min. is required with a 440 V electric source. It takes 4 h to charge from a 240 V source. It can be charged from a 110 V source, but it will take more than 8 h. The charger is designed to run off 110 V, 220 V, and 440 V electrical sources. The 8,000-cell lithium-based battery pack can be swapped in 5 min., allowing the drivers to purchase a 160-mile battery pack and rent a 300-mile battery pack, if desired. The battery pack uses Li-ion technology.

4.3.5 Baker Motors

According to published reports, "The First and Foremost Electric Car" was developed more than a century ago by Baker Motors. Although the interior and exterior features of the car are not very fancy or attractive, it nevertheless provides a reliable and safe ride over short distances. Specific details on the electric motor, recharging time, types of batteries used, and performance parameters are not readily available. These EVs offer cost-effective transportation, particularly within the city limits, involving multiple trips, as long the battery mains fully charged. Although Walter Baker's electric Torpedo demonstrated a top speed of 75 mph in 1902, this EV is suitable only for city driving and has a maximum range of 35 miles.

4.4 Development History of the Latest Electric and Hybrid Electric Vehicles and Their Performance Capabilities and Limitations

On the basis of published reports on EVs and HEVs, it seems that GM, Ford, Toyota, Nissan, and Honda have done significant work on design, research, development, and testing on EVs and HEVs. In addition, these companies have identified suitable batteries for their vehicles. The author will recommend the latest battery technology that will significantly enhance the electrical performance and reliability of the vehicle involved. The need to reduce the hydrocarbon-based pollutants produced by the ICE and the dependence on costly foreign oil have forced the automobile manufacturers to look for EVs and HEVs. The automobile mass market is eager to adopt the HEV technology that fits into the average consumer's already squeezed budget. Studies performed by the author on affordability and battery-recharging infrastructure indicate some major hurdles. First, the current price projections for HEVs range from $35,000 to $45,000, which is too high for most consumers. The second hurdle is how the current electric grid infrastructure will handle a significant increase in automobile batteries that require daily recharging. Third, the 21st century's most advanced and high-capacity battery technologies are still very expensive. Fourth, the battery pack costs vary from $6,000 to $10,000 depending on the type of batteries and their longevity and reliability requirements. Currently, the battery pack suppliers predict the operating life between 7 and 10 years. High costs and long recharging times are major shortcomings of EVs and HEVs. Consumers must undertake a brief trade-off study to understand the benefits and drawbacks associated with all-electric and hybrid electric cars to determine whether buying such a vehicle is best suited and most cost-effective in the long run [2].

4.4.1 GM Chevy Volt

The Chevy Volt is GM's answer to the electric car challenge. GM designed the initial version of this car in 2007. The Chevy Volt is an extended-range electric vehicle (E-REV) car and was first available in 2011. The E-REV is essentially a plug-in hybrid electric car with an all-electric driving range of 40 miles without filling the gas tank [3]. The Chevy Volt is a four-cylinder vehicle, requires a

4 h recharge time at a 240 V electric source, and uses the four-cylinder gasoline engine, which runs when the battery is depleted. In the Chevy Volt vehicle, the battery is located beneath the rear passenger and driver's seats. An electronic battery-charge monitor system provides the driver with charge status information as well as an audible warning signal at least 15 min. before the battery charge level reaches its predetermined charge threshold. Its estimated retail price is less than $25,000.

The 2011 Chevy Volt has an entirely new power train, patented battery pack, extended driving range, and innovative transmission system. The GM engineers have used unique design concepts to reduce weight and drag significantly to boost the driving range. This five-door vehicle carries a 435 lb., Li-ion battery pack with 16 kilowatt per hour (kWh) capacity, which is located under the seats occupied by the driver and the passengers. The GM designers selected Li-ion chemistry over metal-hydride technology because it packs two to three times the electric power in a smaller package. According to GM engineers, this particular battery pack experiences little loss of charge when not in use and is not as prone to losing its storage capacity after repeated charge and discharge cycles. The battery pack contains 288 prismatic cell elements grouped into nine modules. The battery management system deploys the latest diagnostic circuit technology, which performs 500 tests every 10 seconds, with 85% of those tests ensuring that the battery is working safely and the other 15% tests tracking battery performance parameters and its life. The battery pack also contains thermal-management circuitry and a liquid-cooling system that heats or chills the battery package, depending on the ambient conditions. This thermal-management system permits the battery to operate efficiently at ambient temperatures ranging from –13 to +122°F and extends the battery life. The Chevy Volt carries three liquid-cooling subsystems: one for the battery; one for the traction, generator motor, and electronics; and one for the gasoline ICE.

This vehicle has a remarkable electrical and mechanical performance. The 3,780 lb. vehicle accelerates from 0 to 60 mph in 9 seconds and has a top speed of 100 mph. The low-speed torque is so impressive that it makes the car feel like a sedan with a 250 HP V6 engine. The rechargeable Li-ion battery sends the electrical power through an inverter to an AC 149 HP traction motor to power the Chevy Volt for about 40 miles. The 40-mile range can vary between 25 to 50 miles, depending on the road conditions, the number of passengers, and the operator's driving style. Once the battery is down to a certain level, an 84 HP (62.7 kW), 1.4-liter (L) gasoline engine switches on to spin a 54 kW generator, which powers the traction motor. The regenerative braking capability lets the traction motor act as a generator to recoup up to 0.2 g of braking force and convert it to electricity. Note 1 g of acceleration is equal to 32.2 ft./sec./sec. The traction motor also switches to a generator when the Chevy Volt coasts, which slows the car to generate more electricity. The combination of traction motor and electrical generator offers this car's unique performance, which has not been demonstrated by any all-electric or hybrid electric car to date. Another unique and economical feature of this car is that it can be instructed via a smart phone to recharge itself when the commercial utility rates are the lowest.

The Chevy Volt comes with two charging options. It can use a household source using the plug-in technique, but it will take 10 to 12 h to recharge the battery pack. The car also comes with a $400 option, which allows the driver to use a 240 V charging station. This method takes 4 h to recharge the battery pack with a charging station fee ranging from $0.10 to $0.15 per kWh. Recharging expenses associated with all-electric or hybrid electric vehicles plus the cost of gasoline must be considered when conducting a meaningful, economical trade-off study. The lithium battery pack is the most costly item in the car, and its typical initial price ranges from $4,000 to $6,000. Therefore, its high mechanical integrity is of critical importance in an EV or HEV. This vehicle offers a maximum lifetime exceeding 15 years or more. The 16 kWh Li-ion battery pack was developed by GM scientists, who claim that never fully charging or fully draining the battery extends the life of the battery. That is why the battery pack designed and developed by GM carries an 8-year/100,000-mile warranty. Lowering the center of gravity of the EV and deploying high-strength materials with optimum stiffness for the chassis are needed for the safest and most comfortable ride.

4.4.2 Ford

Ford has designed and developed an all-electric car (the 2011 Ford Focus) as well as hybrid electric cars (the 2012 Ford Focus, 2013 C-Max, and 2013 C-Max Energi). The year given before the model name indicates the year in which car is or will be available. These Ford cars deploy the latest and advanced materials that will provide the highest mechanical integrity and safety. Advanced, high-voltage Li-ion battery packs are used in all-electric cars. According to Ford scientists, the advanced Li-ion battery packs occupy between 25 and 30% less space and are lighter by 50%. This reduction in size and weight allow the Ford engineers to install other electronic components, if needed, to improve car performance in terms of reliability, safety, longevity, and passenger comfort.

4.4.2.1 Ford Focus

Details about this new, all-electric Ford Focus car were announced at the 2011 International Consumer Electronics Show in Last Vegas, Nevada. The Focus is the first all-electric and fuel-free, rechargeable passenger car from Ford Motor Company, and it is one of five new electrified vehicles that Ford will deliver to customers by 2013 in North America and Europe. The Focus was available in 2011 and is designed to offer enough travel range to cover the majority of daily driving habits of Americans. Ford designers claim that this vehicle will offer gas per gallon rates better than GM's all-electric Chevy Volt and other battery-based electric vehicles. Highlights of the Focus and Focus Electric vehicles can be summarized as follows:

- Fuel-free with zero emissions, rechargeable passenger car.
- Offers better mileage than Chevy Volt.
- The Focus is designed to provide outstanding energy efficiency, which will increase the lifetime of both the rechargeable battery and the EV.
- Focus Electric offers a remarkably quiet, comfortable car because of the absence of the gasoline engine.
- Focus Electric will be powered by a high-voltage, light-weight, advanced Li-ion battery pack.
- The battery system uses heated and cooled liquid to maintain the optimum battery pack temperature needed to maximize the battery life and, consequently, the life of the car.
- Focus deploys regenerative braking technology to achieve maximum energy efficiency and the energy of the battery with a fully charged status.
- The battery is designed for a 23 kWh energy level.
- This EV can use a household 110 V source for overnight battery charging or an optional 240 V charging station for a quick charge not to exceed 30 minutes.
- The estimated price for a partially loaded car is about $18,600.

4.4.2.2 Ford Escape

The Ford Escape is an HEV that uses gasoline for long-range driving and electric energy for city touch-and-go driving to reduce driving expenses. This is a five-passenger, plug-in hybrid car most ideal for family use. The highlights of the Escape can be summarized as follows:

- This vehicle comes both with a four-cylinder, 2.5 L, Atkinson-engine, which features an electric traction motor to run the 177 HP motor, or a six-cylinder, 3 L, flex-fuel engine.
- The regenerative braking system captures the kinetic energy produced by braking, which will charge the 330 V battery pack and keeps the battery fully charged.
- Several ultrasonic and electronic sensors are provided for performance improvement, reliability, and passenger safety, and all of these sensors are powered by the battery.
- The Escape hybrid is equipped with a 100 V power outlet for charging a laptop or other portable electronic devices.
- The Escape uses a permanent-magnet AC-synchronous electric motor with an output power rating equal to 94 HP.

- Ford offers a 10-year/150,000-mile limited warranty on the battery pack.
- The car comes with a 330 V, sealed nickel-metal-hydride (Ni-MH) recharge-able battery pack.
- The estimated price of this hybrid vehicle is about $30,000.
- This hybrid car offers 41 miles per gallon (mpg) in the city and 33 mpg on the highway.

Deployment of a basic alternator starter mechanism as shown in Figure 4.2 would offer the most cost-effective design for the induction motor used in the HEV.

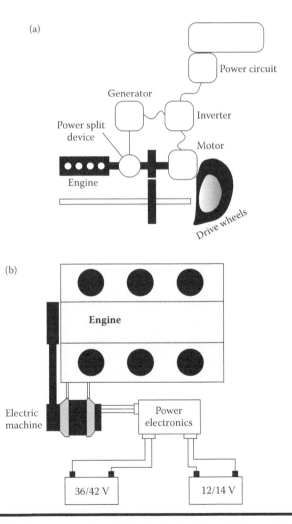

Figure 4.2   Critical components of (a) hybrid electric vehicle and (b) basic alternate starter mechanism.

4.4.2.3 Ford C-Max and Ford C-Max Energi

Ford has developed two impressive, plug-in hybrid electric cars, namely the C-Max and C-Max Energi. Each is capable of providing impressive performance in terms of reliability, comfort, and longevity. Both cars deploy a power-split hybrid architecture capable of operating in a fuel-saving electric mode beyond 47 mph and use advanced Li-ion battery systems. According to Ford, the C-Max Energi is a plug-in hybrid, provides a driving range in excess of 500 miles, and delivers improved charge-sustaining fuel economy over current plug-in hybrid cars. The major benefits of these plug-in hybrid electric cars can be summarized as follows:

- Electric driving range, most desirable for emission-free and silent city driving.
- Reduced dependency on gasoline.
- Significant saving on energy and fuel cost.
- Increased use of electricity from a renewable energy source (namely solar cells) for recharging the Li-ion battery pack while driving, thereby reducing the cost of recharging.
- Use of an advanced Li-ion battery offers up to 30% reduction in battery size and 50% reduction in battery pack weight.
- Ford provides onboard and off-board features and tips for remote-control recharging and preconditioning settings, monitoring battery SOC, and maximizing the energy efficiency needed to extend the electric mode of operation.
- Ford offers the "My Ford Touch" driver-connect technology, which provides useful information such as current fuel level, current battery power level, instant vehicle status information for driver safety, and average and instant miles per gallon.

4.4.3 Nissan

Nissan has developed and marketed its all-electric car, the LEAF. Its affordable price range and impressive gas mileage performance have attracted many customers whose major emphasis is on the affordability factor. The most attractive features of this all-electric car can be summarized as follows:

- 80 kW AC synchronous electric motor.
- Li-ion battery pack with capacity of 24 kWh.
- Li-ion battery pack offers the best energy-to-weight ratio with zero memory effect.
- Regenerative braking system helps to feed back the kinetic energy produced by the application of the brakes to the battery. This means recharging time is reduced.
- Onboard charger with 3.3 kW output power.

- 120 V portable trickle charger cable.
- Zero emission system.
- 100 miles with one charge.
- Regenerative brakes.
- Electronic brake force distribution.
- Photovoltaic (PV) solar panel spoiler.
- 120 V, 1.4 kW battery trickle charging provision for 24 h at home or in public.
- 240 V, 3.3 kW battery charging provision for 8 h at home.
- 480 V, 50–70 W battery charging provision 30 min. at home or at a charging station (optional).
- 107 HP capability of the four-cylinder engine.
- Five-year limited warranty.
- Suggested retail price between $25,000 and $30,000.

4.5 Performance Requirements of Various Rechargeable Batteries

Comprehensive studies have been undertaken by the author on various battery types best suited for EVs and HEVs. Cost, weight, and longevity were given serious consideration during these studies. These studies indicate that batteries run out of electrical energy quickly, especially at higher speeds, no matter what type of battery is deployed. As far as battery types are concerned, the studies recommend either the advanced Li-ion battery or an Ni-MH battery pack. These batteries would be most suitable for all-EV and HEVs because of their unique OCV performance as a function of DOD for both the new and aging vehicles as illustrated in Figure 4.3. Research and development activities undertaken by the scientists at Lawrence Livermore National Laboratory reveal that compared with Li-ion batteries, the zinc cell has demonstrated several advantages. The cell combines the atmospheric oxygen or air with zinc pellets to generate electricity, which can be stored in the battery. When all the zinc is consumed, the only by-product is zinc-oxide, which is completely recyclable. The fuel cell can be refueled in 10 minutes, which allows the cell to be used instantly. In fact, a continuous-feed cell will never have to shut down for refueling. This is the most important advantage of zinc-air fuel cells [4]. Furthermore, zinc is the least expensive option and is widely available.

The United States has only 35% of the world's zinc supply, which currently stands at 1.8 gigatons (1.8×10^9 tons). Zinc-air battery designers predict that 21 months' worth of global zinc production could be used to manufacture 1 billion 10 kWh zinc air cells. By contrast, it would take 180 years' worth of lithium material to make an equivalent amount of Li-ion batteries. Because most of the supply of lithium is located outside of the United States, Li-ion batteries would have to made outside of the United States. In addition, Li-ion batteries require charging

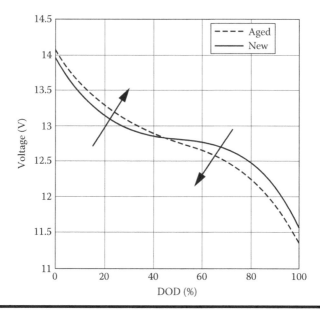

Figure 4.3 **OCV as a function of DOD for both new and old Ni-MH recharge-able batteries.**

time of 8 to 10 h and contain toxic elements that present health hazards. These cells output power between 1 and 3 kW. Most manufacturers of EVs and HEVs use Li-ion battery packs. Regardless of the battery pack selected for a specific EV or HEV vehicle, the battery pack must have the performance characteristics summarized in Table 4.2.

Table 4.2 Principal Battery Performance Requirements for an EV or HEV

Battery Pack Characteristics
High specific energy density (kWh/cm²)
High volumetric energy density (kWh/cm³)
High power density (W/cm²)
High shelf life (better than 10 years)
Stable battery performance at lower temperatures (20°C or lower)
Easy and less expensive charging management
Dormant capacity loss per year at constant temperature (not to exceed 2% per year)
Minimum capacity loss per year at low operating temperatures
High ampere-hour efficiency (ratio of charge to discharge) of the battery

Preliminary studies performed by the author reveal that Li-ion batteries satisfy all the performance characteristics as summarized in Table 4.2.

4.5.1 Battery Pack Energy Requirements

A single gallon of gasoline contains 33 kWh of electrical energy, and about two-thirds (67%) of the energy must be available from the entire battery pack, regardless of battery types. These estimates indicate that an HEV would derive driving energy from the gasoline and roughly two-thirds of its energy from the battery pack.

A high-speed expensive electric car such as a Mercedes would require a battery pack with 392 kW capacity, whereas compact EVs, such as Chevy Volt or Ford Fusion, would require a battery capacity ranging between 16 kW (the Volt) and 24 kW (the Fusion). Most compact or midsize electric cars could require battery packs with an energy capacity ranging from 18 to 35 kW, whereas the full-size all-electric or plug-in HEVs could require battery packs with a capacity ranging from 40 to 55 kW. As far as the peak battery capacity is concerned, a single-pack module could be designed with an electrical power output between 15 and 25 KW, depending on the space allocated for the battery pack. For all-electric or plug-in HEVs, however, the single large-size battery pack should be broken into two to three battery modules with a total capacity equivalent to that of a single battery pack, which may be too large to locate in the vehicle. As mentioned, the battery pack is generally located closest to the vehicle floor and between the two rear wheels. This type of module arrangement will be able to locate other critical components or devices needed for passenger safety and vehicle reliability.

4.5.2 Battery Materials and Associated Costs

Most of the EV and HEV manufacturers are using lithium material for the battery packs. This material is expensive and its availability in the future is questionable. Because this material is toxic, its temporary storage and disposal expenses could be more than 15% of the initial cost of the battery pack. In addition, there is a high drop-off fee close to $1.00/lb. The overall weight of Li-ion battery packs varies somewhere between 450 and 550 lbs. for compact and midsize passenger EVs, depending on the battery pack capacity. The battery pack used in full-size EVs could weigh between 600 and 825 lbs., depending on battery capacity. Even the 16 kWh Li-ion battery pack deployed by the Chevy Volt weighs 435 lbs. (198.1 kg). This particular battery comes with a warranty of eight years/100,000 miles. The cost of an Li-ion battery pack varies from $4,000 to $8,000 depending on battery capacity. Finally, justification of buying an EV is strictly dependent on the life-cycle cost criterion.

4.5.2.1 Materials for Rechargeable Batteries Deployed in EVs and EHVs

On the basis of studies performed by the author on materials best suited for fabrication of rechargeable battery cells, it is found that lithium, nickel, cadmium, polymer, zinc, spinel, and manganese are most suitable for manufacturing the rechargeable battery packs widely used in EVs and HEVs. Although zinc is widely used in the manufacturing of zinc-air fuel cells, it can also be used in the fabrication of rechargeable battery packs for EVs and HEVs.

The zinc-based cells combine atmospheric oxygen with zinc pellets to produce electricity. When all the zinc is consumed, the only by-product is zinc oxide, which is completely recyclable. The cells can be refueled in less than 10 minutes, thereby allowing the cells to be used almost constantly. This type of cell would never require shutdown for refueling purposes. Note that zinc cells offer several advantages over Li-ion batteries. First, zinc is widely available. Second, zinc is cheaper than lithium. Third, the worldwide zinc supply is enough to manufacture more than 1 billion 10 kWh zinc-air cells. By contrast, it would take roughly 180 years of lithium production to make an equivalent amount of Li-ion batteries. Fourth, most supplies of lithium are located outside of the United States, which means that Li-ion batteries would be manufactured outside of the country on the basis of cost-effective criterion.

Li-ion batteries are widely used by the manufacturers of all-electric and hybrid electric vehicles. Li-ion batteries could require charging times of 4 to 10 h depending on the type of car and the battery power density and energy density. Therefore, recharging costs for Li-ion batteries will be much higher for HEVs compared with all-electric cars

Li-ion batteries come in two categories: Li-ion-alkaline batteries and Li-ion-polymer batteries. The Li-ion-alkaline battery packs are relatively cheaper than Li-ion-polymer batteries. The performance of Li-ion-polymer batteries is superior to the performance of conventional Li-ion-alkaline batteries, particularly in terms of energy density. Materials used, specific performance characteristics of various rechargeable battery packs widely used in EVs and HEVs, and parameters of various rechargeable batteries are summarized in Table 4.3.

It is evident from the tabulated values that the optimum values of specific energy, energy density, and life-cycle parameters are possible only with Li-ion-polymer batteries using manganese spinel as the positive electrode. The procurement cost for these rechargeable Li-ion-polymer battery is the highest. It appears that the higher the battery performance, the higher the procurement cost. Li-ion rechargeable batteries are free from memory effects and have demonstrated the lowest self-discharge rates, which offer high reliability as well as high longevity. For these reasons, Li-ion batteries are widely used in military systems, EVs, and aerospace applications in which high performance and longevity are of critical importance.

Table 4.3   Performance Characteristics of Rechargeable Batteries Widely Used for Commercial Transport Vehicles

Critical Cell Parameter	Critical Cell Parameter		
	Lead-Acid	*Ni-MH*	*Lithium-Ion*
Specific energy (Wh/kg)	30–50	65–95	150–194
Energy density (Wh/L)	80–90	300–350	350–475
Nominal cell voltage (V)	2.00	1.25	3.72
Life cycle (to 80% of initial capacity)	200–300	500–1,000	500–1,250
Fast charge time (h)	8–16	1	2–4
Overcharge tolerance	High	Low	Moderate*
Self-discharge per month (%)	5	20	3–5
Operating temperature for discharge cycle only (°C)	−20/+60	−20/+65	−20/+75
Estimated cost ($/Wh)	0.4	0.5	0.8

* Parametric value improvement for Li-ion-polymer battery only.

4.5.2.2 *Impact of Road and Driving Conditions on Battery Charging Times and Costs*

The operating costs for the EVs and HEVs are strictly dependent on the operating conditions, including the surface conditions of roads, weather environments, city or highway driving, plug-in charging stations, and infrastructures available within reasonable intervals. According to a California Energy Commission report, starting in 2012, Californians will buy 12% of all EVs and 24% of all HEVs. The report further indicated that roughly 5,500 plug-in hybrid electric and all-electric vehicles were being driven in the state at the end of 2011. Such a deployment of all-electric and hybrid electric vehicles in California will strain the local grid transmission lines and could damage weak points in the distribution power gridlines and power transformers. Utility power companies worry that EV charging could cause blackouts in the neighborhoods of some early EV adapters and give a bad reputation to the emerging technology. According to the director of EV readiness efforts of the California-based Southern California Edison, a few thousands EVs would not crash the local grids in California, but they could cause temporary power interruptions, which can be easily rectified. The director further stated that the neighborhood electric circuits, power transformers, and distribution transmission lines will be made robust enough to support the additional electric loads created by charging stations.

Battery life is strictly dependent on the battery type, environmental temperature, use of battery for specific applications, and operating conditions. Limited studies undertaken by the author seem to indicate that retention of battery charge or energy density in Li-ion battery packs is dependent on road conditions, climatic environments, and the operational condition of the battery pack. The studies further indicate that a minimum energy density of 8.5 Wh/in^3 in an Li-manganese-dioxide battery, which is generally known as Li-ion battery, can be maintained over a temperature range of 60° to 140°F or 15.5° to 60°C. The battery designers claim that Li-ion batteries can be exposed to a maximum temperature 75°C for a short period of time without affecting the battery's performance, reliability, or life cycle. The EVs or HEVs must be operated at reduced power levels at higher temperatures. Note that charging at elevated temperatures could result in long-term damage to the battery structure. Furthermore, because lithium is a delicate material, charging and discharging must follow specific guidelines for the safety and reliability of the Li-ion battery packs.

Authentic life-cycle costs on rechargeable battery packs for EVs and HEVs are not available in explicitly clear terms. Performance data on prismatic-type Ni-MH batteries are available as a function of cell capacity (Ah), and charging current level are shown in Figure 4.4. The dealers of EVs and HEVs are offering limited warranties, ranging from 5 to 10 years with conditions in fine print. If an EV is involved in a rear-end collision, then the car owner will be involved in complex arguments and negotiations with the insurance companies. Practically all rechargeable battery packs are located between the two rear wheels, which is vulnerable to unlimited damage to an EV or EHV. Furthermore, even with limited damage, the battery pack needs to be replaced. Currently, the retail price for Li-ion battery packs varies from $4,000 to $6,500 depending on the energy density (Wh/in^3). Zinc-air battery packs offer a peak energy density of 17 Wh/in^3 but over a narrow temperature of 40° to 120°F or 5° to 38°C.

Similar reliable performance data on Ni-MH battery packs for EVs and HEVs are not readily available. Limited performance data on cell voltage as a function of cell capacity and current charge level on Ni-Mb battery are shown in Figure 4.4. The OCV for new and aging Ni-MH battery packs as a function of DOD are impressive, as illustrated in Figure 4.5. Published performance data on Ni-MH battery packs seem to indicate that the energy density is about 50% less compared with that of Li-ion batteries. But these battery packs come with robust design features and offer high mechanical integrity. Because reliable life-cycle cost data on zinc-air battery packs are limited, meaningful performance comparison cannot be made with other battery packs. These data will be of significant importance if realistic comparisons of different energy sources are to be made. Sealed lead-acid, Ni-MH, Li-ion-alkaline, and Li-ion-polymer rechargeable batteries are considered for this comparison in terms of power density (kW/kg) to energy density (kWh/kg) ratios (P/E ratios) for various energy storage batteries as illustrated in Table 4.4.

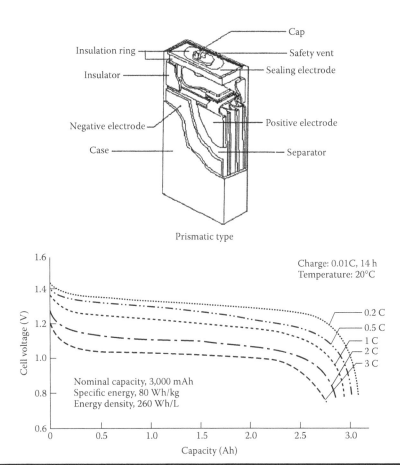

Figure 4.4   Prismatic version of Ni-MH rechargeable battery pack and its cell voltage as a function of battery capacity (Ah) at various charge current levels (V).

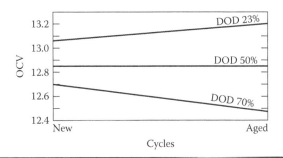

Figure 4.5   OCV of Ni-MH rechargeable battery as a function of DOD and battery cycles. If desired DOD is 70%, then the OCV for a new battery is 12.7 V, and for an aging battery it is about 12.47 V. For a DOD of 23%, the OCV for a new battery is 13.5 V and for an aging battery it is 13.2 V.

Table 4.4 Performance Comparison for Different Energy Storage Devices

Type of Storage Device	P/E	E/P	Remarks
Sealed lead-acid battery	6.0	0.166	Widely used for ICE-based cars
Ni-MH battery	2.7	0.370	Appropriate for small EVs
Li-ion-alkaline	7.0	0.1143	Ideal for EVs
Li-ion-polymer	36.0	0.033	Most suitable for HEVs

Note: P: power; E: energy.

4.6 Materials for Rechargeable Batteries

This section identifies the materials most suitable for rechargeable battery packs for possible applications in EVs and HEVs. While identifying such materials, emphasis has been placed, in particular, on material cost and battery performance, including reliability and longevity. Because nickel and lithium materials are the most ideal materials for automobile rechargeable batteries, these materials will be discussed in detail with an emphasis on reliability and safety under harsh temperature and mechanical environments. Hereafter, discussion limited to Li-ion rechargeable batteries will be given serious consideration.

4.6.1 Materials Requirements for Three Functional Components of the Li-Ion Battery

During the discharge cycle, Li-ions (Li^+) carry the current from the negative to positive electrode, through the nonaqueous electrolyte and separator diaphragm. During the charging cycle, an external electrical power source (namely the charging circuit) applies a higher voltage of the same polarity than the voltage produce by the battery, thereby forcing the current to travel in the reverse direction. The Li-ions then migrate from the positive electrode to the negative electrode, where they become embedded in the porous electrode material in a chemical process known as intercalation.

The three functional components of an Li-ion rechargeable battery are the *anode*, the *cathode*, and the *electrolyte*. In a conventional Li-ion cell, the anode is normally made from carbon, the cathode is made from metallic oxide, and the electrolyte is made from a lithium salt in an organic solvent. The following materials are used for the three distinct components of the lithium cell:

4.6.1.1 Anode

The most commercially available anode material is graphite and this material is not expensive.

4.6.1.2 Cathode

The cathode can be made from one of the three materials described below:

- A layered oxide, such as lithium cobalt oxide (LiCoO)
- A polyanion, such as lithium iron phosphate (LiFePs)
- A spinel, such as lithium manganese oxide (LiMnO)

4.6.1.3 Electrolyte

Electrolyte is the most important element of a rechargeable battery. The electrolyte is generally a mixture of organic carbonates, such as ethylene carbonate or diethyl carbonate containing complexes of Li-ions. These nonaqueous electrolytes generally use noncoordinating anion salts, such as lithium hexafluorophosphate ($LiPF_6$), lithuim hexafluoroarsenate monohydrate ($LiAsF_6$), lithium perchlorate ($LiClO_4$), lithium tetrafluoroborate ($LiBF_4$), and lithium triflate ($LiCF_3SO_3$).

4.6.2 Major Performance Characteristic of Li-Ion Batteries

The terminal voltage, battery capacity, cell life, and safety of an Li-ion rechargeable battery can change dramatically depending on the material choices and their amount. Recently, battery designers and material scientists have investigated novel cell architectures using nanotechnology, which have demonstrated significant improvement in the cell performance. Pure lithium is very reactive. It reacts vigorously with ordinary water to form lithium hydroxide and hydrogen gas. Thus, a nonaqueous electrolyte is generally used in designing a lithium cell, and a sealed container rigidly excludes water from the battery pack.

Li-ion batteries are widely used for aerospace applications and EVs and HEVs because of the advantages summarized thus far. Li-ion batteries are more expensive compared with other batteries with high energy densities, but they are capable of operating over harsh thermal environments and under moderate mechanical conditions. Most important, these batteries have unusually high energy densities over wide temperatures as shown in Table 4.3. Li-ion batteries are fragile and, therefore, they must be handled with care, and they need a protective circuit to limit the peak voltages. These batteries are smaller and lighter than counterpart rechargeable batteries. Li-ion batteries are readily available in the market. These batteries require long recharging times as high as 10 h and thus will need 220 V or 440 V plug-in charging stations at appropriate locations. Long recharging times for Li-ion batteries compared with Ni-MH batteries is the major shortcoming for owners who cannot wait too long or who are in hurry.

4.6.3 Characteristic of Nickel-Metal-Hydride Rechargeable Batteries

Studies performed by the author on alternate rechargeable batteries reveal that Ni-MH batteries could be used for EVs and HEVs. Their specific-energy levels are hardly 20 to 25% compared with Li-ion batteries and, hence, they require more frequent recharging. Because of frequent recharging requirements, charging costs for owners of HEVs will be much higher. Furthermore, HEVs equipped with Ni-MH battery packs require more frequent charging, leading to personal inconvenience as well as higher operating costs. On the other hand, their recharging time is about 1 h compared with 10 h for Li-ion batteries. Because of lower specific-energy levels, the Ni-MH batteries will be bulkier and heavier compared with Li-ion batteries, which could slightly decrease vehicle performance. Important characteristics and shortcomings of Ni-MH rechargeable batteries are summarized in Table 4.3. The procurement cost of Ni-MH rechargeable batteries will be lower compared with Li-ion batteries with the same energy-level ratings. The author has not seen wide use of Ni-MH battery packs in EVs or HEVs manufactured by well-known automobile manufacturers. So far, Toyota's small hybrid-electric car, the Prius, is equipped with an Ni-MH battery, and it has demonstrated a satisfactory performance, according to owners of this particular HEV. This battery has demonstrated impressive OCV as a function of DOD as illustrated by Figure 4.5. Research studies undertaken by the author on potential battery packs seem to indicate that several battery types, namely lithium-manganese, lithium-iron-sulfide, zinc-air, nickel-zinc, and others, are available that could be deployed in EVs or HEVs. But limited experimental performance data on their reliability and longevity are available.

4.6.4 Zinc-Air Rechargeable Fuel Cells for EVs and HEVs

Zinc-air rechargeable fuel cells could trump Li-ion batteries [4]. The zinc-air cell combines the atmospheric air or oxygen with zinc pellets to produce electricity. When all the zinc is consumed, the only by-product is zinc oxide (ZnO), which is completely recyclable. Fuel cell designers and engineers claim that the fuel cell can be fueled in less than 10 minutes, thereby allowing the cell to be used almost instantly. As a matter of fact, a continuous-feed cell would never have to shut down for refueling. Zinc-air fuel cells have the following advantages over the Li-ion rechargeable batteries:

- Currently, the United States has more than 35% of the world's zinc supply.
- Twenty-one months' worth of global zinc production could be used to manufacture more than 1 billion 10 kWh zinc-air cells.
- It would require 180 years' worth of lithium production to produce an equivalent amount of rechargeable Li-ion batteries. This means that the manufacturing speed for zinc-air cells is much faster.

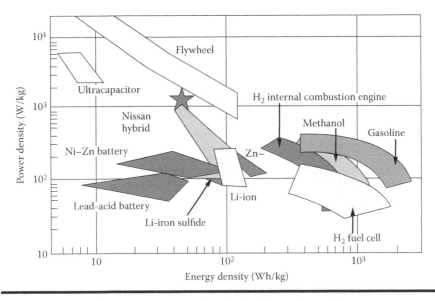

Figure 4.6 Specific power on power density (W/kg) and specific energy or energy density (Wh/kg) for various storage devices and energy conversion techniques.

■ Most supplies of lithium are imported from outside the United States; therefore, the battery designers are at the mercy of foreign suppliers in terms of cost, quality control, and delivery.

■ Lithium batteries contain toxic elements, whereas zinc-air cells contain none.

■ Zinc-air cells are much cheaper than lithium cells, despite the benefits of zinc-air fuel cells or battery packs. Some EVs and HEVs manufacturers are investigating other battery packs, such as lithium-iron-sulfide, lithium-manganese, and nickel-zinc battery packs as shown in Figure 4.6.

4.6.5 *Energy Density Levels for Various Rechargeable Batteries*

For the benefit of readers and owners of EVs and HEVs, this section summarizes the important parameters of rechargeable batteries, which are best suited for EVs and HEVs. Critical performance parameters of these three distinct rechargeable batteries are shown in Table 4.5.

Comprehensive studies performed by the author on various materials, including nickel and lithium, reveal that lithium appears to be the most suitable material, which offers optimum battery performance in terms of specific energy (Wh/kg), energy density (Wh/L), self-discharge rate per month, and longevity. Furthermore, this material is readily available, is free from memory effects, and maintains reasonably good performance over a wide operating temperature range. It requires care in handling and disposal of the material, however, because of its mild toxicity. The

Table 4.5   Critical Performance Parameters of Li-Ion-Based and Zinc-Air Rechargeable Batteries

Parameters	Rechargeable Battery Type for an Electrical Vehicle		
	Zinc-Air	Li-Ion-MnO	Li-Ion-Polymer
Specific energy (Wh/kg)	18.0	8.5	11.5
Battery pack cost (%)	45	100	80
Optimum energy over (°F)	50–100	80–140	80
Charging time (h)	1	4–10	4–8
Operating costs	Minimum	Moderate	Moderate
Estimated weight of battery pack in a small electric vehicle (lbs.)*	421	450	415
Approximate cost (US$/Wh)†	0.39	0.83	0.78

* Estimated values of weight for the Li-ion-MnO battery pack are assumed to be 450 lbs., whereas the weights of zinc-air battery pack and Li-ion-polymer battery packs are interpolated with respect to weight of Li-ion-MnO battery pack. The interpolated weight estimates could be accurate within 15 to 25%, because nobody knows the exact weight of the lithium, manganese oxide, and polymer contents.

† Similarly, the cost of the rechargeable battery packs is based on the estimated cost of the Li-ion-MnO battery pack. In brief, the weight and cost estimates show the trend for the respective battery packs.

level of toxicity does not pose a serious health hazard. Because of its high energy density and lowest self-discharge rate, lithium is widely used for military and aerospace applications for which optimum battery performance, high reliability, and longevity are the most critical design requirements.

The Li-ion battery is sometimes identified as Li-ion-manganese oxide (Li-ion-MnO). Essentially, the battery is made out of lithium and manganese oxide, whereas the Li-ion-polymer batteries are fabricated using lithium and polymer materials. Both the Li-ion-MnO and Li-ion-polymer battery packs are widely used by the EV and HEV manufacturers. The Li-ion-polymer battery packs can be easily damaged if not handled with extreme care. This particular battery pack requires outside walls made from high–mechanical strength material to protect the delicate control of IC circuits and battery structure [5]. The Li-ion battery packs have unique characteristics.

4.6.5.1 Li-Ion Battery Pack Configuration

There is no fixed configuration requirement for the Li-ion battery pack. The battery capacity rating (kWh) strictly determines the pack configuration. Numerous designs are possible for assembling the cells into a battery pack for an EV or HEV. In most cases, 6 to 12 cells are packaged together into a unit called a module. Control and safety circuits are provided in each module to avoid severe structural damage on overcharge because of temperature rise, which can degrade battery performance. Battery performance loss as a function of temperature is shown in Figure 4.7. The modules can then be installed in the battery pack, which is sized to match the space allocated for the battery pack in the vehicle. For example, the Nissan Altra EV is equipped with an Li-ion battery pack consisting of 12 modules each. It carries eight cells and each cell is rated for 10 Ah capacity. In other words, the battery pack rating is 960 Ah. The overall weight of this battery pack is about

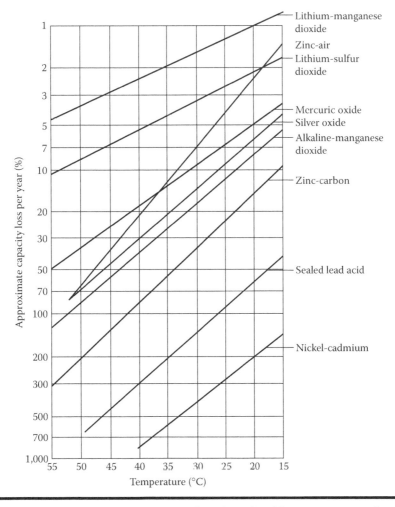

Figure 4.7 Capacity loss per year as a function of ambient temperature for various rechargeable battery packs.

Table 4.6 Typical Weight Breakdown for Each Component in a Li-Ion Battery Pack

Pack Material	Li-Ion Battery Pack Consists of 100 Cells Each Rating 10 Ah	
	Weight of the Component (lbs.)	Weight (%)
Negative electrode (dry)	56	17.2
Cathode (dry)	94	28.8
Container or housing	70	21.6
Electrolyte	44	13.5
Separators	16	5.0
Miscellaneous items	45	14.1
Total	325	100

365 lbs., excluding the weight of control and safety circuits. Studies performed by the author on weight optimization indicate that the principal objective in the development of battery packs for EVs is to maximize the specific energy (kWh/kg) or the electrical energy stored per unit mass. Weight breakdown for each component in an Li-ion battery pack is shown in Table 4.6.

4.6.5.2 Some Unique Problems Associated with Li-Ion Battery Packs

There are several advantages of these battery packs, but there are also some drawbacks of lithium technology, in particular when operating under harsh thermal and mechanical environments. The advantages and disadvantages of this battery technology can be summarized as follows:

- The Li-ion-polymer batteries have higher efficiencies compared with Li-ion-oxide batteries.
- They are most suitable for applications for which a sleeker profile is the principal design requirement, such as EVs or HEVs [5].
- They offer unique design features that allow for battery pack removal or replacement by the consumer with minimum effort.
- These battery packs are very thin.
- These batteries have very light-weight components.
- They meet multipurpose applications, including data capture, data storage, Smart Grid, and other commercial and industrial applications.

- These batteries are somewhat less toxic than Li-ion-oxide batteries.
- They are most suitable for either plug-in HEVs or parallel hybrid-type vehicles.
- These rechargeable batteries are ideal for commercial electronic devices, such as the iPod, iPad, and iPhone.
- These batteries can operate over temperatures ranging from –30° to +60°C with no compromise in reliability and safety except for a slight reduction in power capability, particularly at –30°C.
- When operating this battery, once you start drawing power from the battery at cooler temperatures, it heats up very nicely but shows no degradation in battery performance.
- Industrial users noticed that there are no serious problems at low temperatures, although they have observed some manageable problems above 40°C (104°F). If air cooling is not used, one could damage the Li-ion batteries, leading to some safety problems. Li-ion batteries should not be exposed to some environmental conditions, such as desert conditions or direct sunlight, to avoid damage to the heat-sensitive materials used in the battery. Li-ion batteries should not be located near heat-generating sources, such as a running ICE, exhaust tail pipe, or a military system operating at high temperatures.
- The greatest advantage of Li-ion technology is that as the operating or environmental temperature goes up, the internal resistance of the battery goes down. This leads to significant improvements in the battery, but it does so at the expense of reliability. It is important to install an accurate temperature sensor at the point closest to the highest temperature on the battery with a red warning symbol or red flag at 40°C.
- Studies performed by the author on the reliability and longevity of the Li-ion batteries indicate that the lifetime of these batteries is seriously affected at temperatures exceeding 40°C (104°F). The studies further indicate that when you go above 40°C, one has to consider the long-term impact on the battery life.
- In the case of transportation applications, use of full-power applications are allowed under safe temperature limits, but as the battery temperature approaches even 35°C, the electric car operator should start to limit the output power.
- Battery designers say that drivers of EVs and HEVs can continue to use full-rated power, if necessary, as long as the battery temperature stays below 35°C. This type of EV or HEV operation will have no impact on the long-term life of the vehicle.
- In some applications, a battery pack is needed to go in a sterilizer to meet a specific performance requirement. In such cases, the battery could be exposed to high temperatures exceeding 45°C for a short period of time, which will not affect the battery's performance, reliability, or life cycle.
- Some Li-ion battery suppliers mention that their maximum specified temperature range is 75°C, but they recommend battery operations at

reduced power levels, preferably at 75% of the full power rating of the battery pack.

■ Charging and discharging of the Li-ion battery packs already installed in the EV or HEV must follow specific guidelines for the safety of the battery. Li-ion battery owners are warned that charging at elevated temperatures could result in long-term battery damage, which increases with the timing at elevated temperatures and the SOC. Customers are advised to reduce the battery voltage during charging at elevated temperatures.

■ Battery owners should limit the current draw at higher temperatures to ensure the long-term life of the battery pack.

■ Li-ion batteries can meet the requirements for low-energy consumers and high-energy consumers, including owners of EVs and HEVs.

■ A battery-based power system composed of about 40 MW of rechargeable batteries worldwide is used to help regulate frequency and spinning reserves, which will lead to a stable electrical power grid.

■ Minimum charging costs for EVs and HEVs can be achieved by avoiding using charging stations during peak energy consumption times. For example, if you are paying $0.11 per kWh during regular operations when the utility company really needs the electrical power, the utility company will be willing to pay four, five, or six times that amount for power at that time to avoid having to operate another power plant to serve the energy peak.

■ Currently, the cost for Li-ion battery packs is very high, ranging from $4,500 to $6,500. Battery suppliers predict that when multiple industries accept the potential benefits of Li-ion technology, including the 10-year life span, the battery pack prices will come down (perhaps within the next four to six years).

■ Battery scientists and designers believe that for many applications for which space is at a premium, and considering the Li-ion's advantage in its charge density, the premium for Li-ion batteries is acceptable.

■ Currently, the weight of Li-ion battery packs varies approximately from 400 to 550 lbs. for EVs and from 500 to 750 lbs. for HEVs, depending on the power capacity of the batteries. The weight of these vehicles is a critical issue for the car manufacturers, and they are willing to pay a price premium for the rechargeable batteries with reduced weight and slim design.

■ Battery designers believe that the Li-ion batteries will last considerably longer under moderate ambient temperature inside the vehicle. The battery designers further believe that if the application has a lot of cycling, and if you are comparing just cost and life cycle to a lead-acid battery, the Li-ion battery will certainly be more cost-effective over a period of 10 years or so. The lead-acid batteries like to be cycled and, according to the battery designers, also will not last long in high temperatures. Customer requirements demand anywhere from 8,000 to 10,000 cycles and upward. According to the EV designers, these cycle estimates are compatible with a 10-year use cycle for transportation vehicles.

- Battery cost-reduction techniques require a fine-balancing approach, which will involve balancing electronics and monitoring electronics. Both of these techniques will ultimately lead to a potential growth area for all battery management. Note that the Li-ion battery packs seriously address these balancing issues and other potential techniques that can be used for balancing. To explain the balancing concept, one has to assume that the current into every cell is the same. In theory, the balancing techniques are based on the cell voltage. If we consider the effect of temperature on cell voltage, however, the expression for the cell output voltage becomes complicated and requires computational theory. Battery designers think that active balancing concepts will not be included in first-generation Li-ion battery packs. More advanced techniques, such as an active-balancing technique, will be involved in the design of second- and third-generation lithium-based battery packs.

- Currently, Li-ion, Li-polymer, and Ni-MH rechargeable batteries are manufactured for deployment in EVs and HEVs. Toyota Prius uses the Ni-MH rechargeable battery, which is automatically charged by an ICE whenever it runs. Furthermore, the cost of the Ni-MH battery pack is slightly lower than the Li-ion battery pack. Reliable cost data on the Ni-MH battery packs are not available. No external charging station is required for charging a battery pack in this car, but the charging provision is provided from a 110 V outlet in the customer's garage.

- Various types of lithium chemistries for rechargeable batteries have been investigated for possible applications in future EVs and HEVs. Comprehensive research and development activities are being pursued, in particular, on lithium-phosphate, lithium-iron phosphate, lithium-sulfide, and lithium-manganese rechargeable batteries with a maximum emphasis on cost- and weight-reduction techniques. Comprehensive examination of important properties of cathode materials must be undertaken to conclude whether a particular material selected for the cathode offers minimum cost and other desirable characteristics, such as antirunaway, anticorrosion, and nontoxicity properties. Studies performed by various material scientists reveal that lithium-iron-phosphate material offers optimum safety because of the absence of thermal runaway. In general, battery cells tend to experience capacity loss at elevated temperatures and suffer from thermal-runaway problems under extreme temperature rise.

- State-of-the-art algorithms are necessary to monitor vital electrical performance parameters, temperatures, and structural reliability parameters of each cell and module to ensure the reliability of the battery pack as well as the safety of the passengers in the vehicle. Second-generation control chips must be used to synchronize current and voltage measurements with an accuracy of +/-3 mV. Specification requirements for self-diagnosis for reliability and maintainability must be defined. Safety and maintenance experts recommend the use of battery management integrated circuits for everything and

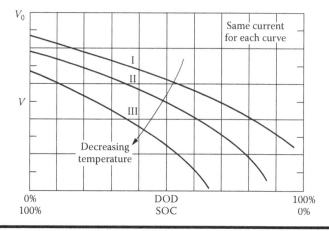

Figure 4.8 OCV of rechargeable battery as a function of DOD and SOC, both expressed in percentage.

precision analog electronics to manage and control the battery packs being deployed in EVs and HEVs. OCV as a function of DOD and SOC must be maintained, as shown in Figure 4.8, if optimum reliability, consistent performance under various load conditions, and optimum safety in EVs or HEVs are the principal operating requirements.

4.6.6 Design Concept Incorporating the Smart Grid Technology

Smart Grid technologies promise a major overhaul of the aging energy infrastructure. Charging stations must have facilities to charge the batteries installed in EVs and plug-in hybrid electric vehicles (PHEVs) with minimum inconvenience to the customers. Customer input via a long-term, two-way communication will be beneficial to the owners of EVs and HEVs. The ability to successfully charge for servicing outside the customer's home is strictly dependent on the state-of-the-art charging infrastructure. One of the most critical aspects of the Smart Grid is the emergence of the EVs and PHEVs. These mobile loads and charging energy sources will have to be integral parts of the Smart Grid system [6], but not at the same location. Imagine several EV owners at a business meeting or at a party who want to charge their vehicles. In addition to the sudden impact on the local transformer, the EVs and PHEVs need to connect and communicate with a Smart Grid so that the vehicles can be charged within a minimum time frame and in an orderly manner.

The success of EVs and PHEVs, which can be collectively called plug-in electric vehicles (PEVs), heavily depends on the effective charging infrastructures. The locations of these charging infrastructures are identified in the Smart Grid technology handbook, which is provided to every PEV owner [6]. The U.S. federal government wants to have more than 1 million PEVs on the road by 2015. The EV's battery charging requirements will increase the household's electricity consumption as high as 50% or more. The electricity-consumption requirement could be easily satisfied by the existing power transformers and distributing transmission lines.

4.6.6.1 Charging-Load Impact on the Utility Gridlines

On the basis of the rechargeable batteries installed in EVs and HEVs, the electrical load effects on the home and the local transformer will be quite different. As far as battery ratings are concerned, an all-electric small sedan will have a battery rating between 22 and 25 kWh. Conversely, both the Ford Fusion HEV and Nissan LEAF HEV use a 24 kWh battery. As far as electric consumption is concerned, because of climatic variation, the daily use of energy consumed in the charging battery at home could differ from 30 kWh/day in Arizona to 20 kWh/day in the Detroit area. EV charging levels, types of batteries involved, and the charging expenses for different EVs and HEVs are summarized in Table 4.7.

Table 4.7 Electric Vehicles Charging Levels, Types of Batteries Involved, and Charging Expenses Involved for Various Electric and Hybrid Electric Vehicles

Vehicle Type	Charging Level or Voltage (V)	Phase	Charging Current (A) and Energy (kWh)	Charging Time (h)	Typical Charging Expense (US$)*
GM Chevy Volt (EV)	AC 110 V	Single	16 and 18	2–3	1.80
GM Chevy Volt (HEV)	AC 240 V	Single	15 and 36	<1	3.60
Maxima-OptaMotive	AC 420 V	Single	40 and 88	2–3	7.68
E-Tracer (HEV)	AC 240 V	Single	80 and 92	1.5	28.10

* The charging expenses assume the commercial utility charge of $0.10 to $0.11/kWh of electricity. So the total electricity consumed is equal to the product of voltage and current. Total charging expenses is the product of electrical energy consumed and the cost per unit of electricity. In the case of the Maxima, the utility voltage is AC 110 V, which is updated to AC 420 V using a step-up transformer because the Li-ion battery used requires a charging voltage 420 V to reduce the charging time to 1.5 h. All battery-based EVs are HEVs except the GM Chevy Volt, which is only EV.

4.6.6.2 Typical Charging Rates for Rechargeable Battery Packs and Electrical Load

Charging rates for all-electric vehicles are different than those for PEVs and, furthermore, the charging rates are a function of line voltage available at the charging locations and the EV charger rating. The chargers are rated at 3.3 kWh or 6.6 kWh. The low-rating chargers generally are used for small electric cars, whereas the high-rating chargers (6.6 kWh) are used for PHEVs or high-power all-electric vehicles. For example, the 18 kWh battery will take 6 to 7 h to recharge when using the 3.3 kWh charger. The Volt has an 8 kWh battery, which can be recharged using a household 110 V source drawing only 12 amperes, which can take 5 to 7 h, or a 240 V commercial plug-in source drawing 40 amperes, which could take less than 1 h of charging time.

After 2015, practically all PEVs will be able to communicate with the Smart Grid systems. For example, GM's Chevy Volt can take advantage of built-in communication capabilities, such as the OnStar subsystem, which is an integral part of Smart Grid system. The National Institute of Standards and Technology (NIST) is taking great interest in this technology and is directing the consolidation of all standard activities. The NIST Framework and Roadmap for Smart Grid Operating Standards (see Figure 4.9) identifies the following priorities [6]:

- Demand response
- Consumer energy efficiency
- Energy storage availability
- Wide-area situational awareness
- Electric transportation

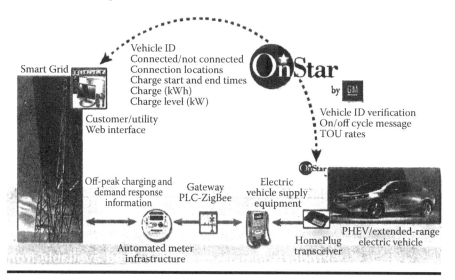

Figure 4.9 **The National Institute of Standards and Technology Framework and Roadmap can be integrated with the Smart Grid system concept.**

- Advanced metering infrastructure
- Distribution grid management
- Cybersecurity capability
- Network communication capability

4.6.7 Materials and Their Properties Best Suited for Rechargeable Batteries

Studies performed by the author on various types of rechargeable batteries for EVs and HEVs reveal that lithium-based batteries are widely used by various manufacturers. The most widely used rechargeable batteries include Li-ion-manganese-dioxide, Li-ion-iron phosphate, Li-ion-sulfide, and Li-ion-polymer. Performance capabilities and other important characteristics of these batteries are described in great detail in this section, with a particular emphasis on energy density, self-discharge rate, longevity, and power density.

High-cathode material cost, density of materials used in various battery components, and the number of cells and modules used determine the battery cost and total battery weight, regardless of battery type. Estimated material cost and density of various materials used by the critical elements of the battery are summarized in Table 4.8. The overall battery weight strictly depends on the number of cells used; the density of the materials used for cathode, electrode, and electrolyte; the number of separators deployed; and the container or the enclosure. Weight contributions by electrode, cathode, electrolyte, and container come to 82% of the total weight of the high-power rechargeable battery pack currently used in the Nissan Altra EV. The total weight of this particular high-power Li-ion battery pack is just under 800 lbs. Comprehensive weight analysis indicates that the cathode contributes roughly 50% of the total material cost for the high-energy battery cell.

4.6.7.1 Major Material Costs for a 100 Ah High-Energy Rechargeable Battery Pack

The material cost of a battery pack is strictly dependent on the energy level, electrical load, and battery capacity requirements. Suitable materials are required for fabrication and protection for the active cathode, anode, electrolyte, separator, cell, and pack container or housing. The latest material cost estimates for a typical high-energy rechargeable battery pack with 100 Ah capacity are summarized in Table 4.9 [7].

4.6.7.2 Estimated Costs for Battery Packs Widely Used in All-Electric and Hybrid Electric Vehicles

Li-ion and Ni-MH are the only two batteries currently used for all-electric and hybrid electric vehicles. Battery pack weight and cost are the most critical

Table 4.8   Estimated Material Cost and Density of Various Materials Used

Component	Base Element	Density (g/cc)	(lbs./in.³)	Material Cost/ Pound
Anode (–electrode)	Graphite	2.20	0.21	$7.00–$16.00 (depending on the grade)
Cathode	Cobalt	8.90	0.868	$10.00–$12.00
Cathode (+electrode)	Nickel	8.90	0.863	$3.00–$4.00
Lithium core	Lithium carbonate	5.44	0.528	$1.00–$2.50
Cathode (+electrode)	Manganese	7.44	0.725	$0.10–$0.92
Electrolyte	Li-salt (LiPF$_6$)	5.22	0.506	$14.00–$35.00
Module housing	Polybutylene plastic	0.92	0.089	$1.00–$2.00
Pack housing/frame	Aluminum	2.70	0.263	$1.60–$2.00

Note: The material price quotations are just approximate values. Price quotations will be accurate within ±10% depending on the location, quality of material, and amount of material purchased. Platinum is hardly used in any component deployed by the rechargeable battery pack because it is not very expensive. However, it has a very high density, which will significantly increase the battery pack. To convert the density of the material from g/cc to lbs./in.³, multiply by a factor of 0.097.

Table 4.9   Estimated Material Costs for a High-Energy Rechargeable 100 Ah Battery Pack

Component	Material Cost (US$/lb.)	Cost of the Component (%)
Active cathode	25.00	50.2
Electrolyte	27.30	22.2
Graphite	13.60	10.9
Separator material	81.80	7.0
Container material (Al)	9.10	0.8
Copper	6.80	1.5
Can and vents materials	8.20	2.2
Miscellaneous materials	17.70	5.2
Total	Material cost/lb. × 100% weight of the material	

parameters that dictate the retail price of EVs and HEVs. As far as the battery pack price is concerned, it varies from somewhere around $3,550 for a small all-electric vehicle to more than $6,000 for a full-size, fully equipped HEV. The battery pack base price of Li-ion battery packs could be as high as $1,000 per kWh, which means that the retail price of the Li-ion battery pack for the GM's PHEV (the Volt) will be somewhere between $13,000 and $15,000, depending on the battery pack supplier. Higher cathode cost (51%), electrolyte cost (22%), and anode cost (11%) are the principal reasons for higher battery pack costs. In other words, 84% of the total material costs are due to the cost of the cathode, anode, and electrolyte.

Most economists feel that the high cost of a battery pack appears to be the most negative factor in owning an all-electric or hybrid electric vehicle. In addition, recharging and recycling costs are relatively high. Furthermore, tax experts and economists believe that depreciation on the new battery, if an owner buys one, may be questionable, because it is an integral part of the car. They further believe that with a lower cost battery pack, there is a reduced investment in recovering fuel-cost savings, less weight to lug around, and less risk from uncertainty of future gasoline prices. Toyota's Prius took a long time to reach a breakeven point despite its moderate cost of close to $16,000. Economists further believe that high base prices of EVs and HEVs will discourage most customers from purchasing these cars, except for affluent buyers. To sell these vehicles, however, we need a breakthrough in the battery technology that will deliver three things—namely higher energy density, lower rechargeable battery cost, and higher overall performance than currently available from lithium-based batteries. The next section offers an affordability example based on several factors involved in conducting a cost-effective analysis of the EVs or HEVs.

4.6.7.2.1 Affordability Example for an EV or HEV

The GM Volt (a PHEV) was considered in a cost-effective analysis [8]. On the basis of published reports, this particular vehicle would achieve 250 Wh/mol. or 402 Wh/L when operating on electricity. The gasoline will cost about $3.00 per gallon or $0.79 per liter, and the electricity consumed will cost about $0.11 per kilowatt-hour. Under these costs, the HEV would cover about 150,000 miles, or 241,000 meters, over an assumed life of 12 years for the rechargeable battery pack.

4.6.7.2.2 Results of Analysis

This example offers a lifetime savings of $4,875, approximately, ignoring the battery charging costs. By discounting this amount at 10% over the 12-year duration to cover the cost of borrowing the money to buy this vehicle, one can realize a net savings of roughly $3,000 in fuel costs over the life of the vehicle This saving is achieved by running the vehicle on a domestic current source. The high costs of cathode and electrolyte materials are the principal reasons for the high cost of rechargeable battery packs.

4.6.8 Impact of Component Costs on the Procurement Cost of Battery Packs

The author feels that component costs as well as rechargeable battery pack costs will go down as the sale of EVs and HEVs increase in the future. On the basis of the information gathered from a few automobile dealers and rechargeable battery pack suppliers, the author expects that procurement costs for lithium-based rechargeable battery packs will go down from current prices by 15 to 20% within three to four years.

4.6.8.1 Estimated Current and Future Component Costs

On the basis of the cost data obtained from the component suppliers, the author feels that the procurement costs for rechargeable lithium-based battery packs will come down by more than 20% by 2015 [8]. The author believe that when the Li-ion battery technology is fully matured, the component designs are optimized and frozen, and the significant increase in sales of EVs and HEVs occurs, the prices of lithium-based rechargeable battery packs will most likely see a significant price reduction. The author's summary of current and future estimated component material costs for 100 Ah high-energy, rechargeable Li-ion battery packs and the relevant data are shown in Table 4.10. Cost estimates for a rechargeable battery pack are a function of the following factors: the ampere-hour rating of the rechargeable pack, the output capacity of the battery, the cost of material used in the battery components, the types of materials used for major battery elements (such as the cathode), the costs and type of electrolyte, the separator, the total number of cells and modules involved, the type of housing or container used, and the quality control and reliability assurance specifications that need to be satisfied. On the basis of the data presented in Table 4.10, it appears that the cathode cost currently is the highest, and will remain so, because of the shortage and high cost of cobalt material as illustrated in Table 4.10.

4.6.8.2 Material Cost Estimates

Actual material cost estimates for each battery component are not readily available from a battery supplier or from an automobile manufacture. Only the rough cost estimates for the battery's major components are avoidable, which might have significant errors between 10 and 15%. These cost estimates at least will demonstrate realistic trends in the component costs as well as in the overall battery pack cost. A comprehensive review of published technical papers and reports have allowed the author to summarize the current and future material costs for each discrete component used in a high-energy Li-ion battery pack, as shown in Table 4.10. Limited information on the materials used and their cost in each discrete component used in Ni-MH and Li-air battery packs is

Table 4.10 Current and Future Material Costs for the Components Used in High-Energy Battery with Lithium-Based Battery Packs Capacity Rating of 100 Ah (US$/lb.)

Component and Material	Current Cost and Percentage of Total Cost		Future Cost and Percentage of Total Material	
	Material Cost/Pound (US$/lb.)	*Material Cost (%)*	*Material Cost/Pound (US$/lb.)*	*Material Cost (%)*
Cathode (Co)	25.0	50.2	9.1	48.1
Electrolyte (lithium salt)	27.3	22.1	9.1	19.6
Separator (polyethylene)	81.8	7.0	18.8	4.2
Anode (graphite)	13.6	10.9	6.8	14.5
Housing/can (plastic/ aluminum)	8.2	2.1	7.2	2.6
Low-loss lines (copper)	6.6	1.5	4.6	1.5
Container (aluminum)	9.1	0.8	6.8	1.6
Miscellaneous items (plastics/metals)	17.7	5.3	8.6	6.8
Total percentage of material cost	100		100	

available. A preliminary market survey made by the author on Li-ion battery packs widely used in EVs and HEVs reveals that battery pack costs could range between $3,000 and $4,500 for small EVs and between $6,500 and $11,500 for HEVs, depending on the battery energy capacity and the configuration of the cells within a module.

The material prices summarized in this section are the prices for the materials only and do not include other expenses, such as incoming inspection charges, fees for compliance with quality control, verification of the quality and quantity of the material, cost of refining if needed, and return of the material shipment if the material did not meet the purchasing specifications.

An examination of the tabulated parameters reveals that about 83% of the material costs are involved in buying materials for the cathode, anode, and electrolyte. Attempts must be made to use alternate materials with lower costs, provided that the performance, longevity, and reliability of the rechargeable battery pack are not compromised. It is interesting to see that materials purchased with lower costs in 2015 will still represent the same percentage of the materials for the same

Table 4.11   Material Prices for Cathode (2000)

Base Element	Price (US$/lb.)*	Cathode Material	Price (US$/lb.)*
Co	18.50	$LiCoO_2$	25.40
Ni	3.20	$LiNi_{0.8}Co_{0.2}O_2$	30.45
Mn†	0.25†	$LiMn_2O_4$	27.44

* The material prices quotations are based on small quantities, but the prices will go down if an order is placed for a large quantity.
† The manganese price quotation is based on the Mn content in the raw ore. This means that the price of refined metal would be higher depending on the purity specification.

capacity rechargeable battery pack, but the reduction in the fabrication of the battery pack will be between 15 and 20%.

4.6.8.2.1 Critical Materials and Their Impact on the Affordability of Li-Ion Rechargeable Batteries for EVs and HEVs

Studies undertaken by the author on the most critical materials for the lithium-based rechargeable batteries indicate that the most suitable materials are lithium metal for the battery core, cobalt for the cathode, and lithium-phosphate-fluorine compound salt for the electrolyte. The materials for the cathode and electrolyte are the most expensive. Therefore, a reduction in the price of lithium-based battery packs is strictly dependent on the raw material costs used for the cathode and electrolyte elements of the rechargeable battery packs. A significant reduction in price of Li-ion-based rechargeable battery packs can be realized if cheaper materials for the electrolyte and cathode are available [9].

A preliminary review on cathode material costs indicates that three distinct materials, namely cobalt, nickel, and manganese, could be used in the fabrication of the cathode. The 2000 procurement costs of these materials, based on small quantities, are summarized in Table 4.11.

4.7 Critical Role of Rare Earth Materials in the Development of EVs and HEVs

Rare earth materials (REMs) have been widely used in commercial, space, and military products. It is important to mention that the use of certain rare earth materials such as cobalt, nickel, cerium, neodymium, terbium, and dysprosium has not accelerated the development of EVs and HEVs, but it has provided significant improvements in component performance and reduction in weight and size. The higher costs of rare earth materials have contributed to the higher costs of EVs and

HEVs. Recent studies on rare earth materials by the author indicate that the rare earth elements (REEs) or materials are classified in two distinct categories, namely heavy rare earth materials and light rare earth materials. These materials are widely deployed in the design of Ni-MH rechargeable battery packs, catalytic converters, electric motors, liquid crystal display (LCD) screens, headlight glass, hybrid electric motors and generators, and optical temperatures, which currently are being used in various components associated with EVs and HEVs.

4.7.1 Identification of Various Rare Earth Materials Used in EVs and HEVs

REEs are widely used in various components of the EVs and HEVs. In particular, the use of REEs in Ni-MH rechargeable battery packs, catalytic converters, and hybrid electric motor generators has led to significant improvements in efficiency and reliability and has considerably reduced the weight and size of the components involved. The use of REEs in various EV and HEV components is identified in Table 4.12.

The procurement costs of these materials are very high compared with other conventional materials, because these elements are usually found scattered in small fragments among rocks and must be separated, processed, refined, and subjected to quality control inspections. Several complex processes are involved, which make the procurement costs of these materials very high. Furthermore, a large number of rare earth materials are used to manufacture the critical components used in EVs and HEVs. Therefore, the retail prices of these vehicles are higher compared with standard gasoline-based automobiles because of the unusually high costs of rare earth materials. For example, the price of REM cerium material has gone up from $4.55 to $31.82/lb. in the last three to five years. Similarly, the price per pound of other rare earth materials has gone up. This is due to the fact that China produces 97% of the world's supply market and has restricted the export of these materials, thereby leading to higher price hikes for the REMs. The deployment of these materials has realized a significant reduction in the weight and size of the essential components used in EVs and HEVs, in addition to remarkable performance improvements in many components and a significant reduction in harmful exhaust gases.

4.7.2 Impact of Future Rare Earth Materials on the Performance of EVs and HEVs

This section predicts the impact of future rare earth materials on the performance of EVs and HEVs in terms of weight and size reduction, on performance improvements of specific components or subsystems, and on charging times and costs. According to the market survey, only a few suppliers of rare earth materials are

Table 4.12 Identification of Rare Earth Materials Widely Used in Electric and Hybrid Electric Vehicle Components and Their Specific Contributions to Component Performance Reliability and Longevity

EV/HEV Component	Rare Earth Material(s) Used	Specific Benefit or Improvement for REMs
Nickel-metal-hydride battery	Cerium, lanthanum	Minimum voltage depression
Catalytic converter	Cerium, lanthanum	Improvement in efficiency and reduction in weight and size
Diesel fuel additive	Cerium, lanthanum	Accelerates the combustion process and reduction in toxic contaminants
LCD screen	Cerium, europium, yttrium	Offers high-quality screen parameters
Glass and mirror polishing powder	Cerium	Highly efficient polishing agent with no damage to surface
Component sensors	Yttrium	Offers improved sensitivity, accuracy
Headlight glass	Neodymium	Intensity enhanced
Hybrid electric motor and generator	Praseodymium, dysprosium, terbium	Improves motor and generator output
Motor magnets	Neodymium, terbium, praseodymium	Offers high magnetic density

in business, and they are reluctant to share the cost and benefit information on these materials. The author feels that other benefits of these materials need to be researched thoroughly. These unknown benefits could not only increase the sales of rare earth materials but also identify the applications of these materials in the components deployed in EVs and HEVs.

Note that important characteristics of some of the REM-based magnetic materials critical in the design of AC induction motors and generators are not fully known, such as retentivity, coercivity, and B-H characteristics; therefore, further research is required on these materials. Table 4.13 identifies a few rare earth materials that provide a specific benefit to a particular EV or HEV component or subsystem.

The data presented in Table 4.12 indicate the bare minimum benefits of rare earth materials deployed in first-generation EV and HEVs. The author hopes that aggressive research and development activities on the potential rare earth materials

Table 4.13   Specific Benefits of a Particular Rare Earth Material in the Design of the Next Generation of Electric and Hybrid Electric Vehicles

Rare Earth Material	Density (g/cm³)	Benefits and Applications for Particular Component
Ce	6.77	Motor, generator, catalytic converter
Dy	8.54	Motor, generator, permanent magnet
Eu	5.25	LCD screen
Ho	8.78	Monitoring sensor
La	6.16	Catalytic converter, Ni-MH battery pack
Nd	7.08	Magnetic motor, generator
Pr	6.63	AC induction motor, generator
Sm	7.54	Microwave amplifier, AC induction motor
Tb	8.23	Compact high-intensity magnet, motor
Tl	6.63	Cryogenic electronics

summarized in Table 4.13 will bring greater benefits for the components and sub-systems to be deployed in the next generation of EVs and HEVs that will hit the market after 2015.

4.7.3 Costs Associated with Refining, Processing, and Quality Control Inspection of Rare Earth Materials

Deployment of rare earth materials offers several advantages, which are not available with other materials, such as aluminum, copper, and iron. But the conventional materials do not offer other benefits. Difficulties in mining, processing, and refining of REMs and their outstanding benefits can be summarized as follows [9]:

■ REMs have become a hot commodity because they are used in a variety of commercial, military, and space applications.
■ Mining of REMs is very difficult and expensive.
■ Large amounts of dirt have to be dug out from the mines, because these elements are usually found scattered in small fragments among rocks and then must be separated and processed.
■ Material separation from the dirt and rocks, processing, refining, and quality control inspection takes a long time, leading to high material costs.
■ The entire process to get the finished product requires hundreds of gallons of water per minute, consumes huge amounts of electricity, needs toxic materials

for its refining procedure, and occasionally requires removal of radioactive dirt.

■ China controls 97% of the world's supply of these materials, and there is no supply chain for these materials. Under these circumstances, the costs of these materials are expected to remain high in the near future. Most of the mines are located in southern Tibet, which has been illegally occupied by the Chinese military since 1955.

■ The supply squeeze continues to limit the export of REMs and is a violation of the World Trade Organization's regulations.

■ The U.S. Geological Survey has identified deposits of REMs in Mountain Pass and Music Valley, both located in Southern California. Both mining facilities will take a couple of years to develop before they can produce the REMs capable of meeting the country's needs.

■ Currently, there are few companies in the United States that can process the REMs needed for rechargeable batteries and magnets for hybrid electric motors and generators. Samarium-cobalt magnets are best suited for motor, generator, and microwave traveling wave tubes (TWTs), for which high-efficiency and minimum weight and size are the principal design requirements.

■ Molycorp, a California-based company, is planning to expand its operations to produce selected REMs.

■ Currently, two HEVs, namely Nissan LEAF and Chevrolet Volt, are using rare earth magnets, which offer high magnetic density with minimum size and weight. Both of these EVs use more REMs compared with other HEVs. That is why the retail price of these vehicles is relatively higher than other vehicles with identical electrical and mechanical performance levels. In addition, the use of praseodymium-cooled (PrC) rare earth materials could significantly improve the performance of motors and generators but at the expense of significant system cost and complexity. That is why the author recommends PrC for electronic motors and generators used in Toyota HEVs. Toyota's engineers are exploring ways to use alternate materials to reduce the price of the car.

■ The manufacturers of the latest EVs and HEVs are planning to use an appropriate earth material from these REMs, including neodymium, praseodymium, dysprosium, and terbium, for hybrid electric motors and generators. REM selection will be made strictly on the basis of significant performance improvement and considerable reduction in vehicle emissions. Some automobile manufacturers are investigating potential substitute materials, such as copper, aluminum, and iron, to replace REMs because of their high cost and the Chinese monopoly on REMs.

■ Higher REM costs are due to initial geological surveying costs required to identify the mine location or sites where geological scientists expect to find deposits of REMs. In brief, development of a new mine for REMs requires the following:
   1. Comprehensive surveying efforts
   2. Cost-effective exploration techniques

3. Construction permit from the authorities concerned
4. Experienced manpower to extract the materials with minimum efforts
5. Cost involved in material processing, refining, and quality control inspection

■ Recently, REM costs have significantly increased. For example, the price for cerium materials, which are widely used for catalytic converters, has jumped from \$4.55 to \$31.82/lb. Retail prices of REMs are expected to remain high because of the shortage of these materials and the monopoly by the Chinese.

4.8 Conclusion

Comprehensive studies performed by the author reveal that four distinct rechargeable battery packs are best suited to operate EVs and HEVs, namely lithium-manganese battery pack, lithium-iron-phosphate, Li-ion-polymer battery pack, and sealed Ni-MH. Performance capabilities and limitations of each rechargeable battery pack have been discussed, with an emphasis on reliability, longevity, cost, and complexity. Typical battery life and current retail battery price for various rechargeable battery packs are provided for the benefit of buyers of these EVs and HEVs. According to Volt dealers, the Li-ion battery pack used by the Chevy Volt weighs around 525 lbs. (maximum), has a maximum battery life exceeding 15 years, and has a maximum retail price close to \$6,000. The battery pack's weight, price, and longevity are strictly dependent on battery pack capacity and longevity requirements. The chronological development history of early EVs and the associated chargeable battery packs is described. EVs and HEVs by various companies are identified with an emphasis on the battery type used, critical performance parameters, and the retail price of the battery pack. The rechargeable battery pack comes with several cells enclosed in modules, a thermal management circuitry, a liquid cooling subsystem, a light-weight enclosure to protect the battery under harsh operating environments, and electronic sensors to monitor temperatures at strategic locations on the battery surfaces to ensure the reliability and structural integrity of the battery pack. Most EVs and HEVs come with recharging provisions either using the 110 V power outlet in a garage or a 240V/440 V outlet at a charging station. The chapter provided approximate charging times and charges and electricity used for various EVs and HEVs.

Performance characteristics and rechargeable battery requirements used for commercial transportations are summarized with an emphasis on energy density, specific energy, estimated life cycles, charging times, self-discharge per month, and estimated cost (dollars per Wh). Brief studies performed on self-discharge rates of the batteries seem to indicate that the lowest self-discharge rate offers enhanced reliability, high longevity, and sustained battery performance under variable electric load conditions. Improved reliability and optimum electric performance are possible only from Li-ion-polymer rechargeable battery packs. But this particular battery pack can take 6 to 10 h for recharging depending on the battery capacity. Ni-MH battery packs suffer from low specific-energy levels, but their recharging time is 1 h. Furthermore, these battery packs require frequent recharging compared with Li-ion battery packs. As far as toxicity problems are concerned, lithium batteries contain some toxic elements and, hence, adequate ventilation is recommended during the maintenance cycle. Battery designers and material scientists believe that attempts must be made to consider the deployment of zinc-air battery packs in EV and HEV applications because of their low costs and lack of toxicity problems. Despite their advantages, manufacturers currently are avoiding the use of such rechargeable battery packs.

Material requirements for important battery components, such as the anode, cathode, electrolyte, and battery housing, are identified. Important characteristics of these materials are briefly summarized with an emphasis on thermal and mechanical properties. The cathode is the most expensive component of the battery pack and its cost represents about 50% of the total battery cost. It is made from LiCoO. The next most costly component is the electrolyte and the least costly item is the anode, which is made from graphite. Material costs of other battery components, including the separator, housings for cell and modules, and the container, are provided for the customers' benefits.

Safety and reliability of the rechargeable batteries are strictly dependent on the engine and battery surface temperatures. To ensure safe surface temperatures, thermal sensors are provided at strategic locations on the engine and battery pack, which constantly monitor the temperatures. State-of-the-art algorithms are necessary to monitor vital electrical performance parameters, temperature levels, and critical structural parameters of each cell and module to ensure the safety and reliability of the battery pack. Second-generation control chips can be used to synchronize current and voltage measurements. Specifications requirements for self-diagnosis of reliability and maintainability must be defined, if reliability and safety are the principal design parameters.

The author has considered various battery pack configurations for Li-ion batteries. The space allocated for the battery pack in the vehicle determines the optimum configuration for a specific EV or HEV. In most cases, 6 to 12 cells are packages together in a unit called a module. Control and safety electronic circuits are provided in each module to avoid severe structural overcharge or overdischarge. A battery pack typically can accommodate 8 to 12 modules, depending on the battery capacity and the space available inside the vehicle. The modules then can be installed in the battery pack housing, which is sized to match the space allocated for the pack in the vehicle. The battery pack is always located between the two rear wheels with a low center of gravity, which yields optimum mechanical stability under most driving conditions. NIST has helped to consolidate all standard activities. The NIST Framework and Roadmap are integrated with the Smart Grid system, which provides the public charging locations and their infrastructures and available facilities to owners of EVs and HEVs.

References

1. Editor, "Technology report on all-electric vehicles progress to shock the automobile market," *Electronic Design* (May 2010), pp. 36–44.
2. Roger Allan, "Electric and hybrid vehicles technologies charge ahead," *Electronic Design* (March 2010), pp. 27–33.
3. Reporter notebook, "Are you a one-volt family?" *Machine Design* (November 18, 2010), p. 22.
4. Reporter notebook, "Zinc could trump lithium for vehicle batteries," *Machine Design* (October 7, 2010), p. 26.
5. Jim Harrison and Paul Shea, "Energy-saving forum: Batteries and battery controller ICs," *Electronic Products* (February 2010), pp. 24–28.
6. R. Frank, contributing editor, "Electric vehicles: The Smart Grid's moving target," *Electronic Design* (June 17, 2010), pp. 65–68.
7. Ronald Jorgen, *Electric and Hybrid Electric Vehicles,* Warrendale, PA: Society of Automotive Engineers Inc. (September 2, 1985), p. 142.
8. Editor, "Winners and losers," *IEEE Spectrum* (January 2010), p. 38.
9. Tiffany Hsu, "High-tech's ace in the hole," *Los Angeles Times* (February 20, 2011), p. 1.

Chapter 5

Low-Power Rechargeable Batteries for Commercial, Space, and Medical Applications

5.1 Introduction

This chapter is dedicated to low-power secondary (rechargeable) and primary batteries [1] that are widely used in commercial, medical, and precision industrial applications. This chapter discusses critical design aspects, performance capabilities, and limitations of micro- and nanobatteries with an emphasis on cost, reliability, and longevity. Small low-power batteries are best suited for the detection, sensing, and monitoring devices and sensors widely used in commercial, industrial, and domestic applications, such as flashlights, portable radios, small musical instruments, video cameras, precision infrared cameras, laptops, global positioning system and general navigation devices, camcorders, iPads, iPhones, electronic toys, and host of electronic components widely used for domestic applications. Most low-power batteries that are specially deployed in household applications, such as perimeter security, temperature, and humidity sensors, and health-monitoring and portable diagnostic devices, such as electrocardiogram (EKG), electroensuflogram (ESG), and other sensors used in car and nose examinations, require enclosures to safeguard the mechanical structural integrity of the battery. Rechargeable batteries reveal that lithium-ion (Li-ion) rechargeable battery packs would meet such requirements. The studies further indicate that these batteries would be best suited for radios and security devices operating in remote locations where the ambient temperature could be as low as −40°C. These batteries are most ideal for medical devices that need battery packs capable of operating after temperature exposure to steam sterilization. Both of these operating requirements cannot be satisfied by other conventional battery packs. In battlefield environments, medical devices and security sensors are required to operate under high temperature and explosive environments. Therefore, the batteries used for such applications must meet performance requirements for operating under harsh thermal and mechanical environments. This chapter identifies the battery types, materials required, and design parameter requirements that are most satisfactory in operating under specific operating environment. Performance requirements and design aspects of low-power rechargeable batteries for various applications will be discussed in great detail. Since 1960, rapid changes from vacuum tubes to transistors to microcircuits to nanocircuits have resulted in major improvements in energy density, efficiency, shelf life, and reliability of batteries. This progress is slow, however, compared with advances in electronic circuits and devices.

Furthermore, growth in the new electronic systems and components is relatively slow, and the electronic circuits and devices are getting more complex, depending on the development of new devices and systems that need batteries with new performance requirements. Performance requirements for low-power batteries can vary from application to application. Older batteries or electrochemical system such as carbon zinc (C-Zn), zinc-air (Zn-air), nickel cadmium (Ni-Cd), and lead acid continue to get better and better, and such batteries can maintain the market for the electronic devices that can use them. The most popular lithium cells or primary lithium-based batteries are growing in use as new electronic and electro-optical devices are designed to operate requiring higher voltage, lighter weight, more compact size, and superior shelf life. Market surveys reveal that lithium-manganese dioxide ($Li-MnO_2$) batteries have dominated the commercial market. Studies undertaken by the author indicate that at least 16 manufacturers produce many battery sizes and configurations, ranging from high rate D cells to 50 milliampere-hour (mAh) thin, flat batteries.

Strict environmental regulations continue to affect battery deployment and its disposal. These regulations force the battery industry to consider secondary batteries, which can be used multiple times. Use of secondary batteries will not only eliminate disposal problems but also will yield maximum economy. The author will not describe low-power mercury-based primary batteries, because mercury is considered very harmful to general health and these batteries have been banned.

The author will discuss existing and emerging low-power batteries with a particular emphasis on energy density, shelf life, life cycle, and reliability. Both the energy density and life cycle are application dependent and will change with discharge rate, duty cycle, and operating voltage of the device in use. The author will describe the performance advantages and limitations of microelectromechanical system (MEMS)-based and nanotechnology-based primary batteries, because these batteries offer minimum size, weight, and power consumption. In other words, ultra-low-power battery designs incorporating MEMS and nanotechnologies are best suited for such applications as airborne and space systems, where power consumption, weight, and space limitation are the principal requirements.

Advances in primary battery technology have been linked to electronic sensors and electro-optical devices, where minimum power consumption is the principal design requirements. Circuit developments such as the transistors operating at microwave frequencies and microelectronics with stringent performance requirements have placed a major emphasis on batteries with higher energy density, longer shelf life, and freedom from current leakage. Normally long research and development time is required to achieve significant performance improvement for the batteries. Studies performed on microbatteries by the author seem to indicate that the ultimate size and energy density of the battery are limited by the system chemistry, materials used, and enclosure required to meet specific performance requirements under operating environments.

Advances in material science, packaging, and electronic miniaturization have required significant design changes in older batteries such as alkaline manganese, lead acid, and Ni-Cd. Newly developed systems such primary and secondary (rechargeable) lithium, Zn-air, and Ni-MH have been commercialized to meet higher energy density and longer shelf-life requirements [2]. Leading design engineers and material scientists believe that Li-ion batteries using solid polymer as electrolyte offer higher safety, reliability, and durability over other lithium-based batteries. But most of these battery types in general are not suitable for low-power applications.

The author will present various sections dealing with commercial battery types that are designed and developed specially for low-power applications. Low-power batteries often are characterized by energy and power per unit weight, but in many applications, energy and power per unit volume is more critical, especially for portable electronic devices. The energy available from a specific battery source strictly depends on the rate at which electrical power can be withdrawn from the source. Furthermore, as the current is increased, the amount of energy delivered is decreased. This characteristic can be displayed as a log-log plot of energy delivered versus the power. This particular plot is known as a Ragone plot. Battery power and energy are also affected by the construction technique used, the cell size, the duty cycle employed, and the materials used to produce the batteries. Because of these factors, the battery manufacturers define the capacity of a given battery or cell in ampere-hours or watt-hours when discharged at a specific discharge rate to a specific voltage cutoff. The discharge rate, frequency, and cutoff voltage are selected to simulate a specific application, such as a video camera, a smoke detector, or a security alarm. Sometimes battery capacity is specified in terms of the rate and is expressed in ampere-hours or watt-hours, which could create confusion. Most primary batteries or cells are often rated at a current that is 1/100th of the capacity expressed in ampere-hours and is written as C/100, whereas the secondary cells or batteries are rated at C/20, which means that the secondary cell is rated at a current that is 1/20th of the capacity expressed in ampere-hours.

5.2 Low-Power Battery Configurations

Low-power battery design configurations determine how the battery will provide the most efficient function, how much it will cost at a retail price, and how it will meet the minimum weight and size requirements. The weight requirement may demand the low-density material in the cell, whereas the size specification has to consider the optimum design configuration capable of yielding the smallest size with a desired form-factor.

Most primary cells use aqueous electrolytes and employ single thick electrodes arranged in parallel or in concentric configurations [1]. Specific cell constructions of this type include *cylindrical*, *bobbin*, *button*, and *coin* configurations. Most small secondary batteries deploy a *wound* or *jelly roll* construction, in which long thin electrodes are wound into a cylinder and placed in a metal container using a low-density material. This design configuration yields high power density, but it does so with decreased energy density and a higher cost of construction. Because the construction uses low-conductivity electrolytes, many lithium primary batteries are required to employ *wound* construction to provide higher discharge rates. Increasing numbers of both primary and secondary low-power batteries are also being made in prismatic and thin, flat constructions. These form-factors allow for the most cost-effective construction and provide better utilization of the space allocated in the devices, which often leads to lower energy density.

5.2.1 Low-Power Batteries Using Cylindrical Configuration

Commercially available low-power primary AA batteries known as dry cells mostly come with cylindrical configurations and typically are rated at 1.5 volts (V). These batteries are designed to provide hundreds of nanoamperes to a few millamperes. They generally use lead or aqueous alkaline electrolyte. These cells are best suited for analog cameras, small electronic devices, electronic toys, and flashlights. Specifications of such batteries are summarized in Table 5.1.

5.2.2 Carbon-Zinc Primary Low-Power Batteries and Their Characteristics

Battery market surveys indicate that C-Zn primary batteries are widely used worldwide for low-power battery applications. The survey further indicates that shelf-life performance has improved 750% and current leakage has reduced significantly during the last 90 years. This particular battery is available in two distinct design configurations, namely the standard version, which uses natural manganese

Table 5.1 Characteristics and Physical Parameters of Low-Power Batteries

Description	Electrolyte Used	Output	Diameter	Length
Sony New Ultra, AA	Lead	1.5 V	11/16 in.	1 13/16 in.
Energizer, AA	Alkaline	1.5 V	11/16 in.	1 13/16 in.
Energizer, AAA	Alkaline	1.5 V	11/16 in.	1 13/16 in.
HomeLife (thin), AAA	Alkaline	1.5 V	9/16 in.	1 11/16 in.

Note: The physical dimensions are accurate within approximately 5%.

dioxide (MnO_2) as the cathode and ammonium chloride as electrolyte, and the premium version, which uses electrolytic MnO_2 as the cathode and zinc chloride as the electrolyte. The premium version cell offers enhanced capacity and ultra-high reliability.

The majority of these C-Zn cells are available in D, C, A, and AAA sizes. These cells are manufactured in Europe, United States, China, and other countries. The United States favors the production of the alkaline version of these cells over the C-Zn versions. As far as worldwide application is concerned, both the standard and premium versions are widely used to power electronic devices where cost, performance, and availability are the prime considerations. Because of worldwide demand, these low-power batteries are being manufactured by more than 250 companies. Chinese companies are producing close to 7 billion C-Zn batteries per year because of high demand due to very low cost. Commercial publications indicate that production of older batteries such as C-Zn will slow down as newer low-cost batteries are developed and become available in abundant quantities.

Because of environmental controls, cadmium, mercury, and lead contents are being eliminated from the zinc cans. Some U.S. states such as California have restricted the sale of C-Zn and alkaline batteries containing mercury because of health hazards. The elimination of mercury from the batteries makes the recycling process safe and economical. Removal of the mercury, however, resulted in poor performance in terms of high current and voltage, which are vital in cutting applications. Manufacturers have made some progress in correcting this defect, and alkaline batteries can now be used specifically for cutting applications. In summary, C-Zn batteries are not best suited for electrical or electronic devices, such as tapes, disk players, flash units, automatic cameras, and toys, because of their inability to provide adequate power for satisfactory performance.

5.2.3 Performance Capabilities and Limitations of Alkaline Manganese Batteries

The reduction of power requirements for some advanced miniaturized motors, displays, and electronic devices has contributed to a trend in favor of the growth and production of alkaline batteries. Significant performance improvement in rechargeable alkaline batteries has significantly reduced the need to purchase new batteries when their output power is drastically reduced.

Recognition of the health hazards of mercury and elimination of mercury from the zinc anode have resulted in significant growth in demand for alkaline batteries. The elimination of mercury has simplified the collection, disposal, and recycling processes, leading to significant reduction in the cost and complexity involved in these processes. The elimination of shortcomings associated with mercury-based cells has allowed the battery manufactures to sell worldwide *green* versions of alkaline batteries, which are best suited for low-power applications, including electronic toys, automatic cameras, flash units, and so on.

A performance survey of low-power batteries indicated that major improvements in the performance of alkaline-based batteries are due to the introduction of plastic technology. In the original fabrication of the alkaline battery, cardboard was used to insulate the tube over a bare cell and a thin steel outer jacket was inscribed with a label. Replacement of the old label with a plastic label has resulted in an increase of close to 20% in the internal volume or space needed for active materials, which are responsible for battery performance.

5.2.4 History of Primary Lithium-Based Batteries and Their Performance Parameters

Market surveys indicate that $Li-MnO_2$ batteries are widely manufactured by more than 16 companies worldwide for various electronic and electrical devices and sensors. These cells are available in a wide range of sizes and shapes and their typical characteristics are shown in Table 5.2.

Originally, lithium-based batteries were manufactured in large quantities and were accepted for various applications. Users were impressed with the electrical

Table 5.2 Characteristics of a Lithium-Manganese-Dioxide Cell

Cell Configuration	Cell Characteristics			
	Voltage (V)	Capacity (mAh/cell)	Rated Current (mA)	Energy Density (Wh/L)
Coin (circular)	3	30–1,000	0.5–7.2	500
Cylindrical wound	3	150–1,300	20–1,200	500
Cylindrical bobbin	3	65–5,000	4–10	620
Prismatic	3	1,150–1,200	18–20	490
Cylindrical D cell	3	10,000	2,500	575
Flat	3 or 6	150–1,500	20–135	300

performance of these batteries. Eventually, however, both the sale and growth of these batteries tapered off strictly due to safety concerns. Some users said that the basic reason for the decline was the battery's operating voltage of 3 V. Most of these batteries have been designed with coin-shaped configurations.

In the meantime, new designs of lithium batteries were developed, tested, and evaluated, and newly developed batteries demonstrated significant improvement in battery performance parameters, including energy density, shelf life, longevity, reliability, and consistent electrical performance under high and low operating temperatures with no evidence of failure mechanisms.

Lithium-iodine (Li-I_2) batteries were specifically designed and developed for medical applications. These batteries consistently demonstrated the best performance and suitability, particularly for pacemakers, over a period exceeding 25 years. This particular battery is high in energy density but low in power level. Li-I_2 is a low-conductivity solid-state electrolyte, which limits the current to a few microamperes. According to the manufactures, an operational life ranging from 7 to 12 years for this battery has been demonstrated in the field. The battery suppliers claim that these batteries could be used in other applications, such as watches and memory-retention devices.

High-power implantable batteries have been developed for possible applications in pacemakers, automatic defibrillation devices, and other medical diagnostic sensors. Such cells will be described in great detail in Section 5.4, Batteries for Medical Applications.

Based on the sale and widespread use of Li-MnO_2 batteries, they appear to enjoy the highest popularity. These batteries are produced in large quantities, are widely used worldwide, and are sold at reasonable price. The Li-MnO_2 batteries are produced by more than 18 companies worldwide, and these cells are available in various sizes and configurations. Among the AA-size rechargeable batteries shown in Table 5.3, the Li-MnO_2 cells have demonstrated high specific energy and power and the lowest self-discharge rate (SDR) per month. The lowest discharge rate plays a critical role in longevity, shelf life, and reliability of a cell.

On the basis of published technical papers and reports, the lithium-carbon-fluorine (LiCF$_x$) device, which was once thought to dominate the lithium primary battery category, was produced in fewer configurations and by fewer manufacturers, despite the fact that it is the demonstrated best choice for long-term, high-reliability, and high-temperature operations. This product uses fluorine, which is a pale yellow, flammable, irritating toxic gas and requires careful handling. This particular battery has demonstrated a real-time storage exceeding 15 years at room temperature with a capacity loss of less than 5% over that time. Despite such remarkable performance history, this particular cell has not enjoyed wide acceptance by a large number of suppliers. The original cell has demonstrated a voltage of 2.5 V. Furthermore, reliable cost data on this particular cell are missing. A recently filed patent indicates, however, that the design modification of this device with certain changes in the electrolyte, separator, and seal of the CF$_x$ coin cells has

Table 5.3   Typical Characteristics of AA-Size Rechargeable (Secondary) Batteries

Battery Type	Voltage (V)	Capacity (mAh)	Cycles	Power Loss per Month	
				Density (Wh/kg)	Percentage (%)
Ni-Cd	1.2	1,000	1,000	60	15
Ni-MH	1.2	1,200	500	65	20
Li-ion-CoO$_2$	3.6	500	1,200	8	8
Li-ion-Polymer	2.5	450	200	110	1
Li-MnO$_2$	3.0	800	200	130	1

Note: The tabulated parametric values summarized in this table indicate the estimated values of cell performance parameters and could have an error of ±5%.

extended operation at temperatures leading to 125°C. No coin cell has demonstrated an operation at 125°C without compromising performance and reliability. In summary, the product's higher cost, the operating voltage of 2.5 V, and interest shown by a limited number of suppliers in this product have prevented acceptance of this device worldwide.

Another lithium-based cell, known as the lithium-ferrous-sulfide (Li-FeS$_2$) AA cell, has a voltage rating of 1.5 V and was manufactured mostly in Canada, Europe, and Japan until 1992. The device was made available in some U.S. retail stores in 1993. The cell has a *wound* construction format, a higher voltage rating, and high-voltage cutoff characteristics. This cell is best suited for photoelectronic applications, such as camera flash units, cellular phones, disk players, and small computers.

Lithium cells using liquid cathodes, such as sulfur dioxide and thionyl chloride, are capable of yielding higher energy density, a higher capacity rating, and significantly improved low-temperature performance compared with cells using solid cathodes. This is the fundamental advantage of using liquid cathodes. Because the cathode material is also the electrolyte, the packing efficiency is higher for cells using solid electrodes, but the liquid cathode offers much improved electrochemical kinetics.

Despite improved electrical performance, the major disadvantage of this device is that an isolating layer is required on the lithium to prevent shelf discharge, which could lead to a large voltage drop during the first several seconds of use. Another problem with this cell is the safety concern associated with the high reactivity of the liquid cathode if the protective layer is disrupted or damaged, or if the lithium becomes molten above 180°C. Because of these drawbacks, the commercial market for this device is limited to small low-area cells. Larger cells are widely used in military applications. The use of thionyl-chloride batteries for military applications has been restricted or discouraged, but lithium-sulfur-dioxide batteries are widely used for military applications. On the basis of the market survey, there are limited commercial applications for lithium-based sulfur-dioxide batteries for these reasons.

5.2.5 Nickel-Metal-Hydride, Nickel-Cadmium, and Lithium-Ion Rechargeable Batteries

The most popular small and low-power rechargeable batteries include Li-ion, Ni-MH, and Ni-Cd batteries. On the basis of published reports, it appears that Ni-MH and Li-ion batteries are experiencing rapid growth because of their high demand. In addition, their demand is strictly based on relatively lower costs, high longevity, and availability without any restriction. The small rechargeable battery market is experiencing a growth rate better than 20% because of the explosion in use of cellular phones, portable computers, mini-notebooks, and entertainment devices shown in Figure 5.1.

Although Ni-Cd rechargeable batteries can be used in cases in which power requirements are low to moderate and weight and size are small, research studies performed by the author reveal that both batteries suffer from memory effect. The studies further indicate that the memory effect is higher for Ni-Cd rechargeable batteries than that for Ni-MH batteries.

The memory effect is strictly due to voltage depression. During the discharge cycle, the current is interrupted at three distinct values of the depth of discharge (DOD), as denoted by points 1, 2, and 3 in Figure 5.2. Certain electrical load patterns indicating power removal and charging are responsible for memory effect. The load patterns involve repeated, narrow bands of change in the state of charge (SOC), which causes the voltage depression in step fashion at points 1, 2, and 3, as illustrated in Figure 5.2. The memory effect is created by the increased concentrations of certain chemicals. The electrochemistry theory indicates that a few deep discharges will eliminate the voltage depression, but nobody can predict the number of deep discharges.

Furthermore, the number of deep discharges is dependent on the discharge level, battery capacity, and performance boundary. One boundary, which allows the maximum allowable DOD, is shown in Figure 5.2. This boundary is necessary to prevent the loss of rechargeable battery life. The vertical line indicates the allowable DOD. The electrochemistry theory states that another performance boundary is due to the minimum allowable inverter voltage, which is represented by open-circuit-voltage (OCV) swings, if the battery is connected to power electronics. Decreases in OCV occur strictly because of voltage depression. At point B, all available battery energy has been used. At point A, the battery is completely turned off, and the actual battery SOC value is much higher than the allowable value. Consequently, the battery energy has not been used completely. The memory effect and its influence on usable energy from the Ni-MH rechargeable battery are fully evident from Figure 5.2.

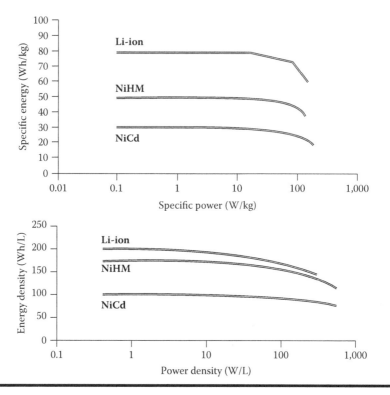

Figure 5.1 Energy density and specific power capabilities of the most widely used rechargeable batteries.

Ni-MH batteries are widely used in electric vehicles (EVs) and hybrid electric vehicles (HEVs) because of their lower costs, enhanced reliability, and improved longevity. Ni-Cd batteries are used for limited applications because of their memory loss and toxicity problems. These batteries can be used where foolproof ventilation is provided. In an Ni-Cd cell, the anode is made from cadmium and the cathode from Ni-oxides. Typical electrical characteristics of an Ni-Cd battery can be summarized as follows:

- Cell voltage: 1.2 V
- Energy density: 33 Wh/L
- Specific energy: 60 Wh/kg

Because of low OCV and memory effect and environmental problems due to toxicity, Ni-Cd batteries are not widely acceptable.

Similar curves for an Ni-Cd battery can be generated showing the maximum allowable DOD, minimum inverter voltage, loss of DOD, and voltage swings causing the memory loss in an Ni-Cd battery. The memory effect of this particular

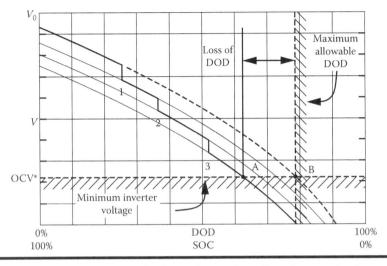

Figure 5.2 **Memory effect and voltage depression effect frequently observed in Ni-MH batteries.**

rechargeable battery can be more severe. Studies performed by the author on memory effects indicate that Li-ion batteries do not exhibit such an effect.

5.2.5.1 Peculiarities in Rechargeable Batteries

Electrochemical devices such as rechargeable batteries sometimes exhibit peculiarities, such as voltage depression, memory effect, and voltage recovery, which are strictly dependent on the DOD and SOC values (Figure 5.2). Not all rechargeable batteries exhibit these peculiarities. Furthermore, these peculiarities are often noticed when the batteries are supplying power to high-power electronic devices, HEVs, or plug-in HEVs. Normally, such peculiarities are not observed in small low-power rechargeable batteries.

5.2.5.2 Design Considerations for Small Low-Power Rechargeable Batteries

Many combinations of electrode material and electrolyte can yield a cell. A few hundred possible cells are required to design a high-power battery. An Ni-Cd cell can be used in the design of a low-power rechargeable battery to power small electronic devices such as a minicomputer, musical recorders, and flash lamps for cameras. For a compact rechargeable battery, two components of the battery contribute to the mass and volume. First, the chemicals and materials used in the development of the battery strictly determine the physical parameters. Second, the mass and size of the packaging materials affect these parameters.

One can write the expressions for mass or weight and for volume of the battery as follows:

$$\text{Total mass} = [\text{mass of chemicals used}] + [\text{mass of structural materials}] \quad (5.1)$$

$$\text{Total volume} = [\text{volume of chemicals}] + [\text{volume of structural materials}] \quad (5.2)$$

The structural materials include the materials used for the separator, electrode, and battery enclosure.

Practically all battery designs include the materials needed for separators to separate the electrodes. The separator material provides the mechanical strength and maintains the desired separation distance. A high-conductivity grid may be embedded in an electrode to reduce the internal resistance of the battery needed to reduce the voltage drop within the battery. The cell voltage is determined by a pair of electrodes that are connected along with the electrolyte. The cell energy can be determined from the overall cell reaction. In the case of a lead-acid battery, the chemical reaction equation can be written as follows:

$$[2PbSO_4 + 2H_2O] = [PbO_2 + 2HSO_4^- + 2H^+ + Pb] \quad (5.3)$$

This chemical reaction equation merely identifies the products generated after the chemical reaction. But the electrochemist uses the knowledge of the specific energy or energy per mass of each chemical species to determine the overall cell or battery energy. The cell capacity and the energy are determined by multiplying the mass of the reactants. These values are called theoretical values and are the mass of reactants excluding the mass of the packaging materials used in the fabrication of the battery.

5.2.5.3 Frequent Mathematical Expressions Used in the Design of Batteries

SOC and its associated quantity, DOD, are strictly based on the battery's energy. In other words, the quantity SOC can be written as follows:

$$SOC = [(\text{actual battery energy})/(\text{fully charged battery energy})] \quad (5.4)$$

This equation, regardless of battery type and capacity, can be rewritten as follows:

$$SOC = [100\% - DOD] \quad (5.5)$$

Rapid charge and discharge are the most desirable features of the battery, regardless of its type and energy capacity.

As far as battery OCV and efficiency are concerned, voltage drop in a cell is of paramount importance, because it can affect both of these performance parameters. The expression for the voltage drop in a battery can be written as follows:

$$\Delta V = [I \, L \, R \, / \, A] = [R \, J \, L] \qquad (5.6)$$

where I is the cell current, L is the distance between electrodes, R is the cell resistance per unit length, A is the cell plate area, and J is the current density.

Examination of the electrical parameters indicates that smaller separation between the electrodes reduces both the battery length and its volume or size. The space between the electrodes is filled with the electrolyte, which has two functions: to provide a conduction path for ions, which for a Li-ion battery is Li^+, and to impose a high resistance to electrons. Electrons are forced to flow in the external load circuit or to a device connected across the battery.

An alternate method is to use a separator material between the electrodes, which are pressed against both sides of the separator. There are two functions of a separator: to maintain a specific spacing between the electrodes and to prevent short circuits. A separator comes in three forms, namely liquid, solid, and semisolid. Studies performed on various separators by the author indicate that a solid separator serves dual functions, namely a mechanical spacer and an ion-conducting electrolyte.

5.2.5.4 Contributing Factors to Battery Weight

For a Li-ion battery or Li-ion-polymer battery, the electrodes serve as a housing and interstices (small spaces for holes) for lithium atoms and ions. The added weight, however, will reduce the specific energy (Wh/kg), which is the most vital performance parameter of a battery. The added weight comes from the following sources:

- Electrolyte
- Separator
- Electrodes
- Housing or casing
- Ionic conducting chemicals (such as Li^+ ion for a Li-ion battery)
- Pressure-release device
- Current collectors
- Safety electronics
- Additives to enhance conductivity

5.3 Batteries for Miniaturized Electronic System Applications

The current practice to deploy miniaturized electronic circuits and devices [2] in space and airborne system applications requires the most compact, lightweight, and reliable micro- and nanotechnology-based batteries. Research studies conducted by the author reveal that the miniaturized batteries will be best suited for high-performance supersonic jet fighters, space-based surveillance and reconnaissance systems, armed unmanned air vehicles (UAVs), counter-improvised explosive devices (IEDs), and microcontroller and embedded-system applications. These studies further indicate that the secondary batteries designed to meet the performance requirements for embedded-system applications [3] will unquestionably meet the battery requirements for these applications.

Federal, defense, and commercial organizations are pressing for a major emphasis on electronic miniaturization [2] using the MEMS and nanotechnologies. In other words, nanotechnology and MEMS are the most ideal applications for military and aerospace applications, in cases in which size, weight, power consumption, longevity, and reliability are the principal design requirements.

Defense scientists and planners believe that UAVs and robotic devices and systems will benefit the most from the electronic miniaturization technology. In particular, the UAV drones conducting surveillance and reconnaissance missions will observe significant improvement in mission accomplishment because of the effective implementation of this technology. Miniaturization technology involving MEMS and nanoelectronics is considered a key in the design and development of micro- and nanobatteries. Furthermore, deployment of carbon nanotubes (CNTs) and graphene nanoribbons can play critical role in the design and development of battery management systems.

Comprehensive studies undertaken by the author on the materials used to produce various low-power batteries reveal that alkaline batteries are best suited for embedded systems. An alkaline battery uses zinc (Zn) as an anode (negative) and MnO_2 as a cathode. The manufacturers typically fabricate these batteries using MnO_2 and Zn powder with a caustic alkali of potassium hydroxide (KOH) as an electrolyte. This battery technology is best suited for low-power batteries (Figure 5.3) widely used in many standard applications, such as portable radio, medical diagnosis equipment, smoke detectors, portable audio devices, and high-energy flashlights. The nominal voltage of a rechargeable alkaline cell or battery is 1.5 V, with a discharge voltage of 0.9 V.

Among the primary and rechargeable batteries best suited for embedded-system applications, alkaline, Zn-carbon, and lithium batteries are widely used. In a Zn-carbon cell, Zn is used for the anode and MnO_2 is used for the cathode. These batteries are used in cases in which cost-effective performance is the principal performance requirement. The nominal voltage and the discharge voltage of this battery are identical to those of the alkaline battery. In the case of lithium batteries, lithium is used for the anode and carbon monofluoride gel or MnO_2 is used for the cathode. Two types of lithium batteries are available: BR type and CR type. The BR lithium battery uses carbon monofluoride as the cathode, and the CR lithium battery uses MnO_2 as the cathode. But lithium is used as the anode for both types. Lithium batteries come in various form-factors, but the coin-cell

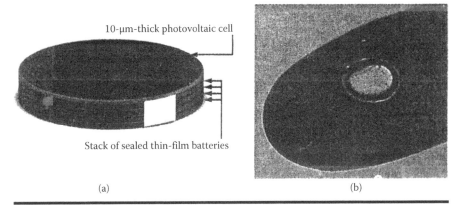

10-µm-thick photovoltaic cell

Stack of sealed thin-film batteries

(a)

(b)

Figure 5.3 **(a) Isometric and (b) plan views of a low-power battery using thin-film technology. Typical performance parameters and physical dimensions are as follows: diameter, 2.5 cm; thickness, 0.16 mm; battery capacity, 80 mAh per discharge; thickness of stack containing four thin-film microbatteries, 0.64 mm; photovoltaic cell thickness, 10 µm.**

battery configuration is the most popular for several applications. The most widely used rechargeable (secondary) and primary batteries in embedded-system applications are shown in Table 5.4.

5.3.1 Brief Description of Rechargeable Batteries Best Suited for Embedded-System Applications

Because the embedded system is generally compact, the weight, size, and power consumption for each of the subsystem element must be such that all of the elements can be accommodated within the size, weight, and power consumption specifications or provisions. Furthermore, the temperature of an embedded system could be very high, owing to the operation of several miniaturized electrical, electronic, and mechanical devices in a given space. On the basis of a trade-off study, the author is recommending three distinct battery types: one for a simple embedded-system application, one for an embedded-system application with medium complexity, and one for an embedded-system application with the most complexity.

5.3.1.1 Characteristics of an Alkaline Battery for a Simple Embedded-System Application

Alkaline batteries are the most ideal option for a simple embedded-system application. This battery can be fabricated using Zn for the anode (negative electrode), MnO_2 for the cathode (positive electrode), and Zn powder with a caustic alkali of KOH for the electrolyte. This particular battery offers long shelf life, supports

Table 5.4 Performance Characteristics of Batteries Best Suited for Embedded-System Applications

| Battery Type | Anode Material (–) | Cathode Material (+) | Nominal | | Energy Advantages |
			Voltage (V)	Density (mJ/kg)	
Alkaline	Zinc	Manganese dioxide	1.5	0.51	Long shelf life
Zinc-carbon	Zinc	Manganese dioxide	1.5	0.13	Low-current device
Lithium (BR)	Lithium	Carbon monofluoride	3.0	1.3	Wide temperature operation
Lithium (CR)	Lithium	Manganese dioxide	3.0	1.0	Stable volume performance
Zinc-air	Zinc	Oxygen	1.4	1.68	High energy density
Lithium-thionyl chloride	Lithium	Sulfur-oxygen chloride	3.6	1.04	Low self-discharge rate and 22-year longevity

Source: Oland, K., *Electronic Design News*, 36–39, 2010.

The values of nominal voltage and energy density shown in this table are estimated values and could have errors of ±5%.

high- to medium-drain applications, and deploys the least expensive fabrication method. Its nominal voltage and energy density are relatively low and are shown in Table 5.4.

5.3.1.2 Performance Characteristics of a Battery Best Suited for the Least Complex Embedded-System Application

Zn-carbon batteries seem to be the most attractive option for the least complex embedded-system applications. This particular battery can be fabricated using Zn for the anode, MnO_2 for the cathode, and Zn powder for the electrolyte. Low operational cost, least complex fabrication, and ultra-low current-consumption characteristics make this device the most suitable for the least complex embedded-system applications. The battery offers a nominal voltage of 1.5 V and an energy density of 0.13 MJ/kg [3].

5.3.1.3 Characteristics of a Battery Best Suited for the Most Complex Embedded-System Application

Studies performed by the author indicate that BR-type lithium batteries appear to be the most ideal battery for the most complex embedded-system applications. Because the embedded system is so complex, it can have several system elements, such as a microcontroller algorithm, temperature-monitoring device, pulse-shaping and -forming network, current-leveling device, and so on. When several elements are operating in a complex embedded system, a wide temperature range becomes the most pressing requirement. In addition, battery performance requirements may become more stringent. Battery manufacturers generally fabricate the devices using carbon monofluoride and a lithium alloy. This battery composition offers excellent temperature characteristics over a wide range, long service intervals, low power characteristics, and low self-discharge characteristics, which are most desirable for embedded applications. Such applications include heat-cost allocations, water and heat meters, electronic toll-collection systems, and tire-pressure-monitoring systems. Lithium (BR) batteries appear to be the most qualified option for this embedded-system application. This battery provides a nominal voltage of 3 V and energy density better than 1.3 MJ/kg. Because of higher energy density and higher nominal voltage, this particular battery could meet the power requirements of several system elements. This battery is capable of operating with the utmost reliability over a wide temperature range and providing a low-pulse current.

Lithium (CR) battery designs use lithium for the anode and MnO_2 for the cathode. MnO_2 reduces the internal impedance of the battery. This particular device is best suited for applications in which higher pulse currents and stable voltage during discharge are the principal specification requirements. Both BR and CR lithium batteries offer a nominal voltage of 3 V and an energy density in excess of 1 MJ/kg. These batteries are widely used for radio-frequency identification (RFID), remote key entry, and watches for which reliable performance over a wide temperature range and excellent pulse capability are the system requirements.

The lithium-thionyl-chloride battery can be used for embedded-system applications [3], if an ultra-low SDR and longevity greater than 20 years are the principal design requirements. This battery can be fabricated using lithium as the anode and sulfur-oxygen chlorine as the cathode. This battery offers the highest nominal voltage of 3.6 V, moderate energy density of 1.69 MJ/kg, and extremely low SDR, leading to a shelf life in excess of 20 years. The flat discharge profile of this battery over time offers a constant output voltage throughout the entire service life. Preliminary research studies on the fabrication materials indicate that this battery technology is more costly than other lithium chemistries. Despite slightly higher fabrication costs, this particular battery is in great demand for industrial and military-electronic applications, where higher efficiency and reliable operation over a period of 20 years or more are the principal requirements [4]. As mentioned, these batteries have nominal voltages greater than 1.4 V and discharge voltages of 0.9 V.

Energy capacity can be expressed in MJ/kg or in milliampere-hours. According to battery designers, joules is more convenient way of comparing batteries with different chemistries. One can easily convert the battery capacity from milliampere-hours using the following expression and inserting relevant electrical parameters:

$$E = [(C \, V_T) \, (3.6)] \tag{5.7}$$

where C is the battery capacity in milliampere-hours and V_T is the terminal voltage.

5.3.2 Battery Suitability and Unique Performance Requirements for Aerospace Applications

Battery suitability for a specific application requires a meaningful assessment of critical parameters by the engineer. Some of the most common parameters to be used in the analysis are nominal voltage, energy capacity, energy density, SDR, and other dynamic considerations. The nominal voltage is the voltage as measured across the battery terminals. Battery chemistries rely on electrochemical reactions to provide electrical energy. Some space and military applications [5] define specific requirements for the weight and size of the battery. The size-to-energy ratio is known as energy density. As a general rule, the higher the energy density, the higher the cost of the battery technology. Battery designers constantly struggle to find the optimum balance between cost and energy density.

Regardless of battery type, battery energy does not last forever. Even if the battery is sitting idle on the shelf, electrochemical reactions are taking place that diminish the battery's energy capacity. This process is known as SDR. The operating life is called the service life (SL). The SL for alkaline batteries is typically between 7 to 10 years. The service life for lithium (BR and CR) batteries varies from 10 to 15 years. But the service life for a lithium-thionyl-chloride battery is more than 20 years. SDRs and other deteriorative mechanisms that affect battery life can highly depend on the environmental temperature and duty-cycle characteristics. Furthermore, fluctuating duty-cycle requirements can have an adverse effect on the ultimate discharge characteristics of a battery. In addition, dynamic physical parameters can also affect battery performance. Variation in temperature, output impedance, duty cycle, and energy delivery can affect battery-loading conditions. If some of these conditions are first-order variations, appropriate corrections in dynamic physical parameters are necessary.

Many electronic systems have high dynamic bandwidth with respect to the power demand. For example, a wireless sensing system involving an advanced-metering infrastructure-class gas or water meter can have dormant power consumption on the order of a few microwatts (10^{-6} watts [W]) and an active peak consumption of a few watts. In other words, dynamic-system power-demand bandwidth can be microwatts during the low-duty-cycle sleep mode and can be watts during active-duty and high-duty-cycle radio-transmit mode. This operational situation creates additional power delivery requirements that the battery must be able to accommodate alone or with another energy-storage device. In some cases, additional design issues must be considered, such as charging capacitor size, charging scheme, battery-leakage specifications, and battery-discharge profiles.

5.3.2.1 Potential Applications of Lithium, Alkaline, and Zinc-Air Batteries

As discussed, BR and CR lithium batteries are most ideal for applications in which longevity, fluctuating duty-cycle requirements, and high dynamic bandwidth are the major performance requirements. Dynamic physical parameters include variations in temperature, output impedance, duty cycle, and battery energy capacity, which affect the battery-loading conditions and ultimately shape the battery-selection process. These lithium-based batteries are most suitable to operate over high-dynamic-bandwidth physical parameters.

Alkaline batteries play an important role in bidirectional wireless-sensor mode in a home-security application featuring a glass-breakage-detection system with a bidirectional communication link that uses a transmitter and receiver. This system includes a microcontroller device with an integrated direct current (DC)/DC converter, a sub-1 gigahertz (GHz) radio transceiver, a piezoelectric shock sensor, and a single AA-alkaline battery with a nominal voltage of 1.4 V. The home-security sensor monitors the physical condition of the glass window and periodically reports the status of the window and the alkaline battery to the main control panel. The radio communication between the sensor and the control panel uses a transmitter-receiver-acknowledgment protocol, which essentially reduces the number of redundant messages the sensor sends to the control panel. Most of the time, the sensor is operating in a low-power mode to maximize the battery's life and its energy. The glass-breakage-sensor states are summarized in Table 5.5.

Table 5.5 Various Operating States and Function and Description of Various Sensor Elements for the System Described

Sensor Mode	Frequency	Description and Function of Sensor Element
Measurement	Event-driven	Shock sensor interfaced to system I/O
Transmit	Once per minute	Transmit sensor and battery status to panel
Receive	Once per minute	Receives acknowledgment from control panel
Sleep		Maintains low-power operation in sleep mode and I/O function

Note: I/O: input/output.

5.3.2.1.1 System Application Involving AA-Alkaline Batteries

One can make four basic assumptions about the system under consideration. First, the piezoelectric sensor is self-powered and generates a 3 V pulse signal if a glass window breaks. If the glass breaks, the 3 V pulse signal triggers an external interrupt that "wakes up" the microcontroller device. Second, the microcontroller core is regulated to 1.8 V by an internal regulator. The random-access memory power-management unit and the real-time clock can operate at a voltage as low as 0.9 V so that the microcontroller can operate from a single AA-alkaline battery. Third, the power amplifier in the transceiver's transmitter provides higher output power at higher efficiency when its voltage rail approaches the maximum-rated power rail. Fourth, an internal 1.8 V regulator regulates the low-noise amplifier, receiver chain, phased-locked loop (PLL), and synthesizer in the radio. The minimum operating voltage is 1.8 V.

5.3.2.1.2 System Application Using AAA-Alkaline Batteries

It will be interesting to find out how the same system performs using AAA-alkaline batteries with nominal voltage of 1.5 V. Using the same assumptions, it will become clear that dynamically adjusting the AAA-alkaline voltage of 1.5 V will optimize the power efficiency and system performance. One can obtain maximum transmitter efficiency when the transceiver is operating at 3 V. The AAA-alkaline battery has only a nominal voltage of 1.5 V. So one has to achieve 3 V operation with the integrated DC/DC boost converter, which has an efficiency of approximately 90%. Internal regulation limits the receiver chain to 1.8 V, however. This means that supplying 3 V during the receiver transaction would cause the internal low-dropout regulator to reduce efficiency close to 60%. Therefore, it would be better to adjust dynamically the output voltage of the DC/DC converter from 3 to 1.8 V and to increase the efficiency during the sensor's receiving transaction. This illustrates that a battery voltage of 1.8 V is most suitable for this particular case. The energy requirements for each system element using AAA-alkaline batteries with dynamic voltage regulation are summarized in Table 5.6.

5.3.2.1.3 System Application Using Lithium Coin-Cell Batteries

One should compare system performance using a lithium coin-cell battery featuring a fixed voltage rail with an alkaline battery using the dynamic-switching technique. Preliminary system designs indicate that the switching loss is close to zero using the coin cell because the switch-mode supply is not in use and the terminal voltage for the lithium coin cell is 3 V. Because of the cell capacity, one does not need to increase the size of the call to meet the peak current demand. Resizing of the cell would increase the size, weight, and cost of the cell.

Table 5.6   Energy Requirements for Each Element Using AAA-Alkaline Batteries with Dynamic Voltage Regulation

Mode	Frequency	Duration Second (A)	Current Voltage (V)	Voltage Loss (%)	Switching (J)	Energy
Sleep	60	0.955	0.6×10^{-6}	1.5	0	$52/10^6$
Process	1	100×10^{-6}	4×10^{-3}	1.8	10	$0.8/10^6$
Transmit	1	15×10^{-3}	27×10^{-3}	3.0	10	$1.4/10^3$
Receive	1	30×10^{-3}	18×10^{-3}	1.8	10	$1.1/10^3$

The energy requirements of each element of the wireless-sensor application [6] using the 3 V lithium coin cell are summarized in Table 5.7. The sleep duration is one second minus the sum of all other transactions. The processing, receiving, and transmitting functions occur once per minute. Table 5.7 shows the requirements for a lithium coin cell (CR2450) with a terminal voltage of 3 V, capacity of 620 mAh, and an approximate cost of $0.62. The CR2450 battery will not last more than 4.33 years, and the battery will experience a gradual loss in capacity as it ages.

Next, the energy requirements for each element of the wireless-sensing application using AAA-alkaline cells with nominal voltage of 1.5 V will be determined. This particular cell costs about $0.25. The cell capacity is approximately 1,125 mAh. As discussed, the cell experiences reduction in capacity as it ages.

This AAA-alkaline battery will last approximately 4.65 years [3]. This duration indicates 16% higher efficiency, which will result in a 7% increase in battery service life with a 60% decrease in battery cost. This particular battery clearly offers impressive improvements in efficiency, cost, and service life by using more modern dynamic concepts of energy conversion. These gains or improvements are strictly dependent on the duty cycle of the functions that derive the maximum advantage from the high-efficiency power supply. As the duty cycle increases, so too do the benefits of using an alkaline battery source with a switched-mode power supply.

Table 5.7   Energy Requirements for Each Element Using Lithium-Coin Battery with Fixed Voltage Regulation

Mode	Frequency	Duration (sec)	Current (A)	Voltage (V)	Switching Loss (%)	Energy (J)
Sleep	60	0.955	0.6×10^{-6}	3	0	$0.1/10^3$
Process	1	0.1×10^3	4×10^{-3}	3	0	$1.2/10^6$
Transmit	1	15×10^{-3}	27×10^{-3}	3	0	$1.2/10^3$
Receive	1	30×10^{-3}	18×10^{-3}	3	0	1.6×10^3

The DC/DC converter's output voltage will be 3.3 V, which could be higher than that of the lithium-based coin battery. This higher voltage will provide higher output power and improved dynamic range.

Growth in battery technologies and chip-level power-management techniques has demonstrated significant improvement in system performance and reduction in system component costs. Published technical papers indicate that aggressive research and development activities, technological evolution, and rapid innovations in the fields of chemistry, material science, electrical and electronics engineering, and manufacturing have resulted in batteries that are thousands times more sophisticated in design and function than when Volta invented the original versions. In brief, 21st-century system designers and architectures have much greater options when selecting an appropriate battery or cell to support their next embedded-system designs.

System designers or architects must select a battery or cell for their embedded-system applications that should be able to provide optimum efficiency and reliable performance with minimum power requirement, cost, and complexity [3].

5.4 Batteries for Medical Applications

Studies performed by the author indicate that low-power, compact batteries are given great attention for use in medical applications that are needed to collect patients' health-related parameters. Low-power batteries are generally considered to cover ratings of a few microwatts for watches and cardiac pacemakers up to 10 to 20 W to meet the power requirements for notebook computers and laptops. Low-power Li-ion rechargeable batteries for medical applications have been designed and developed by various manufacturing companies that are capable of meeting low-power electrical performance requirements, including the physical requirements such as weight, size, and geometrical configuration. An extended storage period is the principal requirement for rechargeable batteries intended for medical diagnostic applications.

Currently, batteries with specific geometric configuration and size are being used for different medical applications. For example, a transdermal delivery system known as *Smart Relief* requires a coin-size battery to provide a current to generate an ion transfer between a patch and the skin to administer a migraine drug using a coin-size microprocessor and this battery. The migraine patch provides relief for migraine headaches.

For some medical treatments, a battery is required to provide consistently reliable performance over a long duration. For medical applications, battery longevity ranging from two to five years is the principal requirement for optimum economy.

One of the most notable drug-delivery methods is the insulin pump using microelectronic circuits and microfluidic MEMS technology. This miniaturized insulin pump can be mounted on a disposal skin patch to provide continuous insulin infusion to the diabetic patient. A battery for this application must be compact, ultra-lightweight, and most reliable. The battery must provide the required current and voltage with high reliability and minimum fluctuations. A single AA- or AAA-alkaline battery will able to power this pump for a period ranging from one to three months. The lithium-based coin microbatteries are best suited to operate this pump continuously for longer periods. Li-ion batteries, despite their higher costs, can be used to operate the pump even for several months because of their higher energy capacity, longevity, reliability, and safety features.

The U.S. Food and Drug Administration (FDA) is evaluating the latest rapid, affordable, high-sensitivity electronic diagnostic testing system. This system can be used in doctors' offices, other patient care centers, and eventually at home. This compact, lightweight diagnostic equipment can be used to screen for breast cancer and heart disease. Lithium-based batteries using solid electrolytes and other low-power batteries can power this particular system.

Scientists at Xen Biosciences and Cambridge Consultants are jointly developing a compact, modem-size device that deploys time-resolved fluorescence (TRF) spectroscopy technology. This device can be used to perform tests for up to 20 different ailments, enabling the physician to make rapid decisions with minimum cost and complexity. According to the device designer, the power requirements will be less than 100 milliwatts, which can be satisfied by low-power Li-ion batteries

A pocket-size or palm-size cardiopulmonary resuscitation (CPR) device was developed a few decades ago to initiate the critical rescue steps needed to revive someone from sudden cardiac arrest (SCA). This device has been recognized as the most effective life-saving device and has been approved by the FDA for over-the-counter sales. A low-power, coin-shaped Li-ion battery, which has demonstrated utmost safety, reliability, and operation over wide temperatures, will be able to power this pocket-size SCA device.

According to health care providers, powerful signal processing, more accurate and reliable information, and a portable low-power source with regulated output voltage will enable doctors to accurately monitor patients' vital signs and to make medical decisions. Deployment of analog front-end loaded, high-speed integrated circuit devices with micro- or nanobatteries would yield portable, accurate, and reliable EKG and electroencephalography (EEG) machines. These machines would be capable of rapidly monitoring vital medical parameters for the physician's timely review and medical decision making. The following paragraphs identify and describe potential battery technologies that are best suited for medical applications. An emphasis will be placed on performance capabilities, shelf life, and reliability aspects.

Miniature hearing devices are widely used by the elderly. According to audiologists, miniature hearing devices can be placed deep inside the ear canal as a semi-permanent device. The person can wear this device for four to six months at a time. If needed, the audiologist can remove the device, replace the tiny lithium-based coin battery, and replace it in the ear canal. The hearing device contains a min-iaturized electronics-based, high-gain voltage amplifier that requires less than 100 microamperes at a typical voltage close to 100 mV. Because both the current and voltage requirements are extremely low-power, lithium-based coin batteries are best suited for hearing aid devices.

5.4.1 Recently Developed Batteries for Specific Medical Applications

Thus far, batteries for general medical applications have been described. Now low-power batteries for specialized medical fields and special medical applications will be described with a particular emphasis on cardiac rhythm management (CRM) systems [7]. Studies performed by medical experts reveal that three distinct types of devices are capable of treating cardiac diseases: pacemakers, cardioverter defibril-lators, and left-ventricular assist devices [7]. Batteries also have been developed to meet the power requirements of an artificial heart.

Cardiac pacemakers are generally employed when the cardiac rhythm is either abnormal or too slow. To rectify this problem, doctors prescribe implanted pace-makers that detect the slow heart rate and send impulses to stimulate the muscle using microelectronic circuits. The life of these devices developed before 1973 and incorporating zinc-mercuric-oxide (Zn-HgO) batteries was only between 12 and 18 months. When Li-I_2 batteries became available around 1975, the battery life was extended to more than 10 years. The life of devices with batteries developed after 2008 could be more than 15 years.

5.4.1.1 Performance Characteristics of Li-I_2 Batteries

The cathode of Li-I_2 batteries contains a mixture of iodine and polyvinylpyridine (PVP). The latest fabrication approach for this battery involves pouring a molten cathode material into the cell, which ultimately forms a separator layer. The energy density of this battery is high because of the high energy density of I_2 and because the separator and the electrolyte are not added to the cell structure. This device sup-plies the current in the microampere range, which meets the current requirements of the implanted pacemakers over several decades and does so with ultra-high reli-ability. The high safety and ultra-high reliability of this cell make it most suitable for implanted pacemaker applications. Published literature indicates that the per-formance of these medical devices has improved significantly by using titanium casing and special radiofrequency (RF) filters.

Implementation of special filters eliminates the interference in the band-pass region of the pacemaker's sense amplifier coming from power lines, cellular phones, microwave ovens, and other RF sources nearby. The introduction of programmable pacemakers eliminated the need for repeated surgical operations, because the operating parameters of the pacemaker could be set by external signals. Subsequently, advanced pacemakers were developed with rate-response capability, which allows for the detection of body movements affecting the pacemaker rate. Pacemakers can be used as an anti-atrium fibrillation device. Advance pacemaker batteries using advanced technology can last more than 10 years. Modern pacemakers act as microcomputers, which can store the patient's heart data directly in the device's memory and can monitor the patient's heart activity.

Excellent batteries have been designed to power implantable cardioverter defibrillators (ICDs), which are capable of detecting and treating episodes of ventricular fibrillation. This device contains a high-voltage capacitor, which provides a stimulus to the heart when the regular heart pace is lost because of ventricular problems. Batteries capable of delivering high current pulses are needed to power ICDs.

Moderate power batteries have been developed for total artificial hearts (TAHs). The TAH is a mechanical heart pump that essentially replaces the patient's natural heart. According to the TAH designer, the device contains two chambers, each of which is capable of pumping more than 7 L of blood per minute, which is equal to the heart's natural pumping rate. The TAH device uses an implantable Li-ion battery pack, which is recharged through the skin of the patient.

According to medical experts, all implantable batteries must have the following characteristics:

- Ultra-high reliability
- High safety
- Overall performance with high predictability
- High energy density
- Minimum weight and size
- Low self-discharge
- High cycle life and a safe charging process
- Clear-cut indication of end of life

Heart-related battery searches indicate that the following battery types are widely used to treat various cardiac diseases:

- Li-I_2: best suited for implantable pacemakers
- Li-ion: recommended for TAHs
- Lithium-silver-vanadium oxide (Li-AgVO): capable of delivering high current pulses and, therefore, most ideal for ICDs
- Li-MnO_2: most desirable for cardiac pacemakers
- LiCF$_x$: best suited for ICDs

Discharge characteristics of Li-I_2 and PVP batteries widely used in implantable pacemakers are shown in Figure 5.4. The relationship between the battery's

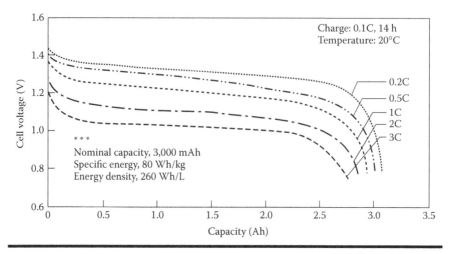

Figure 5.4 Typical discharge curves at various charging rates of a cylindrical 3 Ah cell. (***: electrical characteristics of a 3 Ah cell using cylindrical configuration.)

OCV and the battery's capacity (mAh) is also shown in Figure 5.4. Discharge characteristics of a Ni-MH battery under long-life test conditions at various temperatures are shown in Figure 5.5. The tests are based on applying four 10-second, 2 A pulses every 30 days with a background load of 100 kilo-ohms applied continuously.

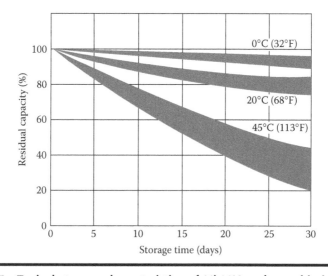

Figure 5.5 Typical storage characteristics of Ni-MH rechargeable batteries at various temperatures. (Courtesy of Duracell.)

5.4.2 Microbattery and Smart Nanobattery Technologies Incorporating Lithium Metal for Medical and Military Applications

Portability and reliability are the most critical requirements for medical and military applications. In some applications, such as battlefield deployment, a maintenance-free requirement is an additional performance specification. Medical and military users want batteries that incorporate the latest microelectronics and lightweight material technologies in addition to meeting state-of-the-art electrical performance requirements. In brief, portability applications are constantly changing and gaining design complexity with the main intent of providing users with more features. These new features always provide design challenges for the portable battery design.

During the past couple of decades, sealed lead-acid (SLA) batteries were widely used in hospitals and clinical medical-diagnostic equipment because of their unique performance and lifetime cost savings. These 12 V, maintenance-free lithium-iron-phosphate ($LiFePO_4$) SLA batteries provide a long runtime, close to 40 Ah current capacity, a fast charge time, and a long shelf life; they are lightweight (13.2 lbs.) and have compact packaging not exceeding 7.8 in. × 7.0 in. × 4.9 in. Note that the lithium-based cathode offers cost-effective battery performance, improved reliability, and a long service life.

Recently, new batteries that are better than SLA batteries have been developed and currently have been deployed for portable infusion pumps, ventilators, wheelchairs, and workstation medical carts carrying portable diagnostic equipment capable of providing the on-the-spot vital information doctors need. These batteries incorporate microelectronics technology and the latest materials, leading to a compact, smaller, and lighter package compared with SLA batteries. Both of these battery designs cannot be characterized as microbatteries or smart nanobatteries.

Microbattery and smart nanobattery technologies have been developed by various companies under the Small Business Innovative Research (SBIR) programs. These batteries are specifically designed and developed for medical diagnostics, computer memory chips, defense systems, and space sensor applications, for which weight, size, and power consumption are of critical importance. Published technical articles claim that various nanotechnology firms are working to perfect nanobatteries under defense organization research and development funding.

A smart nanobattery has been developed and soon will be available for a critical computer memory application. In other words, a research and development company has perfected the design of a small footprint, multicell, 3 V, lithium chemistry, micro-arrayed battery with a minimum shelf life exceeding 20 years and uninterrupted power output during this period. This battery package has potential applications in airborne missiles, UAV drones, and space sensors, for which uninterrupted power availability over long durations are the critical requirements. Battery scientists and designers believe that deployment of multiple cells and silicon-based membrane design configurations will be most ideal for providing constant power to flashlights, which are best suited for defense perimeter security applications during nights regardless of weather conditions.

5.4.2.1 Smart Lithium-Ion Batteries

It is absolutely necessary to follow the mandatory safety precautions during the manufacturing of lithium metal–based batteries. Typical safety devices used during the fabrication, testing, and evaluation of these batteries include the following:

- Tear-away tabs (excess internal pressure)
- Shutdown separators (overtemperature condition)
- Vents (excess internal pressure)
- Thermal interrupt (overcurrent or overcharging conditions)

Typically, these batteries are permanently and irreversibly disabled if safety devices are not used. Smart Li-ion batteries do not require a large number of safety devices for safe operation, thereby yielding a substantial cost reduction in the manufacturing of such batteries.

Cost is a major factor for any battery pack used in an EV or HEV. A preliminary cost prediction study indicates that Li-ion batteries have a 50% higher cost over Ni-MH batteries, and a 40% higher cost over Ni-Cd batteries. Higher cost projections are valid only for the high-power batteries required to operate EVs or HEVs.

5.4.3 Low-Power Zinc-Air, Nickel-Metal-Hydride, and Nickel-Cadmium Rechargeable Batteries

Studies undertaken by the author on batteries indicate that lithium, Zn-air, Ni-MH, and Ni-Cd secondary batteries are best suited for applications requiring low power levels. Because of inferior performance capabilities and operational limitations associated with of Zn-air and Ni-Cd batteries, the author will devote more time to discussing the performance capabilities and limitations of Ni-MH batteries. The following sections briefly summarize the performance capabilities and limitations of these rechargeable batteries [8].

5.4.3.1 Zinc-Air Rechargeable Batteries

Zn-air batteries with effective air management have the lowest SDRs. The discharge rate for a Zn-air cell is less than 5% per month. The presence of the air-management system in a Zn-air cell does not qualify it as a real small battery. As a matter of fact, the small Zn-air batteries are designed for portable computers and small electric cars. In the case of EV applications, the batteries incorporate filters to remove the carbon dioxide from the incoming air and include a small pump to recirculate the electrolyte. Removal of carbon dioxide from the air is highly desirable because the gas tends to react with the KOH electrolyte to form potassium carbonate, which eventually plugs up the oxygen (O_2) electrode. To reduce the weight size of the battery, it will be necessary to eliminate the filter or select a new filter design. Another weak point of this battery is its sensitivity to electrical abuse. This battery can be quickly charged, and if it is discharged below 0.9 V, approximately, the battery will be permanently damaged. Its nominal operating voltage is between 1.0 and 1.2 V.

To rectify these defects, the designers have integrated miniaturized built-in monitoring devices and charging circuits in the Zn-air batteries. The first type monitors the voltage of each cell, puts out an alert signal as cutoff approaches, and shuts down the battery when cutoff is reached. The second type charges the cells at a constant voltage of about 2 V with a current-limiting circuit. The charging process ends when the current tapers off to a specified value.

5.4.3.2 Nickel-Cadmium Rechargeable Batteries

Because of the toxicity of cadmium, deployment of Ni-Cd rechargeable batteries has been restricted to applications for which proper ventilation and other safety precautions have been taken into account. It is of critical importance to point out that during the overall chemical reaction, water (H_2O) is consumed on discharge and reformed on charge. Therefore, there is no net reaction involving H_2O. This means that the concentration and the conductivity of the solution do not change. These favorable features are counterbalanced by reduced power density, faster self-discharge, and lower tolerance to overcharge.

In the case of the Ni-Cd battery, a smooth solid-state process occurs because of hydrogen-ion (H^+) uptake and release. This process eliminates the negative changes occurring during the life of the cadmium electrode in its crystallography, surface morphology, mechanical structural integrity, and electrical conductivity. Again, there is no net chemical reaction involving H_2O, so that the concentration and the conductivity of the solution do change. These favorable features are counterbalanced through a reduced power density, faster self-discharge, and reduced tolerance to overcharge. Finally, toxicity of cadmium and disposal problems have prevented this battery from wide acceptance for various applications. Major advantages and limitations of nickel-cadmium rechargeable batteries can be summarized as follows:

Advantages

- *Simple storage and transportation*: Most airlines accept these batteries without any restrictions or special conditions.
- *The batteries are guaranteed against abuse*: Ni-Cd batteries come with mechanical integrity and are best suited for applications involving severe operating environments.

- *Good value*: Market surveys indicate that Ni-Cd batteries are economically priced and require a fast and simple charge, even after prolonged storage.
- *Long operating life*: Ni-Cd batteries can tolerate a high number of charge and discharge cycles with no compromise in performance. A minimum of five-year storage is possible.
- *Excellent load impedance*: Ni-Cd battery material allows recharging even at low temperatures.
- *Ni-Cd batteries are capable of meeting wide performance requirements*: These batteries are available in a wide range of sizes and performance options.

Limitations

- *Environmentally unfriendly*: Ni-Cd batteries contain toxic materials.
- *Relatively high self-discharge*: These batteries suffer from high self-discharge rates and, thus, need recharging after prolong use or storage.
- *Low energy density*: Ni-Cd batteries are larger and heavier than alternatives. These batteries suffer from memory effect and must periodically be exercised.

5.4.3.2.1 Conclusions

Track records of various applications reveal that Ni-Cd is the only rechargeable battery type that performs with high reliability under rigorous working conditions [8]. These batteries are best suited for power tools. Ni-Cd batteries do not like to sit in chargers for extended durations and do not prefer to work for short periods. These batteries remain a popular choice for two-way radios, emergency medical equipment, and power tools widely used in various commercial and industrial applications. Ni-Cd batteries are ready to provide full energy after a fast and simple charge, even after prolonged use or storage.

5.4.3.3 Nickel-Metal-Hydride Rechargeable Batteries

Lithium-based, Ni-Cd, and Ni-MH batteries were developed and evaluated for low-power applications. The anode of this battery is a metal hydride. The cathode is nickel hydroxide. The electrolyte is KOH. This battery offers impressive power and energy performance with minimum weight and size.

Studies performed by the author on Ni-MH batteries indicate that these devices suffer from voltage depression, leading to a memory effect. The studies further indicate that the memory effect is less severe in Ni-MH batteries than in Ni-Cd batteries. Well-designed Li-ion batteries, however, are free from memory effects. When the current is interrupted, partial voltage recovery may occur in some batteries. Voltage recovery is strictly dependent on the complexity of the electrochemical devices, such as batteries. During the charge and discharge cycles, the hydrogen ion moves back and forth between the two electrodes as illustrated in Figure 5.6.

Actual determination of SOC in an Ni-MH battery is dependent on a chemical reaction between the SOC and OCV. This is true if one knows both the parameters of SOC and OCV. Determination of the critical value of OCV is again dependent on SOC, age of the battery, and the adverse effects of voltage depression. From comparing the OCV with the DOD curves for new and old batteries, one can determine the critical value of OCV. These curves indicate that as the battery ages, if the desired DOD is about 23%, the critical value of OCV increases. The longevity of low-power batteries is strictly dependent on the following conditions:

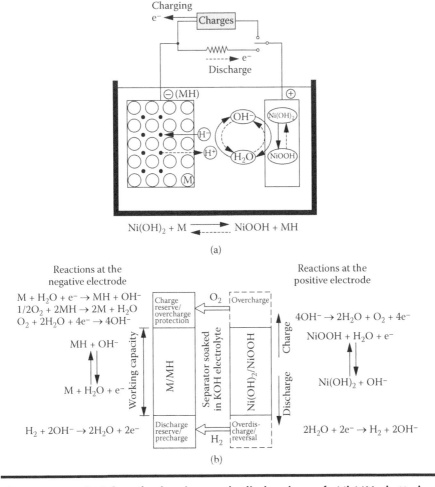

Figure 5.6 Principle of charging and discharging of Ni-MH batteries. (a) Movements of hydrogen ions during charge and discharge cycles of an Ni-MH cell. (b) Scheme of chemical reactions occurring within the Ni-MH cell.

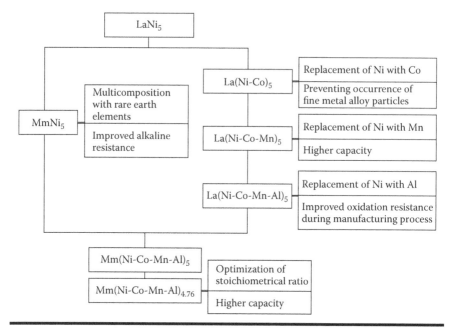

Figure 5.7   Progressive optimization steps required for an LaNi$_5$ alloy (AB$_5$) used in the construction of a sealed Ni-MH battery. (Mm: Mischmetal; Mn: manganese.)

- Charge the battery early and often.
- Reduce the charge to about 40% and store it in cooler place when the battery is not used for several weeks.
- Store the dormant battery in a refrigerator (never below –40°C).
- Never deep cycle with SOC near zero.
- Avoid exposure to high temperatures, which may result in rapid degradation in battery performance.
- Ensure that the shelf life of a battery starts at the manufacturing time and ends with the storage period until it is ready for deployment.
- Follow progressive optimization steps, if optimum performance from a thin prismatic-sealed Ni-MH battery using a rare earth–based lanthanum-nickel (LaNi$_5$) alloy is desired. The progressive optimization steps as shown in Figure 5.7 are necessary for this battery to meet the demanding discharge rates at low- and high-temperature operations.

5.4.3.3.1 Role of Rare Earth Materials for Progressive Optimization

The alloys recently investigated in the fabrication of Ni-MH batteries include the AB$_5$ (LaNi$_5$) type and AB$_2$ (zirconium vanadium [ZrV$_2$]) type. During the evaluation of these alloys, the material scientists failed to explore critical performance characteristics of the alloys other than the discharge capacity and progressive optimization steps of the battery. The investigators, however, have done a wonderful job in determining the charge capacity as a function of terminal voltage. Data presented in Table 5.8 indicate that even a slight change in terminal voltage can bring significant improvement in the discharge capacity of an Ni-H-M battery using appropriate rare earth materials.

Reduction in discharge capacity in Ni-Cd batteries is severe down to 1,000 mAh even at terminal voltage of 1.19 V. The battery discharge capacity is reduced to less than 1,325 mAh at terminal voltage close to 1 V.

The symbol "Mm" shown in Figure 5.7 denotes a Mischmetal, which is a naturally occurring mixture of rare earth elements, such as lanthanum (La), neodymium (Nd), praseodymium (Pr), and cerium (Ce). Rare earth material scientists claim that the material composition $Mm(Ni-Co-Mn-Al)_{4.76}$ offers a discharge capacity close to 330 mAh/g, which is 10% higher than the alloy $LaNi_5$ designated as AB_5. A comprehensive review of various material compositions indicates that the $LaNi_5$ (AB_5) alloy is cheaper and easier to use and, hence, is best suited particularly for sealed Ni-MH batteries.

Material scientists have investigated other compositions, but they suffer from poor corrosion resistance, shorter life cycle, and lower operating voltage compared with the alloy $LaNi_5$. In summary, deployment of a specific rare earth material in the fabrication of a battery significantly improves its capacity. The overall performance of Ni-MH batteries strictly depends on the right kind of alloy for the negative electrode (anode), the cathode formulation, and the unique

Table 5.8   Ni-MH Battery Discharge Capacity as a Function of Terminal Voltage Variations

Battery Terminal Voltage (V)	Battery Discharge Capacity (mAh)
1.30	200
1.26	400
1.25	600
1.25	800
1.25	1,000
1.22	1,200
1.20	1,400
1.19	1,600
1.15	1,800
1.00	1,885

properties of the separator. Both rare earth–based alloys, namely AB_5 and AB_2, have undergone important research developments over the past several years to improve their material characteristics in terms of high capacity and energy, wide temperature range, long cycle life, high electrochemical activity, enhanced reliability, and long shelf life. (See Chapter 4 for specific details on properties of potential rare earth materials.)

5.4.3.3.2 Ni-MH Cell Construction and Its Performance Characteristics

Market surveys indicate that thin prismatic-sealed Ni-MH batteries are widely used in several applications, where portability, low-power consumption, compact packaging, and minimum room-temperature storage loss are the principal procurement specifications. The cathode of the Ni-MH battery produces a cobalt oxide (CoO), which has much higher conductivity than nickel oxide hydro oxide (NiOOH). The addition of yttrium oxide (Y_2O_3) increases the utilization of the positive ions. O_2 that evolves at the positive electrode on overcharge could oxidize the separator element. Therefore, a chemically stable separator material is needed to avoid the oxidation of the separator. The suffocated polypropylene is the most suitable separator material to avoid oxidation.

The construction of sealed Ni-MH batteries will have design specifications similar to those for Ni-Cd batteries. The electrodes will have high surface areas capable of meeting high charging rates. In addition, the electrodes are flat and rectangular and are considered most ideal for cylindrical design configuration as shown in Figure 5.8. Cylindrical configuration is preferred because of design simplicity and lower construction costs. The positive electrode is made of a felt or foam substrate into which the active material is impregnated. The negative electrode is a porous structure and is made from perforated nickel-foil supporting the plastic-bonded hydrogen storage alloy. The electrodes in a prismatic Ni-MH battery are spirally wound similar to Ni-Cd batteries, and they are stuffed into a nickel-plated steel can as shown in Figure 5.9. The cell top contains a releasable safety vent.

Prismatic batteries have thin design structures and are best suited for powering tiny devices. According to battery designers, for a given space, these batteries deliver up to 20% more capacity. Discharge curves as a function of charge rates for a cylindrical, prismatic 3 Ah Ni-MH battery cell are shown in Figure 5.4. Typical values of nominal capacity, specific energy, energy density, and ambient temperature are shown in the left corner of the figure. It is evident from these curves that the cell loses little capacity at the expense of mean voltage. These batteries will have similar exothermal nature of the discharge reaction of Ni-Cd rechargeable batteries. Under these circumstances, the heating due to current flowing in the cell is partly balanced. Repeated discharges at higher discharge currents will reduce the battery cycle. According to the Ni-MH battery designers, best performance results can be achieved with discharge rates of 0.2C and 0.5C.

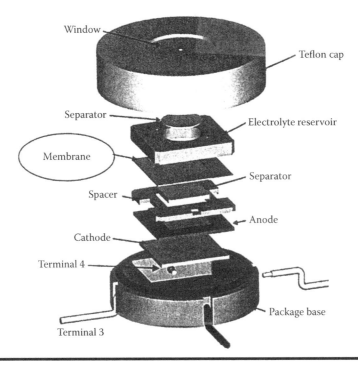

Figure 5.8   Typical design configuration of a reserve battery incorporating nanotube membrane technology.

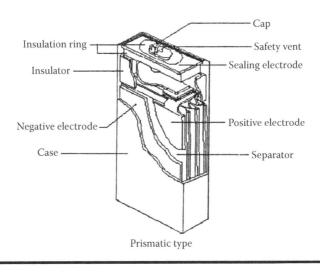

Figure 5.9   Structural details of a thin prismatic Ni-MH battery. (Courtesy of Panasonic.)

Table 5.9   Impact of Temperature on Discharge Capacity of a Sealed Ni-MH Battery

Temperature (°C)	Discharge Capacity (%)
−10	65
−20	41
0	89
10	97
20	99
30	98
40	96
50	90

Note: The data presented in this table indicate the estimated values, which could have errors within ±5% of the values shown. The accuracy of the discharge capacity strictly depends on the ability of the rare earth–based alloy used.

The effect of temperature on discharge capacity of a sealed Ni-MH battery can be visualized from the data presented in Table 5.9. The temperature range outside 0° to 40°C should be used, if optimum cell performance is required. The drop of capacity at low temperatures is strictly due to the increase of the internal resistance of the cell.

The discharge curves of a cylindrical Ni-MH cell as a function of cell voltage and charging rates are shown in Figure 5.4. These curves indicate that the cell loses a small capacity by increasing the rate, which is at the expense cell voltage. The discharge is of an exothermic nature in Ni-MH batteries, and the discharge is of an endothermic nature in Ni-Cd batteries, because the current flowing in the cell is partly balanced. Studies performed by the author reveal that even though the Ni-MH battery is capable of sustaining high discharge currents, repeated discharges at these currents tend to reduce the cycle life of the battery. The studies further indicate that the most satisfactory results are possible at charging rates ranging from 0.2C to 0.5C and using the values of specific energy, energy density, nominal capacity, ambient temperature, and charging rate specified in Figure 5.4.

5.4.3.3.3 Charging of Ni-MH Batteries

Because of the sensitivity of Ni-MH batteries to charging conditions, charging is a critical step in determining the electrical performance parameters and overall life of these batteries. It is extremely important to select the most appropriate charging rate and temperature range.

Generally, Ni-MH batteries are charged at an appropriate and constant current. The charging current levels have to be limited to avoid overheating and incomplete O_2 recombination. Furthermore, the voltage and temperature profiles of Ni-MH batteries, which can be obtained from the suppliers or manufacturers, should be consulted for optimum charging efficiency. The shapes of the charge curves at different rates along with the corresponding temperatures are available in the standard *Handbook of Batteries*. One will see that the voltage raises more sharply around 80% charge because of more significant O_2 evolution, and it tends to level off or to reach a maximum value. In comparison with the charge of Ni-Cd batteries, the voltage drop, if present, is not evident. This is the fundamental difference between the charging process of Ni-MH and Ni-Cd batteries. As discussed, the charging process is exothermic (formed with the evolution of heat or giving heat during the process) in Ni-MH batteries, whereas it is endothermic (formed with absorption of heat) in Ni-Cd batteries. Therefore, the temperature of the Ni-MH batteries will rise more quickly during the charge cycle compared with Ni-Cd batteries under the same charge input. After about 80% charge, however, simultaneously with O_2 evolution, the temperature of both batteries will increase sharply because of the exothermic O_2 recombination reaction. This increase in temperature causes the voltage to drop as the battery reaches its full charge status and enters the overcharge zone.

On the basis of these considerations, the charge temperature has to be controlled to obtain a high discharge capacity. To maintain reliable performance, the temperature should not exceed 30°C, approximately, because an increasing O_2 evolution takes place at high temperatures. Battery charging experts believe that charge efficiency is very high within the temperature range of 10° to 30°C. This indicates that charging at temperatures below 10°C should be avoided, if high efficiency is desired.

Voltage drop and temperature rise should be taken as indicators of the end of the charge. Note that the charging method is strictly dependent on the conditions of the specific charge. A charge control is absolutely necessary when dealing with fast charges. In cases where the temperature and pressure might suddenly reach high values, a vent must be provided. The methods for charge control are summarized as follows:

- *Input charge*: Using the current and time parameters, the charge (Ah) can be calculated. However, this method is applicable only to slow charges less than 0.3C to avoid overcharges that could occur.
- *Voltage drop*: This method is applicable only when the charge is slow and in cases in which the voltage levels off at a charge close to 0.1C.
- *Temperature cutoff*: This particular method stops the charging process when the ambient temperature reaches a preset value indicating overcharging.

▪ *Rate of temperature increase*: This method measures the rate of temperature rise as a function of time and stops the charging process when a predetermined value is reached and is expressed as 1°C/min. This method is preferred to stop high-rate charges to ensure longer cycle life. In addition, this method can sense the starting of overcharge earlier than the voltage drop method. The Ni-MH battery would suffer if repeatedly overcharged. This is the most important criterion to follow to avoid overcharging as well as to protect the battery from complete destruction.

5.5 Selection Criteria for Primary and Secondary (Rechargeable) Batteries for Specific Applications

A common individual will not be able to select a particular battery type for his or her specific application, unless he or she has reasonably good knowledge of the various batteries available in the market. Furthermore, the individual may not know the approximate values of the critical performance parameters, such as reliability, life cycle, and SDR during storage. Batteries are available everywhere, but selection of the right battery for a specific application is of critical importance [8].

The author has conducted research studies on various primary and secondary batteries that currently are available in various stores. The author has heavily emphasized the most essential characteristics, such as cycle life, self-discharge, typical cell voltage, approximate battery cost, and cost per cycle. Table 5.10 provides comparison data of six different batteries for the benefit of readers. The data presented in this table are accurate within ±5%.

5.5.1 How to Select a Battery for a Particular Application

It is sometimes difficult for a common individual to select the right battery for a particular application. But when the most critical parameters are known (such as cost, maintenance requirement, and life cycle as shown in Table 5.10), the customer can select the right battery. This section identifies particular batteries for specific applications and briefly discusses their performance specifications to match their operating requirements. For example, in the case of mobile or cell phones, compact size, light weight, and high energy density are the principal requirements. Other important characteristics may include service life, load characteristics, maintenance requirements, self-discharge, cost, safety, and reliability. For this particular application, Ni-Cd rechargeable batteries are recommended. In the 21st century, however, the most popular designs such as lithium-based batteries are also available for various applications, but their procurement costs are much higher. Because lithium-based batteries have the longest life regardless of application [9], the author

Table 5.10   Currently Available Battery Types and Their Typical Performance Specifications

Characteristic	Battery Type				
	Lead-Acid	Ni-Cd	Ni-MH	Li-Ion	Li-Ion-Polymer
Energy density (Wh/kg)	30–55	45–85	60–120	110–160	100–135
Fast charge time (h)	8–16	1	2–4	2–4	2–4
Cycle life (to 80% charge)	200–300	1,500	300–500	500–1,000	300–500
Cell voltage (V)	2	1.25	1.25	3	3.6
Self-discharge/month (%)	5	20	30	10	–10
Operating temperature (°C)	–20/+60	–40/60	–20/60	–20/60	0/60
Maintenance required (mos.)	3–6	1–2	2–3	Not required	Not required
Estimated battery cost (US$) (for voltage rating)	25 (6 V)	50 (7.2 V)	60 (7.2 V)	100 (72 V)	100 (7.2 V)
Cost per cycle (cents)	10	4	12	14	29

will focus on various lithium-based primary batteries, and their characteristics are shown in Table 5.11 [9]. These batteries offer the longest shelf life and maximum operational reliability.

The performance comparisons between various lithium-based batteries have been made on the basis of energy density, power density, operating temperature range, shelf life, safety and reliability, environmental impact, and price and performance. Under the price and performance category, $LiCF_x$ batteries enjoy significant advantages over all other batteries summarized in Table 5.11. These battery parameters will certainly help a customer to select the right kind of lithium-based battery capable of meeting his or her power requirements.

For power tool applications, Ni-Cd batteries are preferred because of their reliable and safe operations under rigorous working conditions. In fact, Ni-Cd is the only battery type that performs well under rigorous working conditions. The same battery remains the most popular choice for two-way radios, emergency medical equipment, and power tools. Its fast charging capability, even after prolonged use, is considered most beneficial for power tools and emergency medical equipment. No other battery chemistry can meet or beat the superior durability and low cost of Ni-Cd batteries.

Table 5.11   Important Characteristics of the Five Primary Lithium-Based Batteries

Characteristics	Lithium-Based Batteries				
	$LiMnO_2$	$LiSO_2$	$LiSOCl_2$	$Li(CF)_n$	$LiCF_x$
Gravimetric energy density (Wh/kg)	300–450	240–320	500–700	360–480	500–700
Volumetric energy density (Wh/L)	500–650	350–450	600–900	500–600	700–1,000
Temperature range (°C)	−20 to 60	−55 to 70	−55 to 150	−20 to 60	−60 to 160
Typical shelf life (years)	5–10	10	15–20	15	16
Safe (high rate discharge)	Yes	No	No	Yes	Yes
Environmental impact	Moderate	High	High	Moderate	Moderate
Price/ performance	Fair	Good	Fair	Poor	Good

Source: Lind, E., *Electronic Products*, 14–18, 2010.

Li-MnO$_2$ batteries offer low-cost, safe, and reliable capabilities and are best suited for consumer electronics, military communications, RFID, transportation, automated meter reading, medical equipment calibration, medical defibrillators, and memory backup devices.

Lithium-sulfur-dioxide primary batteries are most ideal for military, aerospace, and satellite communications. These low-cost batteries are best suited for applications in which high pulse-power capability and a wide operating temperature range (from −55° to +70°C) are the basic performance requirements.

Lithium-thionyl-chloride batteries are most attractive for applications for which design objectives include an ultra-wide operating temperature range (from −55° to +150°C), high volumetric energy density, high pulse-power capability, and long shelf life. These batteries are widely used in commercial and consumer electronics, military communications, transportations, RFID, and memory backup.

Lithium-polycarbonate-monofluoride batteries are widely used in consumer electronics that require low to moderate power, military communications, RFID, transportation, automated meter readings, and medical defibrillators.

$LiCF_x$ batteries offer high energy and power densities, sustained performance over an ultra-wide temperature range (from –60° to 160°C), and long service or shelf life. These devices are best suited for portable electronics, military search and rescue communications, transportation, RFID, and medical defibrillators.

$LiFePO_4$ batteries have been considered for a variety of applications, including grid stabilization, power tools, HEVs, lasers, naval operations, planes, and helicopters. An upgraded version of this battery known as the Super-Phosphate battery has been developed, which has eliminated the shortcomings of the conventional battery. This upgraded version offers higher power and energy densities and is best suited for military applications for which reliability and ultra-high power and energy densities are the principal requirements. Various battery types and the performance capabilities best suited for military system applications are discussed in Chapter 6.

5.6 Conclusion

This chapter is dedicated to low-power rechargeable batteries widely used for commercial, medical, and certain industrial and space applications. Low-power battery types and performance requirements for consumer electronics are briefly summarized. Battery design configurations for various low-power applications are discussed with an emphasis on size, weight, reliability, life cycle, safety, and self-discharge. Performance capabilities and limitations of low-power batteries using cylindrical configurations are clearly identified in terms of their electrolyte requirements and physical dimensions. Performance characteristics of widely used low-power C-Zn batteries are briefly summarized. These batteries come in two versions: a premium version using MnO_2 as the electrolyte and the standard using zinc chloride as the electrolyte. The premium version offers enhanced reliability, high capacity, and significantly improved overall performance. Performance capabilities and limitations of alkaline manganese are identified. Alkaline batteries are widely used because they do not present health hazards and are free of mercury contents. Performance characteristics and cell configurations of low-power $Li-MnO_2$ are discussed in great detail because they are widely used by most customers. These batteries come in various configurations, such as coin, cylindrical wound, cylindrical bobbin, flat, and cylindrical D. Performance capabilities and major benefits of rechargeable Ni-MH, Ni-Cd, and Li-ion batteries are identified with an emphasis on reliability, longevity, DOD, SOS, and OCV characteristics. Chemical reaction equations for various batteries have been developed and the output products are identified. Weight contributions to the battery come from various sources, such as the electrolyte, electrodes, separator, housing, pressure-release device, current collectors, safety electronics, and additives. To reduce the battery weight, one must try to focus on the contributing sources. The author highly recommends the use of miniaturized, lightweight microbatteries or thin-film batteries for electronic circuits and devices deployed in space systems, airborne sensors, air-to-air missiles, and UAVs for which weight, size, and power consumption are of critical importance. Federal, defense, and security organizations are continuously pressing for battery designs involving MEMS and nanotechnologies to achieve significant reductions in the weight and size of these batteries. Characteristics of alkaline batteries using KOH as the electrolyte are identified for embedded-system applications, because of their long shelf life, high reliability, and least construction method. Lithium-thionyl-chloride batteries can also be used for embedded-system applications, if ultra-low self-discharge rate and longevity greater than 20 years are the basic design objectives. The author identified microbatteries and smart nanobatteries for critical computer memory and sensitive medical diagnostic devices. Such batteries are being designed, developed, and evaluated under SBIR programs The scientists and designers involved in the development of these batteries highly recommend the deployment of these batteries for portable medical diagnostic system, computer memory chip, UAV, and space sensor applications for which weight, size, and power consumption are critical. Smart Li-ion batteries are specifically designed and developed for use in EVs and HEVs. Smart Li-ion batteries do not require a large number of safety devices, leading to significant reduction in weight, size, and cost. These batteries are best suited for EVs and HEVs because of high capacity, safety, and longevity characteristics. Typical battery life and the cost of batteries being used in EVs and HEVs are 10 years and $4,500, approximately. Li-I$_2$ batteries with life better than 15 years are most ideal for heart-treating devices such as pacemakers, cardioverter defibrillators, and artificial hearts. LiCF$_x$ batteries are best suited for ICDs. The role of certain rare earth materials was discussed to achieve significant improvement in the performance of specific batteries best suited for medical and space applications. Performance characteristics of prismatic batteries with thin structures, best suited for powering the tiny medical devices, are identified with an emphasis on reliability, life cycle, and safety.

References

1. Robert A. Powers, "Batteries for low power electronics," *IEEE Spectrum* 83, no. 4 (April 1995), pp. 687–693.
2. Courtney E. Howard, "Electronics miniaturization," *Military and Aerospace Electronics* (June 2009), p. 32.
3. Keith Oland, "Selecting the best battery for embedded-system applications," *Electronic Design News* (November 18, 2010), pp. 36–39.
4. John Keller, "Smart power requirements are looking to improve the efficiency and to reduce power-system-development-cost," *Military and Aerospace Electronics* (March 2011), pp. 39–40.
5. Editor, "Technology focus," *Military and Aerospace Electronics* (June 2009), p. 39.
6. Editor, "Engineering feature," *Electronic Design News* (March 24, 2011), pp. 26–27.

7. Pier Paolo Prossini, Rita Mancini et al., "$Li_4Ti_5O_{12}$ as anode in all-solid-state, plastic, lithium-ion batteries for low-power applications," *Solid State Ionics* (September 2001), pp. 185–192.
8. Kerry Lanza, "What is the best type of battery?" *Electronic Products* (March 2011), p. 40.
9. Eric Lind, "Primary lithium batteries," *Electronic Products* (September 2010), pp. 14–18.

Chapter 6

Rechargeable Batteries for Military Applications

6.1 Introduction

High-technology rechargeable batteries are being developed that incorporate saline- and alkaline-based electrolytes. These batteries are best suited for various military applications for which minimum weight, size, and cost as well as high reliability are of critical importance. In the case of certain battlefield applications, portability, efficiency, and collapsibility are additional design requirements. Preliminary studies performed by the author seem to indicate that aluminum-air (Al-air) batteries in general are considered to be the most suitable for most military applications. Military weapon systems and sensors employing state-of-the art electronic and digital components are continuously pushing the boundaries in terms of performance, portability, safety, cooling, and reliability, while operating in harsh environments. The batteries needed for such systems or sensors must integrate advanced technologies, such as microelectrical mechanical system (MEMS) and nanotechnology, in the design of rechargeable batteries to meet performance specifications defined by the armed forces. Batteries incorporating MEMS or nanotechnologies offer the most compact size, low weight, and ultra-high reliability, thereby making them the most ideal option for air-to-air missiles, air-to-ground missiles, and other aerospace applications. Reliability and safety are the most demanding requirements for the batteries deployed in all military weapon systems and electronic sensors. Rechargeable batteries deployed by mission-critical military systems operating under harsh thermal and mechanical environments—such as battlefield tanks, unmanned underwater vehicles (UUWVs), unmanned air vehicles (UAVs) for surveillance and reconnaissance missions, drones equipped with electro-optic sensors and complex missiles, unmanned ground combat vehicles (UGCAs), microcommunication satellites conducting covert surveillance and reconnaissance missions, and other military systems—must meet stringent reliability, safety, and longevity requirements. In the case of weapons systems specifically designed for battlefield applications, the rechargeable batteries must provide optimum reliability, safety, constant output voltage, and minimum battery capacity as required by the military procurement specifications. In the case of digital systems or circuits, the batteries must provide uninterrupted electrical power over the specified period defined by the procurement specifications. In the case of military covert communications equipment, the rechargeable batteries will be designed to protect the authenticity of the mission-critical voice and video data. In the case of battery failure, irrespective of military applications, mission-critical data could be lost; therefore, a backup battery or redundant battery is recommended for the deployment to avoid the catastrophic failure of critical data. In the case of robotic systems for battlefield deployments, batteries should be designed to operate efficiently, reliably, and without interruption under harsh thermal and mechanical environments. In addition, the rechargeable batteries must operate efficiently under shock and vibration conditions as defined by the procurement specifications.

Defense officials and military program managers are moving to a multilayer architecture that permits consolidation of several discrete electronic devices to realize a significant reduction in cost, weight, size, and capacity of rechargeable batteries. If military planners are looking at such an architecture, then unique battery technologies have to be explored to meet the power requirements of several devices from a single power source. Batteries for tactical communications equipment must be capable of adapting automatically and continuously to network changes and should support the time-critical missions as well as provide uninterrupted power with the utmost safety and security [1]. U.S. battery suppliers are considering the development of next-generation rechargeable batteries specifically for military applications. These next-generation batteries will deploy advanced technology and a unique, cost-effective architecture that will offer higher reliability, portability, and safety, while minimizing the cost, size, and weight.

Batteries best suited for counter–improvised explosive devices (C-IEDs) equipment will have special performance requirements, including portability, light weight, compact size, improved reliability, and appropriate capacity. Because C-IED equipment is carried by the soldiers on their back, while detecting buried IEDs in the hostile battlefield zones, portability, minimum weight, highest safety, and compact packaging are the essential design requirements for rechargeable batteries deployed by this equipment.

In the case of UAV and drone applications, rechargeable batteries must meet the most stringent design requirements under severe aerodynamic environments, namely compact packaging, minimum weight, ultra-high reliability, and safety.

Furthermore, such batteries must be free from toxic gases and chemical agents. In brief, the rechargeable these batteries must be free from greenhouse effects, because there is no provision to remove harmful gases from the UAVs or electronic drones.

Rechargeable batteries for military personnel carriers, battlefield tanks, unmanned ground combat vehicles (UGCVs), and UUWVs must have higher efficiency, compact packaging, and ultra-high reliability under harsh thermal and mechanical environments. The operating conditions in these vehicles might experience internal temperatures as high as 125°C, high amplitude vibrations, and unbearable shocks. Therefore, rechargeable batteries required to operate under these harsh operating conditions must include appropriate materials for the anode, cathode, and electrolyte without compromising electrical performance, reliability, and safety.

This chapter describes charge-management controller devices for the rechargeable batteries that are best suited for space-limited and cost-sensitive applications requiring high current and voltage accuracies. Use of microchip technology and algorithms by these charge-management controller devices will provide such features as constant-current and constant-voltage regulation, cell temperature monitoring and preconditioning, advanced safety timers, automatic charge termination, and current charge-status indication.

Cell-balancing techniques are essential for longer battery life and greater levels of safety under these operating environments. Cell balancing must correspond to the manner in which the battery cells are used and to the environment in which they operate. Thermal mismatch of a cell's battery pack is just one of many factors involved. The self-discharge rate for most rechargeable batteries doubles for every 10°C rise in temperature. This in turn affects the state of charge (SOC) in the cell. To ensure minimum temperature effects, every care must be taken to determine where a battery pack will be placed. The closer the battery pack is to the equipment it powers, the greater the risk of overheating the cells that are in physical contact with the equipment. The effect of overheating and its impact on energy storage capability varies from battery to battery. Battery material scientists believe that lithium-ion (Li-ion) batteries can store more energy than conventional nickel-metal-hydride (Ni-MH) batteries and are roughly 30% smaller and 50% lighter. But the Li-ion batteries tend to overheat when overcharged or during deep discharging, thereby requiring protection and safety features for high-cell-count Li-ion battery packs. In summary, cell balancing offers higher efficiency and longer battery life.

Studies undertaken on cell balancing by the author indicate that there are two distinct balancing techniques, namely active and passive. The active technique is complex and is best suited for high-cell-count Li-ion battery packs. The active technique supports high-voltage operations exceeding 300 volts (V) and therefore is best suited for battery packs widely used by electric and hybrid automobiles and UPS vehicles. The studies further indicate that the active-cell technique avoids energy loss, resulting in high battery pack efficiency. In this active method, the charge is transferred from one cell to another with minimum energy loss.

Conversely, the passive-cell technique uses linear technology incorporating integrated circuit elements. Cell balancing is simple compared with the active technique. In this passive technique, however, shunt resistors are used across each cell, in a multicell-series stack, to reduce the imbalance currents when the battery stack is fully charged. Cell imbalance is a function of the quality of several factors, including the battery cell involved, the impedance mismatch between the cells, the number of cycles, battery pack cell configurations, and the size of the battery pack [2]. Cell balancing is particularly important for portable battery packs designed for C-IED applications in the battlefield. The following sections identify the cell-balancing technology used for specific military system applications.

6.2 Potential Battery Types for Various Military System Applications

This section focuses on the next-generation rechargeable batteries best suited for various military systems and sensors deployed by the Air Force, Army, and Navy in various mission applications. The section describes the advanced battery technologies and design architectures best suited for various military system applications. The section also summarizes the design aspects and critical performance requirements of rechargeable batteries that are most ideal for a variety of military system applications and that involve appropriate platforms, such as electronic drones, UUWVs, UAVs, man-pack C-IED systems, battlefield covert communications systems, and robot systems for battlefield applications. Requirements for batteries for UAVs, which are used for intelligence, reconnaissance, and surveillance missions, will be summarized later in this chapter under Section 6.7.2. Some battery suppliers are developing high-technology saline- and alkaline-based batteries using the latest electrolytes specifically for military applications. For battlefield applications, batteries must be light, reliable, portable, nontoxic, and safe. Battery designers are looking at the collapsible aspect of the rechargeable battery. Battery research scientists believe that Al-air rechargeable batteries, carbon nanotube–based Li-ion batteries, and silver-metal-hydride (Ag-MH) batteries are most suitable for military systems applications [3]. The author will discuss the performance capabilities and limitations of these batteries and other rechargeable batteries in subsequent sections.

6.2.1 Aluminum-Air Rechargeable Batteries for Military Applications

Some battery manufacturing companies are exploring design concepts for the development of rechargeable batteries specifically for military applications. Alu

Power Canada Ltd. has been actively engaged in the design and development of a number of saline and alkaline electrolyte Al-air batteries for such applications [3]. These rechargeable batteries have a major advantage over lithium and other rechargeable batteries—that is, the Al-air batteries have demonstrated a very high energy density exceeding 550 Wh/kg. These batteries are collapsible and highly portable, and therefore can be transported with minimum cost. They can be filled with water at the operating site. Another advantage of the alkaline Al-air battery is its high power density greater than 125 W/kg, which makes it attractive for battlefield applications. For example, if a soldier is looking for buried IEDs in a battlefield, the soldier can carry the battery on his or her back because it is very light weight. Salt-water batteries are best suited for moderate power demand and for multiple use because of their ability to recharge the battery between use cycles. A self-contained, man-portable Al-air battery using alkaline electrolyte is most ideal to provide satellite communications in battlefields and remote areas.

These rechargeable batteries will meet cold-start requirements within 30 min. at –40°C, when the battery output is between 5 and 400 watts (W) and the energy density is as low as 430 Wh/kg. Activation of these batteries can be accomplished by emptying the reservoirs on the top of the battery cells that contain the electrolyte. Both modeling and laboratory tests have demonstrated remarkable electrical performance, including reliability and safety. Because of their improved reliability, high portability, and safe performance, these rechargeable batteries are most attractive for battlefield communications and for UAV, micro air vehicle (MAV), and unmanned ground vehicle (UGV) applications.

The Al-air batteries use aluminum anodes and air cathodes. Al-air batteries recently developed by material scientists offer unique battery configurations and have demonstrated much improved battery performance compared with earlier versions. There are two types of portable Al-air rechargeable batteries, namely saline-electrolyte rechargeable batteries and alkaline-electrolyte rechargeable batteries. Saline-based batteries are best suited for multiple applications for which water is readily available and the batteries can be transported dry and collapsed. The alkaline-based batteries are widely used for high-power applications that are activated by adding alkaline solution before actual use. Both types of these batteries are stored in a dry state and have a long storage life. The dry state is the principal requirement for a long storage life. Studies undertaken on energy density of stored reactants indicate that the Al-air battery has energy density second only to the lithium-air battery. The studies further indicate that the reduction in energy density is due to the inability to operate the aluminum-anode and air-cathode electrodes at their thermodynamic potential and because water is consumed in the energy-producing reaction. Despite these limitations, the practical energy density level of the Al-air batteries has been observed in the 200–400 Wh/kg range, which exceeds that of most battery systems. The overall chemical reaction and the

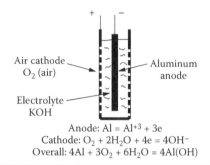

Anode: Al = Al^{+3} + 3e
Cathode: O$_2$ + 2H$_2$O + 4e = 4OH$^-$
Overall: 4Al + 3O$_2$ + 6H$_2$O = 4Al(OH)

Figure 6.1 Chemical reactions from anode, cathode, and electrolyte in a saline regenerating battery.

chemical reaction that occurs at the anode and cathode electrodes are shown in the following chemical equations:

$$\text{Reaction at anode: Al} = [\text{Al}^{+3} + 3e] \tag{6.1}$$

$$\text{Reaction at cathode: } [\text{O}_2 + 2\text{H}_2\text{O} + 4e] = [4\text{OH}^-] \tag{6.2}$$

$$\text{Overall chemical reaction: } [4\text{Al} + 3\text{O}_2 + 6\text{H}_2\text{O}] = [4\text{Al(OH)}_3] \tag{6.3}$$

Specific details on the chemical reactions are illustrated in Figure 6.1.

6.2.1.1 Description of Key Elements of These Batteries

This section identifies the key elements of these batteries for the benefit of readers and design engineers. The aluminum alloy used in the fabrication of the anode electrodes is widely available at a reasonable price in the market with no restriction on the amount to be used. This particular alloy operates at high columbic efficiency compared with other alloys over a wide range of current densities. High columbic efficiency is of paramount importance for these batteries with current density levels to avoid overheating within the cells. Regarding the cathode electrode, a low-cost high-performance air electrode can be manufactured in large quantities at minimum cost. The anode performance after extended discharge is of critical importance. In other words, corrosion rate as a function of anode current density over a wide range must be kept to minimum to avoid degrading battery efficiency and longevity. Advanced aluminum alloys are available in the market and must be used in the design and development of these batteries to keep the corrosion significantly less than 2% over the entire range of anode current densities. The corrosion reaction not only affects efficiency but also produces hydrogen that must be expelled from the battery system. Anode performance as a function of anode current density for various aluminum alloys is shown in Figure 6.2. Anode performance improved over a 30-year period from 1960 to 1990 using various alloys. It is evident from

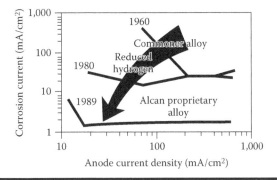

Figure 6.2 Anode performance after extended discharge as a function of corrosion current and current density.

Figure 6.2 that the Alcan proprietary alloy provides the optimum anode performance. Its corrosion current has been kept under 2 mA/cm^2 over the anode current density ranging from 10 to 1,000 mA/cm^2. Further improvements in anode performance are possible by carefully selecting the appropriate alloy through systematic examination of advanced alloys developed over the period from 1990 to 2010, which the author intends to do through comprehensive research.

6.2.1.2 Performance Capabilities, Limitations, and Uses of Saline Batteries

It is evident from Figure 6.1 that saline regenerative batteries generate a reaction product called aluminum hydroxide ($AlOH_3$) dispersed in the salt solution, which is considered safe and environmentally mild and harmless. Under these conditions, the battery can be easily and safely emptied after partial use and can be refilled with the electrolyte for subsequent discharge operation. To maintain the effectiveness and chemical affinity, it is desirable, if possible, to recharge the battery between deployments. It is better to transport a sufficient amount of common salt to provide for multiple deployments. The energy density of the dry battery carrying enough dry salt to achieve a complete discharge could be estimated as high as 600 Wh/kg. These batteries are best suited for field communications equipment, UAVs and UGVs, and field-charging systems for nickel-cadmium (Ni-Cd) and Ni-MH batteries, and lead-acid power packs. Advanced and high-capacity versions of the salt-water battery are required for field recharging of Ni-Cd, Ni-MH, and lead-acid batteries.

6.2.1.3 Performance Capabilities and Uses of Alkaline Batteries

The alkaline electrolyte–based Al-air batteries enjoy higher conductivity than the conventional salt solution electrolyte–based Al-air batteries. In addition, $AlOH_3$

has higher solubility in alkaline electrolytes. Alkaline Al-air batteries are best suited for high-power applications. The gravimetric energy density can be as high as 450 Wh/kg for the alkaline Al-air batteries.

Self-contained, man-portable Al-air alkaline batteries are widely used in battlefield applications and remote military locations. The battery units initially come with dry alkali contained in the cells, and water can be added when electrical power is desired. The modules can provide temporary electrical power for mobile shelters, command and control communications centers, and field medical facilities. The man-portable Al-air batteries can act as power supplies for unmanned vehicles. Specific details on the components used in the self-contained, man-portable Al-air battery are shown in Figure 6.3. Such a battery was designed and developed about 15 years ago for UAVs and UUWVs and is capable of providing a nominal power of 1.6 kW and a peak power of 4 kW. But, the next generation of such a battery with high-quality alloy is expected to have a peak power exceeding 5 kW.

Quiet operation of the battery, long storage life, free from harmful gases and chemical agents, high energy capability, high power capacity, minimum maintenance, ultra-high reliability and safety, and quick recharge (mechanically) cycle are the most remarkable features of the Al-air battery. Because of these exceptional operational features, these rechargeable batteries are best suited for military applications, such as battlefield applications. Because of low acoustic signature, this

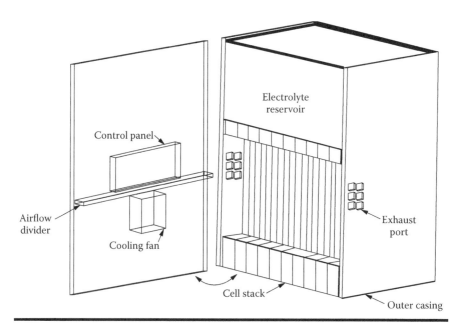

Figure 6.3 Layout of a self-contained, man-portable, aluminum-air rechargeable battery.

battery is most suitable for underwater surveillance and reconnaissance missions. For two or three decades, these self-contained, man-portable batteries have been used to provide power to satellite communications. These batteries are most attractive for providing electrical power under emergency situations. Typical battery performance characteristics of a 300 W, 12 V, 12 h alkaline Al-air battery module comprising 10 cells can be summarized as follows:

- Maximum energy density: 450 Wh/kg
- Maximum power output: 425 W
- Cold-start: within 30 min. at −40°C
- Typical battery output power: 300 W
- Typical battery-rated voltage: 12 V
- Duration for optimum performance capability: 12 h (minimum)
- Number of cells in the battery module: 10

Cold-start at −40°C is of critical importance, particularly in battlefield conditions and remote military locations where satellite communication links have to be established under emergency conditions.

Performance studies performed by the author indicate that battery terminal voltage will decrease slightly as the discharge time increases. The variation of terminal voltage as function of discharge duration for the 300 W, 12 V, 12 h alkaline Al-air battery output is shown in Table 6.1.

The tabulated data reveal that a reduction in battery output voltage occurs after 8 h of operation, excluding the reduction in voltage at start-up. The 0 h reduction is perhaps due to incomplete thermal stabilization of the battery at 55°C (131°F). The terminal voltage will experience a significant reduction at cryogenic temperatures as the operating temperature approaches −40°C. As the discharge hours increase, one should expect the battery output voltage to decrease, which is natural. The battery terminal voltage remains very close to the rated terminal voltage of 12 even after 8.5 h of discharge. If you design the battery architecture for a nominal voltage of 14 V, then one will see battery output voltage greater than 12 V after 12 h of discharge.

Both the saline and alkaline batteries offer unique performance capabilities, including a high energy density level exceeding 500 Wh/kg in the dry, inactivated state; long life; portability due to low weight; and high reliability. Al-air rechargeable batteries using alkaline electrolyte yield much better performance, however, because the alkaline electrolyte (KOH) has higher conductivity than the salt solution (saline), and its by-product $AlOH_3$ has higher solubility in alkaline electrolytes. Because of these unique performance characteristics, Al-air batteries using alkaline electrolytes are best suited for various military applications, such as field communication, UAVs, UGVs, unmanned combat air vehicles (UCAVs), MAVs, and field-charging systems for Ni-Cd and lead-acid batteries.

Table 6.1 Variations in Battery Terminal Voltage as a Function of Discharge Period

Discharge Time (Hour from Start)	Battery Output Voltage (V)
0	11.2
1	13.3
2	12.9
3	13.3
4	13.1
5	13.0
6	12.6
7	12.4
8	12.2
9	11.7
10	11.4
11	10.8
12	9.7

6.2.1.4 Bipolar Silver-Metal-Hydride Batteries for Military Applications

This bipolar Ag-MH rechargeable battery consists of positive and negative electrodes, a positive contact face, a separator, and an insulating border seal. Essentially, the Ag-MH battery uses a wafer cell and a conductive plastic film as the cell face [4]. The wafer cell was developed around 1991 by Electro Energy, Inc. (EEI), Danbury, Connecticut, and the performance evaluation was first reported by Eagle-Picher Industries, in Joplin, Missouri, in 1992.

6.2.1.4.1 Introduction

Battery designers claim that the test results obtained from comprehensive laboratory tests made on a wafer cell covered with a plastic film led to the successful development of the bipolar Ag-MH battery. The cell has slim structural elements, which can be stacked like a deck of cards to form a bipolar Ag-MH battery as illustrated in Figure 6.4. The bipolar plate consists of two opposing cell faces held in compression to achieve a slim bipolar Ag-MH battery. EEI first publicized its performance test results on single-wafer cells and multiwafer cells at a 1993 conference.

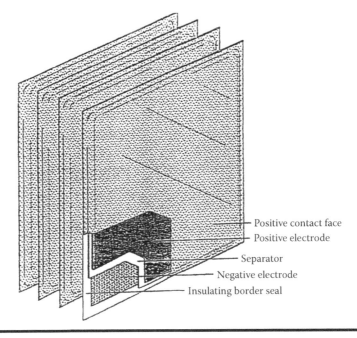

Positive contact face
Positive electrode
Separator
Negative electrode
Insulating border seal

Figure 6.4 Structural details on a bipolar battery cell best suited for some applications.

Experimental test were made on single cells and multicells to obtain performance characteristics using the Ni-MH battery system. The test results indicate their acceptance for a wide range of applications, including electric vehicles (EVs) and hybrid electric vehicles (HEVs). After the initial tests were successful, EEI explored the possibility of replacing the nickel electrode with a silver electrode in this same bipolar battery system. Replacing the nickel electrode with a silver electrode in the next generation of bipolar batteries will exhibit significant improvement in electrical performance, reliability, and cycle life.

6.2.1.4.2 Structural Details and Unique Characteristics of Ag-MH Batteries

Structural details and critical elements of the battery are shown in Figure 6.4. The critical elements or the important components of the battery include positive and negative electrodes, a separator, a contact face, and an insulating border-sealing element. EEI scientists claim that the Ag-MH battery can offer higher energy density than the Ni-MH battery. In addition, the Ag-MH battery offers improved charge retention capability, although it does so at a relatively short cycle life. Because of its unique performance characteristics, this battery has potential applications in critical military and commercial systems, where high energy density, portability, wide operating temperature range, and reliability are the principal design requirements. In commercial applications, however, the cost of the battery may not be attractive. Initial performance characteristics were evaluated by Eagle-Picher Industries in 1992 using a conventional monopolar battery design incorporating sintered silver electrodes opposite either sintered hydride electrodes or pasted foam nickel in a flooded electrolyte medium. In 1994, the same scientists conducted laboratory impedance measurements that indicated the overall impedance of the Ag-MH battery was strictly dominated by the silver electrode. To improve battery performance, sealed prismatic cells were built with a dual-layer separator, which achieved battery conversion efficiency better than 90% at a 10 C charge rate. This was possible with separator layers, which essentially made a large contribution to the impedance of the battery.

Battery scientists at EEI deployed the wafer-cell approach in this bipolar battery design. Use of this approach revealed that the commercial development of a bipolar Ag-MH battery based on stacked wafer-cell design concept would significantly improve the battery performance over conventional monopolar batteries, while also reducing the manufacturing cost. The cost reduction was achieved as a result of the single-step mixing process for the plastic-bonded hydride electrodes, which demonstrated the high design flexibility intrinsic to the polar design approach. This particular design approach would allow for volume production techniques to manufacture the prismatic batteries with minimum cost and complexity. This approach is not limited to cells with square, cylindrical, or annular geometrical configurations.

6.2.1.4.3 Design Analysis

EEI design engineers undertook rigorous design analyses to obtain preliminary design calculations for a bipolar Ag-MH battery to replace a silver-zinc (Ag-Zn) battery with 0.5 kWh capacity, which qualified for an extravehicular mobility unit (EMU) in a manned space flight application. In or about 1992, low-capacity Ag-Zn batteries were deployed in space applications. Typical EMU battery performance requirements can be summarized as follows:

- Battery output voltage: 16 V (minimum), 22 V (maximum)
- Battery capacity: greater than 27 Ah
- Cycle life: 100 cycles
- Wet life: 425 days (minimum)
- Battery outer dimensions (centimeters): 12.7 cm (height) × 26.8 cm (width) × 9.5 cm (length)

NASA engineers specified the battery's outer dimensions so that the battery could be installed in the available space determined by the program manager.

6.2.1.4.4 Description of the EMU Battery Electrical Parameters and Battery Assembly Procedures

According to battery design engineers, 18 wafer cells were required to meet the minimum battery output voltage of 16 V. An electrode size of 25 cm × 10.6 cm (270 cm²) was selected to meet the battery's electrical performance. A silver electrode with 50 porosity provided a capacity of 29 Ah based on a 67% utilization factor. EEI's custom metal-hydride alloys were conservatively rated for 200 mAh/g. The battery cells were not optimized to meet the weight specification requirements for this particular application, but they can be optimized for other applications. Each sealed wafer cell was designed to deliver an energy density of 87 Wh/kg or 274 Wh/L. The high volumetric energy density demonstrates a particular advantage of the deployment of the bipolar battery design approach in a low-voltage battery system such as an Ag-MH battery. The calculated final volumetric energy density for this battery comes to 270 Wh/L.

EEI has used both commercially sintered silver powder with 1-mm-thick electrodes and sintered silver powder with in-house electrodes made from an AB_5-type metal-hydride alloy (also known as lanthanum-nickel alloy) using rare earth material for initial laboratory cell test evaluations. Introduction of rare earth material improves the oxidation resistance during the alloy manufacturing process. The AB_5-type metal-hydride alloy should be widely used in sealed Ni-MH batteries. The use of rare earth metal lanthanum will provide improved electrical performance, enhanced reliability, and ultra-high longevity for the sealed Ni-MH and Ag-MH battery systems.

Physical parameters (weight and size) of various critical elements of a sealed Ag-MH battery cell are summarized in Table 6.2 to demonstrate the extremely

Table 6.2 Physical Dimensions and Approximate Weight of Various Cell Elements

Cell Element	Thickness (cm)	Weight (g)
Hydride electrode	0.234	218
Silver electrode	0.094	88
KOH electrolyte (liquid)	N/A	35
Separator	0.060	14
Conductive plastic film	0.010	4
Wick	0.025	2
Conductive plastic film	0.010	4
Total	**0.433**	**365**

Note: N/A: not available.

portable capability of the bipolar sealed Ag-MH battery. The values of the parameters shown in Table 6.2 are for a sealed Ag-MH battery cell with a capacity not exceeding 30 Ah. These are estimated values and could have errors within +/–5%. These physical parameters indicate that the sealed Ag-MH battery cells are the most compact and highly portable because of their low weight. The cell has a square cross-section of 3 in. × 3 in. (7.62 cm × 7.62 cm) and an electrode size of 25 cm × 11 cm (275 cm²). Because of their compact size and low weight, the cells used for bipolar batteries are best suited for airborne, aerospace, unmanned airborne, and space system applications.

The physical parametric values of the cell elements are used in the laboratory testing phase. During the fabrication process, the wafer cells are compressed between two acrylic compression plates. The current is drawn in a bipolar mode via two nickel foils facing each outer conductive cell face. A conductive plastic film protects the cell face. Barriers and wicking separators and inorganic materials can be used adjacent to the silver electrode to mitigate the cellulosic oxidation. A polyacrylonitrile wick should be used adjacent to the metal-hydride anode, if further mitigation is desired. To enhance overall battery performance, it is desirable to establish a highly conductive interface between the silver electrode and the outer conductive plastic cell face. The standard square anode electrode with a dimension of 6.35 cm has been used for preliminary cell design, which was evaluated by battery design engineers during the initial laboratory evaluation.

The presence of the KOH (35% approximately) electrolyte allows the silver to oxidize and form a resistive surface barrier. The silver foil with 0.001 in. thickness has been hot forged to the porous silver electrode to prevent the KOH electrolyte from penetrating the silver electrode and reaching the interface. A conductive hydrophobic barrier must be placed between the silver and the outer cell to prevent the KOH leakage.

6.2.1.4.5 Initial Evaluation Test Results for the Ag-MH Cell

The initial test results reveal that the wafer-cell cycle life is close to 80 cycles, while the cells are operating in a fully bipolar mode. Oxidation of the silver electrode must be avoided to achieve meaningful test results. Tests must be performed for both modes of operation. A theoretical cell capacity of 4.6 Ah has been demonstrated. Typical utilization of 67% has been demonstrated for the silver electrodes.

In laboratory tests, the cell designers noticed that the silver electrode undergoes two distinct oxidation and reduction reactions during the charge and discharge of the battery, which corresponds to reductions of silver from the +2 oxidation state to the +1 state and ultimately to silver metal. The evaluation tests further indicate that at low rates, these two primary reactions would most likely manifest during the charge and discharge profiles. During the discharge cycle, the reduction of silver oxide (AgO_2) to silver metal increases the conductivity of the silver, thereby resulting in a very flat discharge profile.

Examination of the discharge voltage characteristics of the cell clearly indicates the presence of a dual plateau in the discharge voltage plot, which provides the cell voltage as a function of test duration expressed in hours. This dual plateau is a matter of serious concern at discharges of 16.5 mA/cm². The first plateau occurs at approximately 1.4 V with uniform amplitude, and the second plateau occurs at about 1.1 V. The initial cell capacity is close to 2.75 Ah, which is about 60% of the theoretical capacity, because the silver electrode was not optimized for the capacity. After about 80 cycles in a bipolar mode, the Ag-MH wafer cell has demonstrated and maintained 80% of the cell's capacity.

In summary, these laboratory tests have demonstrated the anticipated performance of bipolar-mode cycling of a single-wafer cell using the Ag-MH battery system. On the basis of the initial experimental test data, battery design calculations, and modeling results obtained by the design engineers and scientists, it can be stated that the energy density performance of the Ag-MH cell in a bipolar configuration is at least comparable to the performance of Ag-Zn in an extended cycle and wet life. Published technical reports indicate that the comprehensive research and development activities have focused on improving the separator performance, enhancing the silver electrode performance, mitigating oxidation to the silver electrode interface, and achieving maximum cycle life. In brief, successful efforts in these areas will bring significant improvement in electrical performance, reliability, safety, and longevity of next-generation Ag-MH batteries. Technical papers published in 1995 have revealed the design and development of low-cost plastic-bonded bipolar Ni-MH rechargeable batteries best suited for military and aerospace applications, for which portability, reliability, longevity, and safety are the principal design requirements.

6.2.1.5 Rechargeable Silver-Zinc Batteries for Military Applications

Battery scientists and design engineers feel that Ag-Zn rechargeable battery technology is fully matured and that these batteries are capable of providing the highest power and energy per unit of volume and mass. Ag-Zn rechargeable batteries are most ideal for military applications for which minimum weight, size, and reliability are of critical importance. As a result, these batteries have been deployed in aerospace, battlefield, and defense system applications where high energy and reliability are the basic performance requirements [4].

6.2.1.5.1 Silver-Zinc Batteries for Medical Applications

In the past, the cost of silver and low production volume have limited the commercial applications of Ag-Zn batteries. Preliminary clinical studies performed on advanced medical devices reveal that extending the cycle and calendar lives, which offer improved performance (including reliability and safety and lower operating costs), could lead to enhancement in a wide range of future medical devices. Clinical studies indicate that the deployment of these batteries could save many lives during critical medical procedures requiring highest safety and reliability.

Medical experts feel that the prime considerations for using rechargeable batteries for medical applications are safety, ease of handling, and reliability. The U.S. Food and Drug Administration (FDA) has imposed safety and reliability requirements for batteries used in medical applications. Departure from these requirements is not acceptable to the FDA and medical authorities. The Ag-Zn battery is one of the safest and easiest systems to work with because it is forgiving of improper handling and accidents. Even though the electrolyte used by this battery is corrosive, the battery does not expel toxic fumes or chemical agents except under very high temperature exposure or fire. Furthermore, reliability of the battery for medical applications is the most critical requirement, because it may be used in a life-saving medical procedure, for example, on a person in cardiac arrest.

6.2.1.5.2 Silver-Zinc Batteries for Military, Defense, and Space Applications

The rechargeable Ag-Zn battery system is an established part of aerospace, defense, and military power supply equipment. Rechargeable Ag-Zn batteries are capable of supplying high energy outputs with precise voltage control when working under the harsh operating environments normally experienced in defense, aerospace, and battlefield applications. This battery technology is fully matured, and according to numbers quoted by designers and suppliers, the cell's reliability exceeds 99.99%. Regarding the weight of the Ag-Zn battery, its weight is approximately one-half that required for equivalent nickel-based batteries (Ni-Cd and Ni-MH batteries) and nearly one-fourth that required for an equivalent lead-acid battery. The weight advantage of this Ag-Zn battery system makes it most suitable for battlefield system applications, such as C-IED applications.

The reliability of this particular battery is beyond any doubt. Because of their ultra-high reliability (cell reliability is better than 99%), Ag-Zn batteries are consistently deployed in manned space flight programs and have been used on every flight since the first NASA-manned suborbital flight of the *Mercury-Redstone* in 1961. This battery requires safe environmental handling to satisfy requirements for a recognized, safe method of disposal.

High mass and volumetric energy density are important and are only possible if the battery is designed for minimum weight and size. Portability of batteries is of critical importance, particularly when the batteries are deployed in the military systems operating in battlefield environments. Deployment of a small battery requires charging and discharging at high rates, leading to high structural stresses on the batteries.

Battery designers at Eagle-Picher Industries have conducted comprehensive tests on thousands of Ag-Zn cells and batteries over the last 45 years. The batteries involved ranged from 0.8 to 800 Ah and covered both low- and high-rate designs. The test data presented in the reports included the critical features such as safety, reliability, disposal, and performance characteristics. These features will be discussed in Sections 6.2.1.5.2.1–6.2.1.5.2.4.

6.2.1.5.2.1 Safety Test — Battery designers conducted safety tests of Ag-Zn cells and batteries by applying internal shorts, external shorts, high temperature, and intentional mechanical damage. These tests indicated similar results in that the reaction was an electrical short. In the case of very small cells, the reaction was benign with a small quantity of heat energy liberated. In the larger cells, the quantity of heat released was sufficient to cause a softening of the plastic cell case, venting of the steam, and leakage of the electrolyte because of extreme heat. Fire or explosion was never noticed. Ag-Zn batteries produce small gases during high-temperature operation; therefore, these batteries must be housed in a vented battery enclosure or case.

6.2.1.5.2.2 Reliability Test — Reliability tests are essential to verify the reliability of the device under such operating environments as temperature, humidity, shock, and vibration. The reliability data represent the overall probability of success or failure relative to a particular set of conditions of service or operation. For any device, regardless of type, reliability is defined not only on the basis of design and manufacture but also on the conditions of service to be performed. For example, a battery may have only a few components, such as an electrode, battery housing, electrolyte, separator, and so on. But a system or subsystem might involve several components. For example, an HEV can have several components, namely a rechargeable battery, alternating current (AC) induction motor, rear and forward mirrors, tires, brakes, fuel tank, steering wheel, air conditioning, and so on. Reliability could be defined in terms of failure rate, or mean time between failure, or mean time to failure, or mode of failure, or probability of failure. Procurement specifications clearly define the quality assurance and reliability goals. The quality assurance paragraph in the specifications will clearly state the following: The desired reliability shall be 99.7%. To make this explicitly clear, the statement should read as follows: For 100 h of operation, the reliability shall be 99.7%.

In brief, reliability represents the overall probability of success (no failures) relative to a particular set of conditions of a specific service. Typically, in the case of military and aerospace programs, cell reliability exceeds 0.9999 probability of success. The conditions of operations in commercial or nondefense applications would be less severe and, therefore, reliability will be higher.

6.2.1.5.2.3 Disposal Requirements — Secondary cells or rechargeable batteries normally do not have disposal requirements, but they are reclaimed at the end of their useful life. This can be accomplished by returning the used batteries to the manufacturer for reclamation service. This service includes shipment of silver-bearing scrap to an independent reclamation contractor for the recovery of precious metal silver. The battery user, if he or she chooses, may use other reclamation services.

The silver content of an Ag-Zn battery is sufficient to justify these reclamation services. When the silver is reclaimed by the independent source or at a government reclamation center, the chemical components of the battery must be disposed of according to Environmental Protection Agency (EPA) laws and regulations that are in effect at that time.

The manufacturer's material data sheets disclose the exact materials contained in the rechargeable Ag-Zn cells or batteries. The material contents are as follows:

- Zinc oxide (ZnO), whose contents include lead, cadmium, and aluminum
- KOH solution
- Silver oxide (AgO_2)
- Mercury

A trace amount of mercury is embedded with ZnO in the Ag-Zn cells, which suppresses outgassing during the battery's normal operation. Disposal procedures are more stringent for the high-power Ag-Zn batteries deployed by military and space programs.

6.2.1.5.2.4 Performance Capabilities and Limitations of Secondary Silver-Zinc Batteries — This section identifies important electrical performance characteristics of rechargeable Ag-Zn batteries with an emphasis on energy density and reliability under harsh operating environments. Electrical performance characteristics of midsize Ag-Zn batteries are shown in Table 6.3. The midsize Ag-Zn rechargeable batteries have a typical range of 40 to 50 Ah. In general, larger cells will exceed these ampere-hour ratings and vice versa. Design modifications can be made to optimize a specific performance parameter, but these modifications could reduce some other performance parameters that could be of significance. Under these circumstances, trade-off studies or computer modeling must be performed to obtain a viable option for optimization of a specific performance parameter. In some cases, a common trade-off must be pursued to reduce the cycle life or calendar life of the battery to increase the energy density.

6.2.1.5.3 Applications of Rechargeable Silver-Zinc Batteries

Ag-Zn rechargeable batteries offer the highest power (W) and energy (Wh) per unit of volume and mass. Because of these outstanding characteristics, these batteries are widely deployed for defense and aerospace applications for which they have proven their high reliability. These batteries recently have proven their suitability for a new generation of sophisticated, portable, advanced medical devices for which

Table 6.3   Typical Electrical Characteristics of Midsize Rechargeable Batteries

	Battery Type		
Characteristic	*Silver-Zinc*	*Nickel-Cadmium*	*Nickel-Metal-Hydride*
Output voltage (V)	1.8	1.25	1.25
Energy density (Wh/kg)	80–110	55–100	60–120
Volumetric energy density (Wh/L)	180–220	85–125	90–135
Power output (kg)	640	165	178
Cycle life (cycles)	50–100	1,000–2,000	350–680
Calendar life (years)	2–3	3–5	2–4
Fast-charging time (h)	1	1	2–3
Cost/cycle (cents)	14	4	12
Loss/month (%)	13	15	20

ultra-high reliability and energy outputs with strict voltage requirements are the principal considerations.

By extending their cycle life and calendar life, significant improvements in performance and lower operating cost are achieved, which may open the doors for a variety of commercial applications. Weight-related trade-off studies undertaken by the author indicate that the weight of Ag-Zn batteries is approximately 50% of the weight of Ni-Cd and Ni-MH batteries and nearly 25% of the weight of lead-acid batteries. Because of high reliability and lower weight, these batteries are most suitable for battlefield, airborne, aerospace, and satellite-based system applications. Their low charge rates as summarized in Table 6.4 could be a hindrance in certain applications.

The data presented in Table 6.4 indicate that no increase in battery output voltage is visible even after 3 h of charging under a low-charging scheme. Some improvement in output voltage is visible after 5 h of charging. It took practically 9 h before the output voltage level approached a constant value of 1.95 V. Thereafter, the output voltage remained more or less constant at 1.95 V. Under a fast charging scheme, full output voltage appeared within approximately 2 h. Typical self-discharge rates of Ag-Zn batteries are summarized in Table 6.5 [5]. Typical reduction in the 12 V battery output voltage as a function of discharge time is shown in Figure 6.5. The battery output voltage is about 7.2 V after a discharge time of 12 min.

Examination of the data presented in Table 6.5 indicates that the discharge rate is about 24% after six months, which comes to roughly 4% per month. This means

Table 6.4 Typical Charge Voltage as a Function of Charge Time

Charging Time (h)	Battery Output Voltage (V)
0	1.50
1	1.51
2	1.51
3	1.52
4	1.53
5	1.54
6	1.55
7	1.58
8	1.68
8.5	1.80
9	1.95
10	1.95
11	1.95
12	1.95

Table 6.5 Typical Discharge Rates of Silver-Zinc as a Function of Days and at 75°F Temperature

Number of Days (Months)	Percentage of Rated Capacity (%)
0	100
30 (1)	97
60 (2)	92
90 (3)	90
120 (4)	88
150 (5)	86
180 (6)	76

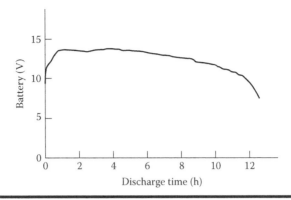

Figure 6.5 **Variations in the output voltage of a 12 V cell as a function of discharge duration. Exact values of the output voltage are shown in Table 6.1.**

that the Ag-Zn battery can be used for a very long period with little deterioration in the electrical performance of the battery, thereby indicating its suitability for long-duration ground and airborne military missions.

6.2.1.5.4 Commercial and Military Applications of Silver-Zinc Rechargeable Batteries

This section will identify the potential commercial and military applications of Ag-Zn batteries. Normally the Ag-Zn and nickel-zinc rechargeable batteries are not light. But they still are portable compared with lead-acid and Ni-MH batteries. As discussed, the weight of Ag-Zn batteries is about 50% of the weight of Ni-Cd and Ni-MH rechargeable batteries and nearly one-fourth of the equivalent lead-acid batteries. Ag-Zn batteries were specifically designed and developed for defense, aerospace, and other military applications, for which reliability and portability are the principal design requirements. These batteries are best suited for commercial and military applications for which high reliability and high energy output with strict voltage control are of paramount importance. A compact version of this battery is developed for a next generation of sophisticated, portable medical equipment.

6.3 Low-Power Batteries for Various Applications

This section of the chapter deals with low-power batteries using advanced technologies such as thin-film technology, microtechnology, and nanotechnology. Batteries incorporating these technologies are best suited for applications that require low power consumption, compact packaging, light weight, reliability, and portability. These batteries are most ideal for space, electronic drone, UAV, missile, and space applications.

6.3.1 Thin-Film Microbatteries Using MEMS Technology

Low-power and miniaturized electronic, electrical, infrared, and electro-optical sensors require miniaturized power sources or batteries to keep their size and weight to a minimum. The miniaturized power sources are nothing but three-dimensional (3D), thin-film microbatteries (MBs). Specific structural details of such a battery are illustrated in Figure 6.1. Deployment of conformal thin-film structures provides distinct advantages over conventional miniaturized structures. Additionally, the planar two-dimensional (2D) thin-film batteries cannot be classified as MBs, because they require large footprints of a few square centimeters to achieve a reasonable battery capacity. Studies performed by the author on miniaturized structures reveal that the maximum energy available from a thin-film battery is about 2 J/cm^3. Commercial thin-film batteries with a 3 cm^2 footprint have a capacity of 0.4 mAh, which comes to about 0.133 mAh/cm. Thus, the 3D, thin-film, Li-ion, MB can meet the power requirements for low-power, miniaturized sensors.

6.3.2 Microbatteries Using Nanotechnology Concepts

Studies performed by the author indicate that it is possible to design MBs using nanotechnology [6]. A battery more powerful than a lithium-based battery has been designed and developed by the Center for Integrated Nanotechnology (CINT) at Sandia National Laboratories [7]. The laboratory scientists claim that they have built the world's smallest rechargeable lithium-based battery, called a nanobattery (NB). The NB has a single tin-oxide (SnO_2) nanowire anode, which has a diameter of 100 nm and a length of 10 mm. This wire is about one seven-thousandth the thickness of a human hair. This battery has a 3 mm long bulk lithium-cobalt oxide and an ionic liquid acting as an electrolyte. The battery designers were expecting to observe changes in the battery's atomic structure during the charge and discharge cycles, so that they could investigate how to improve the battery's energy output and the power densities of Li-ion MBs. The research scientists actually observed the progression of the Li-ions as they traveled along the nanowire to create the high density of mobile dislocations because of lithium penetration of the crystalline lattice, which causes the nanowire to bend. The scientists believe that the penetration of lithium causes the nanowires to sustain lithiation-induced stress without breaking the nanowires, thereby making the wires good candidates for the battery electrodes.

During the laboratory evaluation, the research team discovered that the NB's SnO_2 nanowire rod lengthened during the charging cycle, but it did not increase in diameter, which the investigating team expected. The lengthening could avoid short-circuiting in battery designs. Material scientists believe that lithiation-induced volume expansion, plasticity, and pulverization of electrode materials are the major mechanical defects that can degrade the electrical performance of a battery and reduce the lifetime of high-capacity anodes in Li-ion rechargeable batteries. It is the opinion of the material scientists that observations in structural kinetics and amorphization could have significant implications, in particular for high-energy battery design and to mitigate battery failures.

6.3.3 Critical Design Aspects and Performance Requirements for Thin-Film Microbatteries

Preliminary studies undertaken by the author seem to indicate that 3D, thin-film, Li-ion MBs will be able to meet the power requirements of a miniaturized power source. Critical elements of the 3D, thin-film, MB are described in Chapter 4. The critical elements of this MB include the current collector, graphite anode, cathode, and the hybrid polymer electrolyte (HPE). Structural details of this MB can be found in the cross-sectional view of the device. Cathode thickness and volume determine the maximum energy density and the battery capacity. Studies performed by material scientists indicate that significant gain in the geometrical area and cathode volume are possible with a perforated substrate instead of a conventional substrate. The studies further indicate that the area gain (AG) is strictly dependent on the holes or microchannels in a multichannel plate (MCP) substrate, substrate thickness, and the aspect ratio (height-to-diameter) of the holes, as illustrated by Figure 6.6. Mathematical modeling will indicate that tilting the footprint area with various hole geometries could offer larger AG. In other words, geometry with hexagonal holes separated by constant-thickness walls in a perforated substrate provides the optimum AG, leading to optimum battery capacity and energy density.

The AG as a function of hole diameter (d), spacing between the holes (s), and substrate thickness (t) can be computed using the following mathematical expression:

$$AG = [(22/7)(d)/(d + s)^2] \, [(t - d/2) + 2] \qquad (6.4)$$

All dimensions are expressed in micrometers. Computed values of AG as a function of hole diameter, spacing, and substrate thickness are shown in Figure 6.6. It is evident from the plotted curves in Figure 6.6 that the smaller the hole diameter, the higher will be the AG.

6.4 High-Power Lithium and Thermal Batteries for Military Applications

High-capacity lithium-based and thermal batteries are widely used for various military applications, such as ordinance, aircraft emergency power, and

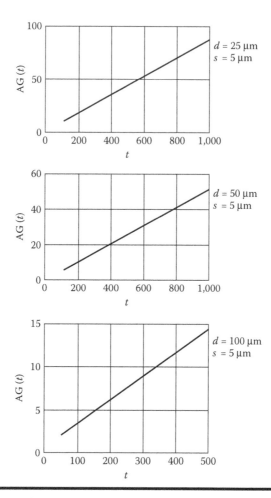

Figure 6.6 **Area gain (AG) as a function of substrate thickness (*t*), hole diameter (*d*), and spacing between holes (*s*). All dimensions shown are expressed in microns.**

hydraulic power in fighter aircraft. Initially, the high-capacity thermal battery systems were developed for –Ag-Zn, seawater, and lead-fluoronic-acid battery replacement [8]. The majority of the thermal battery applications are for ordinance. But later, these batteries were restricted to military aircraft seat ejection systems [6]. Around 1988, Lucas Aerospace Corp. was asked by the U.S. Air Force to develop thermal batteries specifically for aircraft emergency power applications. Around that same time, lithium ferrous sulfide [Li(x)/FeS$_2$] was considered to be the system of choice in developing thermal batteries strictly for ordinance applications because of its long life and high electrical capacity. In the early 1990s, another battery system using lithium aluminum ferrous sulfide

$(LiAl/FeS_2)$ was developed for nonordinance applications, which is currently deployed to supply electricity to emergency electronics and hydraulic systems in fighter aircrafts. Because lithium-based and Li-ion-based rechargeable batteries are widely deployed in several applications, chemistries of these elements or components will be briefly described.

6.4.1 Material Requirements for Cathode, Anode, and Electrolyte Best Suited for High-Power Batteries

The three components of high-power rechargeable batteries are the cathode, anode, and electrolyte. This section briefly describes the materials and their characteristics for cathodes, anodes, and electrolytes with an emphasis on performance capabilities and limitations reliability, and safety.

6.4.1.1 Cathode Materials and Their Chemistries

Lithium has a tendency to intercalate into low-energy sites in a crystal lattice of certain metal oxide materials. Cobalt oxide (CoO_2) was first used by the Li-ion battery and subsequently was used year after year for more than 15 years because of its state-of-the-art chemical characteristics. Since the end of 2000, other intercalation materials, namely manganese oxide (MnO_2), nickel oxide (NiO_2), and vanadium oxide (VO), have been used in the design and development of cathodes because of their high-reversibility characteristic. Intercalation is defined as an ability to insert between the existing chemical elements and essentially is a physical process that involves various chemical changes. Cathode materials capable of producing high energy must be selected. The most popular cathode materials that are best suited and widely used by various battery designers for lithium batteries include lithium cobalt oxide $(LiCoO_2)$, lithium vanadium oxide $(Li_4V_2O_5)$, and lithium manganese oxide $(LiMnO_2)$, because of their maximum specific capacity rating battery of 0.265 Ah/kg, 0.505 Ah/kg, and 0.282 Ah/kg, respectively. As far as operating life is concerned, the MnO_2-based cathode has a shorter life compared with the CoO_2-based cathode. In terms of voltage plateau locations, CoO_2 and NiO_2 have only one voltage close to 3.8 V, whereas the MnO_2 material has a voltage plateau at about 2.8 V. As far as stability is concerned, the NiO_2 cathode is less stable compared with the MnO_2 and CoO_2 cathodes. In summary, plateaus voltage, stability, and specific capacity rating parameters must be carefully evaluated before selection of the material for cathode, anode, and electrolyte for high-capacity batteries. Materials for the cathode, anode, and electrolyte used in high-power batteries must meet some specific electrical, mechanical, chemical, and thermal performance requirements. Some material scientists are exploring the full potential of lithium-fluorine and sulfur-based cathodes for rechargeable batteries in terms of cost-effective performance, safety, and reliability. The following sections summarize the chemistries for anodes, cathodes, and electrolytes used in high-power batteries.

6.4.1.2 Anode Materials and Their Chemistries

The anode materials include lithium metal, carbon structures, some unique materials that can intercalate lithium at potentials close to that of some metal oxide, and metals that can alloy with lithium. The anode materials must retain their mechanical integrity under harsh mechanical environments, including shocks and vibrations. Battery designers believe that dry polymer electrolyte will substantially reduce these hazards. A graphite anode (LiC_6) offers a maximum specific capacity close to 30.375 Ah/kg. As far as safety is concerned, anode carbons with high surface areas offer optimum safety. Preliminary studies seem to indicate that anode materials for lithium batteries must have high specific energy, high energy density, and prolonged cycle life. Hard carbon materials offer a specific energy of 0.425 Ah/kg, an energy density of 0.550 Ah/cm^3, and a cycle life of about 1,000. The graphite material is capable of providing a specific energy of 0.333 Ah/kg, an energy density of about 0.652 Ah/cm^3, and a cycle life of 1,000. Other materials considered seem to yield poor performance in these three distinct areas. If metal alloy electrodes are considered for the rechargeable batteries, then make sure the volume expansion at higher operating temperatures does not pose any reliability or safety problem. The volumetric expansion on charging is about 300% for tin lithium alloy ($SnLi_4$) compared with only 10% for graphite expansion.

6.4.1.3 Electrolytes and Their Chemistries

Brief studies performed by the author seem to reveal that the lithium chemistries prefer to use solid electrolytes or nonaqueous electrolytes for reliability and safety reasons. Electrolytes that tend to generate combustible and toxic products must be avoided. Most organic electrolytes that introduce safety problems are not recommended. Some electrolytes that tend to destroy the graphite anodes should be avoided. Newly developed anode materials may require different types of electrolytes. Research studies performed on lithium salts by various chemical scientists indicate that such materials suffer from stability and hazard problems and, hence, should be avoided. The studies further indicate that organic-based lithium salts would provide improved battery performance, but these salts are relatively costly. The cathode and anode materials selected must demonstrate very good performance with the lithium-based salt that will be used for the electrolyte. Currently, most battery manufacturers recommend lithium hexafluorophosphate ($LiPF_6$) for the electrolyte with minimum water contamination. It seems that this particular electrolyte material offers good overall performance in the absence of water contamination. This electrolyte is best suited for high-power batteries widely used in military system applications with minimum water contamination, for which reliability, safety, and mission success are of critical importance.

Contamination of $LiPF_6$ is possible with excessive water presence. Electrolyte degradation is possible due to water presence, which is evidence from the following chemical equations:

$$LiPF_6 = [Li\ F + PF_5], \text{ with no water contamination} \qquad (6.5)$$

$$LiPF_6 + H_2O = [Li\ F + 2HF + POF_3], \text{ with water contamination} \qquad (6.6)$$

It is evident from Equation 6.6 that the electrolyte $LiPF_6$ is less stable with water contamination. Research scientists have investigated the chemical properties of polymer electrolytes involving two different types of materials, such as a homogeneous dry material and a polymer matrix material. But these two electrolyte materials for high-power rechargeable batteries have not been widely used. The homogeneous dry material is being used, however, for low-power communication system applications. In summary, a specific electrolyte material for high-power batteries does not exist that will meet full performance specifications under a variety of operating environments.

Once the right materials for the anode, cathode, and electrolyte are selected, it becomes easy to develop high-power conventional batteries and thermal batteries to provide emergency power for fighter aircrafts and other military weapon systems. Thermal battery systems were developed as an alternative to the oxygen-methanol or hydrazine emergency systems used to supply emergency power to fighter aircraft. Two distinct thermal batteries were designed, developed, and tested: one battery system designated as TB1 to provide power for electrohydraulic pump (EHP), and the other designated as TB2 to provide electric power to aircraft electronic systems and sensors.

There are three distinct reasons for the deployment of thermal batteries, which can be summarized as follows:

■ Thermal batteries have lower maintenance requirements than the other battery systems, such as oxygen-methanol and hydrazine battery systems.
■ Thermal batteries are environmentally less hazardous than other battery systems.
■ The thermal battery systems can provide the required electrical power with a fast rise time (less than 1.5 sec.) at extreme operating temperatures (ranging from $-40°$ to $+80°C$) over its entire operating life.

6.4.2 Design Requirements for Thermal Batteries for Specific Applications

Emergency systems require direct current (DC) power for EHP and for the DC emergency bus bar. Two batteries are each dedicated to providing power to two different sources. The first thermal battery system (TB1) is used for EHP, which requires a square wave current pulse load capable of simulating the function of the EHP for its entire operating life. The second thermal battery (TB2) provides electric power to the DC bus bar and is required to meet a constant power for its entire operating life. For both thermal batteries, the operating life has to maintain high reliability and the required consistent performance.

6.4.2.1 Design Requirements for TB1 Battery Systems

Battery designers believe that the basic requirement for the TB1 battery is to provide a current capacity in excess of approximately 91,000 A-sec. and a total electrical energy of approximately 764 Wh for 9 min. or 540 sec. with a start time of less than 1.5 sec. at –40°C and less than 1.1 sec. at +80°C [8]. These thermal battery performance requirements are summarized in Table 6.6. It is extremely important to ensure that the battery voltage is required to maintain between 26 and 38 V as shown in Figure 6.3. Because the batteries are used in military and commercial aircraft, the weight of the battery is of critical importance. The total mass or weight of the TB1 is less than 24.50 kg or 50.3 lb. Thus, the energy density for the TB1 battery is 31.18 Wh/kg (764/24.5 = 31.18 Wh/kg).

6.4.2.2 Design Requirements for TB2 Battery Systems

The principal objective of the TB2 battery is to supply DC power for the electronic systems and DC emergency bus bar for its entire operating life. In brief, the TB2

Table 6.6 Performance Requirements for the TB1 Thermal Battery

Performance Characteristics	Value and Units
Minimum voltage	26 VDC
Maximum voltage	38 VDC
Rise time	1.1 sec. from 25° to +80°C
	1.5 sec. from –40° to 25°C
Operating life	9 min.
Battery capacity	91,000 A-sec.
Average power (CW rating)	5.1 kW
Average current	168 A
Maximum surface temperature	325°C

Source: Kauffman, S., and G. Chagnon, "Thermal battery for aircraft emergency power," *Proceedings of the IEEE,* © 1992 IEEE. With permission.

Note: CW: continuous wave; VDC: voltage direct current.

Table 6.7   Performance Requirements for the TB2 Thermal Battery

Performance Characteristics	Value and Units
Minimum voltage	23 VDC
Maximum voltage	31.5 VDC
Rise time	1.1 sec. from 25° to 80°C 1.5 sec. from −40° to +25°C
Operating life	9 min. (540 sec.)
Battery capacity	42,000 A-sec.
Average power (CW power)	2.0 kW
Average current range	64.5–87.0 A
Maximum surface temperature	325°C

thermal battery is required to provide a current capacity of 45,000 A-sec., a total energy equal to 300 Wh for 9 min. (540 sec.) with a starting time of less than 1.5 sec. at −40°C. The performance requirements for the TB2 battery are summarized in Table 6.7. The TB2 [8] battery voltage is required to be between 3.5 and 23 V under a constant power of 2,000 W as illustrated in Figure 6.4. The total mass of this battery is required to be less than 10.4 kg. Thus, the electrical energy requirement for the TB2 battery is 28.85 Wh/kg (300/10.4 = 28.846).

6.4.3 Environmental Requirements for Thermal Battery Systems

Environmental requirements for both thermal battery systems are identical and can be summarized in Table 6.8.

Laboratory tests must be performed under specified environmental conditions for temperature, shock, and vibration to qualify these batteries for fighter aircraft applications. Both thermal batteries must demonstrate full compliance with shock, vibration, and temperature specifications. The most critical of these requirements is the random vibration or nonoperational vibration test for 1.5 h in each of the three mutually perpendicular axes. The principal objective of this vibration test is to simulate the aircraft's structural health and mechanical integrity after aging over a specified duration. In both thermal batteries lithium oxide (LiO) is used in the cathode as a voltage suppressor. This allows for fine-tuning of the battery peak voltages to maximize the actual voltage regulation. In both batteries, a cell diameter of 12.7 cm is used, and the average cell thickness is 0.356 cm. A stainless steel outer cover or casing is used for both battery designs. The battery designs differ only in the number cells used in each design and the parallel sections deployed.

Table 6.8   Specifications for Environmental Test Parameters and Condition of Tests

Environmental Parameter	Value of Parameter and Conditions Attached
Operating temperature range	−40° to +80° C
Operational vibration specifications	10.75 g RMS 10 min./axis × 3 axes
Nonoperational specifications	10.75 g RMS 1.5 h/axis × 3 axes
Shock specifications	30 g, 3 axes, 2.5 m.sec., 1/2 sine/7.5 g RMS, 3 axes, 40 m.sec., 1/2 sine
Bump shock specifications	1,000 shocks × 3 axes, 40 g, 6 m.sec., 1/2 sine

6.4.4 Structural Description of the Batteries and Their Physical Parameters

This section describes the structural details and physical parameters of each thermal battery. The TB1 battery consists of two identical cylindrical units. Each unit has a diameter of 15.4 cm and a length of 26 cm. Each battery unit consists of three voltage sections that are connected in parallel. Each section has 20 cells each of a total of six voltage sections that are connected in parallel. The six parallel sections provide a total cell area of 755 cm². The calculated average current density for the TB1 battery system is 0.224 A/cm².

The design of the TB2 battery system is slightly different than the TB1. The TB2 battery consists of a single cylindrical unit. This unit has a diameter of 15.4 cm and a length of 21.0 cm. The TB2 battery consists of three parallel sections, each containing 16 cells. The total cell area is 377 cm² and is capable of yielding an average current density of 0.120 A/cm².

6.4.5 Actual Values of Performance Parameters Obtained through Laboratory Testing

Measured electrical performance parameters of thermal batteries TB1 and TB2 indicated that the actual values obtained through laboratory testing [8] seem to exceed the specified requirements summarized for TB1 and TB2 battery systems as shown in Tables 6.6 and 6.7, respectively. The highest energy density for TB1 battery system during the tests was 70.2 Wh/kg at the test temperature of +20°C, where the battery mass was approximately 24.3 kg, as shown in Table 6.9. The laboratory tests conducted on the TB2 battery system at the suppliers indicated that the lowest energy density of 39.3 Wh/kg at a test temperature of −40°C was achieved when the battery mass was 9.97 kg, as illustrated by Table 6.10. Actual electrical

Table 6.9 Actual Performance Parameters Obtained by the Battery Designer through Tests on TB1 Battery Systems

Performance Parameters	Value	Test Temperature
Energy density (Wh/kg)	56.6	−40°C
Peak voltage (VDC)	36.3	@ 140 A
Mass (kg)	24.3	N/A
Start time (sec.)	1.3	N/A
Operating life (sec./h)	1,040/0.2889	N/A
Energy density (Wh/kg)	70.0	+20°C
Peak voltage (VDC)	37.0	@ 140 A
Mass (kg)	24.3	N/A
Start time (sec.)	1.06	N/A
Operating life (sec./h)	1,318.5/0.3663	N/A
Energy density (Wh/kg)	67.4	+80°C
Peak voltage (VDC)	37.3	@ 140 A
Mass (kg)	24.3	N/A
Start time (sec.)	1.06	N/A
Operating life (sec./h)	1,210.5/0.3363	N/A
Energy density (Wh/kg)	67.4	+80°C
Peak voltage (VDC)	37.3	@ 140 A
Mass (kg)	24.3	N/A
Start time (sec.)	1.06	N/A
Operating life (sec./h)	1,210.5/0.3363	N/A

Note: N/A: not available.

performance parameters obtained through tests on TB1 and TB2 battery systems are summarized in Tables 6.9 and 6.10, respectively.

6.4.6 Conclusive Remarks on Thermal Battery Systems

Research and application studies performed by the author seem to indicate that thermal batteries are best suited for applications for which high reliability, long

Table 6.10 Actual Performance Parameters Obtained by the Battery Designer through Laboratory Measurements on TB2 Battery Systems

Performance Parameters	Value	Test Temperature
Energy density (Wh/kg)	41.2	−40°C
Peak voltage (VDC)	29.6	@ open circuit
Mass (kg)	9.97	N/A
Start time (sec.)	Not recorded	N/A
Operating life (sec./h)	694.8/0.1930	N/A
Energy density (Wh/kg)	54.2	+20°C
Peak voltage (VDC)	29.0	N/A
Mass (kg)	9.83	N/A
Start time (sec.)	1.19	N/A
Operating life (sec./h)	959.21/0.2664	N/A
Energy density (Wh/kg)	52.0	+80°C
Peak voltage (VDC)	31.3	@ open circuit
Mass (kg)	9.83	N/A
Start time (sec.)	0.07	@ open circuit
Operating life (sec./h)	943.17/0.2620	N/A
Energy density (Wh/kg)	39.3	−40°C
Peak voltage (VDC)	28.3	N/A
Mass (kg)	9.97	N/A
Start time (sec.)	1.39	N/A
Operating life (sec./h)	690.65/0.1918	N/A

Note: N/A: not available.

shelf life, and high energy density are the most demanding requirements. Feasibility studies indicate that thermal batteries with high energy density levels are particularly ideal for emergency power systems and underwater vehicles performing covert surveillance and reconnaissance missions in hostile coastal regions. The thermal battery system is also suitable for C-IED search missions in battlefields in cases in which high energy density over long durations is of critical importance.

6.5 High-Power Rechargeable Batteries for Underwater Vehicles

Rechargeable batteries with high energy density and high power capacity are most suitable for underwater vehicles, torpedo propulsion, and minisubmarine applications. Batteries for underwater applications must meet other stringent performance requirements as well, such as ultra-high reliability and proving a specified electrical power over the long durations needed to complete missions successfully. The rechargeable battery for underwater vehicle propulsion must be able to provide high-rate and long-duration discharge capabilities. French scientists undertook research and development activities on a Volta stack battery design using a flowing electrolyte that was based on sulfuryl chloride (SO_2Cl_2). This particular battery design configuration demonstrated output power much higher than 50 kW with a discharge duration of 20 min. [9]. Next-generation battery systems using flowing electrolyte are expected to generate output power exceeding 650 k, which will be able to meet the power requirements for torpedoes and antitorpedo weapon systems. These weapon systems require batteries with long-duration discharges as long as 30 min. These battery systems will deploy advanced technologies leading to the development of more compact system with optimized specific energy density and long-duration discharge with lower power output. These improvements are possible with lithium-based liquid cathode technology, such as lithium-sulfuryl chloride ($Li\text{-}SO_2Cl_2$) battery systems.

6.5.1 Performance Capability and Design Aspects of Li-SO₂Cl₂ Battery Systems

This particular battery design configuration uses the SO_2Cl_2 as a liquid cathode, which could offer several advantages, such as a very high energy density capability close to 1,410 Wh/kg, discharge rates as high as 200 mA/cm², and discharge durations exceeding 25 min. Long-duration discharge is possible only because of the lower anodic corrosion of $Li\text{-}SO_2Cl_2$ battery systems as opposed to Zn-AgO and Al-AgO battery systems. The longer discharge is also due to the lower level of leakage currents in a Volta stack design configuration because of its low electrolyte conductivity of 13 mS/cm as opposed to 600 mS/cm for the KOH electrolyte. The open-circuit voltage of this battery system is 3.95 V associated with an energy density as high as 1,410 Wh/kg. The overall discharge chemical reaction during the discharge cycle can be given by the following expression:

$$2Li + SO_2Cl_2 = [2LiCl + SO_2] \tag{6.7}$$

The chemical product produced at the carbon electrode as illustrated by Equation 6.7 is insoluble in the electrolyte, and it precipitates in the porous structure of the electrode. The SO_2Cl_2 is very soluble in the electrolyte and participated in the lithium-ion (Li⁺ion) generated through the formation of lithium-aluminum chloride (Li-AlCl₄). Consequently, the composition of the electrolyte is modified throughout the discharge cycle.

As opposed to thionyl-chloride battery systems, the absence of sulfur formation during the discharge cycle simplifies the circulation in the $Li-SO_2Cl_2$ battery system. This simplification leads to improved reliability of the battery with improved optimization of the battery energy density. As a matter of fact, only a pump and a heat exchanger are required to boost the system's reliability. Battery designers are predicting that the use of advanced materials and a liquid cathode will significantly improve the power output performance of the $Li-SO_2Cl_2$ battery systems. Battery designers are forecasting a power level exceeding 650 kW in the near future. The next section identifies various factors for improvements in the electrochemistry of the $Li-SO_2Cl_2$ battery system.

6.5.2 Characteristics of Electrolytes Required to Achieve Improvements in Electrochemistry

SO_2Cl_2 acts both as a solvent and a cathodic material. Furthermore, the electrolyte solution contains Li-AlCl₄ as its sally from approximately 1.5 to 3 mol./L. Chemists believe that slight amounts of free Lewis acid (AlCl₃), ranging from 0.25 to 0.80 mol./L, are generally used to increase rate capabilities at the beginning of the discharge and to partially complex the LiCal produced during the discharge cycle on the carbon cathode terminal. The addition of a bromine solution to the electrolyte will increase its rate capabilities because of a catalytic effect of this compound on the reduction of the SO_2Cl_2 solution, which is evident from Equation 6.7.

6.5.3 Effects of Thermal Characteristics on the Flowing Electrolyte

Thermal characteristics such a specific heat and thermal conductivity of the flowing electrolyte, which is used to cool the battery system, must be measured periodically as a function of depth of discharge (DOD) and temperature. An examination of the electrolyte reveals that the main changes in these characteristics generally occur due to variations in the chemical composition of the electrolyte. The examination of electrolytes further reveals that both the thermal conductivity and specific heat have a tendency to increase the DOD. Only the heat capacity or the specific heat is significantly altered by the temperature, but the decrease of this characteristic as a function of temperature can be compensated for by its increase with DOD, as illustrated by data presented in Table 6.11. These data indicate that the thermal characteristics of the electrolyte are not drastically affected during the discharge cycle. The specific heat and thermal conductivity of the flowing electrolyte are of critical importance, if high reliability, long cycle life, and long DOD durations are the principal design requirements. The cooling efficiency of the electrolyte does not

Table 6.11   Variations in the Thermal Conductivity (W/m·K) of Electrodes Used as a Function of Temperature and Assumed Depth of Discharge (DOD) Values

	Substrate	
Temperature (°C)	SO_2Cl_2 (DOD = 60%)	$LiAlCl_2$ (DOD = 40%)
20	187	176
40	182	172
60	178	167
80	174	163

undergo changes during the discharge. Comprehensive examination of the laboratory tests performed by the battery designer on two distinct electrolytes, namely SO_2Cl_2 and $LiAlCl_4$ (see Table 6.11), seem to indicate that they can expect slight variations in the thermal conductivity of the electrolytes as a function of DOD and temperature of electrolyte. Variations in the specifics of the electrolyte as a function of temperature and DOD are summarized in Table 6.12. The data presented in the Table 6.12 indicate that variations in the specific heat of the SO_2Cl_2 substrate seem to be minimum, which indicates that variations in the battery output power level will be relatively small.

It is evident that the electrolyte SO_2CL_2 offers higher values of thermal conductivity as a function of electrolyte temperature, which will yield both the higher thermal efficiency and battery output power at elevated operating temperatures. In addition, the reliability and the longevity of the rechargeable batteries using a flowing electrolyte such as SO_2CL_2 will be much higher compared with solid or semisolid electrolytes.

Preliminary calculations performed by the author on the variations of specific heat for the SO_2Cl_2 flowing electrolyte seem to reveal that very few changes as

Table 6.12   Variations in Specific Heat Values of the Sulfuryl-Chloride Electrolyte as a Function of Temperature and Depth of Discharge (DOD)

	DOD (%)			
Temperature (°C)	60	40	25	0
20	0.268	0.258	0.971	0.204
40	0.271	0.255	0.210	0.182
60	0.266	0.257	0.224	0.213
80	0.262	0.259	0.233	0.228

a function of temperature and DOD can be expected on the basis of the values shown in Table 6.12. Errors seem to be higher particularly at the lowest temperature. In general, the specific heat values have more errors at the starting temperatures, because the properties of any material or substance are not fully stabilized. In addition, it takes time in the beginning to bring the material completely to its highest thermal stability point.

6.5.4 Output Power Variations as a Function of Discharge Duration in Volta Stack Batteries Using Flowing Electrolytes

Long-duration discharge is possible due to lower anodic corrosion and to the lower level of leakage currents in a Volta stack design configuration. Leakage currents are lower because of a low electrolyte conductivity of less than 12 mS/cm as opposed to 600 mS/cm for KOH electrolytes. Both the long discharge and higher power output are possible using the optimized SO_2Cl_2 electrolyte. These improved performance capabilities of a battery design with an optimized electrolyte are shown in Table 6.13.

Power performance of the 10 kW battery using optimized SO_2CL_2 is significantly improved over the discharge duration, ranging from 2 to 17 min.

Table 6.13   Power Output of Volta Stack Batteries with 20 kW and 10 kW Ratings as a Function of Discharge Durations along with the Performance Capability Using an Optimized Electrolyte

Discharge Duration (min.)	Output of 20 kW Battery	Output of 10 kW Battery	Battery Output Using Optimized Electrolyte
0	0	0	0
2	13.5	6.8	7.8
4	14.1	7.3	8.1
6	14.1	7.5	8.2
8	14.1	7.8	8.3
10	13.9	8.0	11.6
12	13.9	7.9	11.4
14	20.0	9.6	10.8
16	16.2	9.3	8.8
18	10.4	6.1	6.6
20	9.8	5.9	5.2

This confirms the fact that battery output can be significantly improved using a flowing and optimized SO_2Cl_2 electrolyte in a Volta stack battery design configuration.

Leading battery designers believe that battery performance enhancement is possible by circulating the electrolyte solution. The designers further believe that circulation of the electrolyte solution will reduce or eliminate the adverse effects caused by space charge and voltage gradient in the boundary between the cells and the electrolyte and also because of the stratification of the varying electrolyte density during the charging and discharging process. In other words, during the charge and discharge process, a space charge exists in the boundary layer between the electrolyte and electrodes.

Battery scientists recommend that the circulation of the electrolyte could improve battery performance by minimizing the abovementioned negative effects. The performance is strictly dependent on whether the circulation of the electrolyte is by forced circulation or free circulation. Experimental data obtained by battery designers indicate that substantial performance improvements result from forced circulation, which could be an improvement as high as 16% in the power output of the battery. Circulation of the electrolyte is considered to be the most effective optimization technique for the liquid electrolyte. This performance enhancement technique would be valid for any battery system that uses a liquid electrolyte. The impact of the electrolyte optimization is evident from the data presented in the fourth column of Table 6.13. The battery designers believe that raising the temperature of the electrolyte will also lead to additional performance improvements in the battery using a liquid electrolyte.

6.5.5 Impact of Temperature and DOD on the Thermal Conductivity and the Specific Heat of the Electrolytes Used in Thermal Batteries

Research studies performed by the author seem to indicate that both the thermal conductivity and the specific heat could change from the electrolyte circulation. Thus, the electrolyte optimization, which has an impact on the power output, undergoes changes as a function of operating temperature and DOD. The variations in the thermal conductivity of SO_2Cl_2 and $LiAlCl_4$ electrolytes as a function of temperature and DOD are shown in Table 6.14.

Thermal analysis on these two electrolytes seem to reveal that variation in thermal conductivity of the electrolytes is only 0.0225% at a DOD value of 60%, whereas it is roughly 0.028% at a DOD of 24%. The data summarized in Table 6.14 reveal that higher values of DOD will yield better performance of the electrolyte over the temperatures specified in the table. Battery efficiency as well as performance are significantly improved in the temperatures shown.

Table 6.14   Effects of Temperature and Depth of Discharge (DOD) on Thermal Conductivity of Electrolytes

	W/m·K		
Temperature (°C)/DOD	*60%*	*40%*	*24%*
20	187	176	165
40	182	173	161
60	178	169	155
80	174	164	150

Table 6.15   Variations in the Specific Heat of the Electrolyte as a Function of Temperature and Depth of Discharge (DOD)

	DOD (%)			
Temperature (°C)	*60%*	*40%*	*25%*	*0%*
20	0.268	0.258	0.202	0.204
40	0.271	0.255	0.210	0.192
60	0.264	0.257	0.224	0.213
80	0.262	0.259	0.233	0.228

The variations in the specific heat of the electrolytes as a function of temperature and DOD are shown in Table 6.15. The variations in the specific heat of the SO_2Cl_2 electrolyte as a function of temperature and DOD as shown are somewhat small and will have an insignificant impact on battery performance. The change in specific heat in the electrolyte solution is nonlinear, but it is difficult to predict the overall impact of that nonlinearity on battery performance.

6.5.6 Impact of Discharge Duration on the Battery Power Output

The power output of the battery is zero at the beginning of the duration and remains low during the discharge duration not exceeding the first 2 min., as illustrated by the data presented in Table 6.16. The power output remains fairly constant, particularly in the case of a 20 kW battery over the discharge duration ranging from 3 to 13 min., as shown in Table 6.16.

These data indicate that the battery power output remains fairly constant over the discharge duration ranging from 2 to 12 min., which represents about 60% of the duration in each case. The battery loses output power by approximately 50% between the last 2 min.

Table 6.16   Effects of Discharge Duration on Battery Power Output Levels (kW)

Discharge Duration Rating (min.)	20 kW Battery Rating	10 kW Battery Rating
0	0	0
2	13.5	6.8
4	14.0	7.3
6	14.0	7.5
8	14.0	7.8
10	14.0	8.1
12	13.9	7.9
14	20.0	9.8
16	16.2	9.2
18	10.3	5.9
20	9.8	1.5

6.5.7 Electrolyte Conductivity and Optimization of Electrolyte

Optimization of SO_2Cl_2 is strictly dependent on the conductivity of the electrolyte. The lower the conductivity of the electrolyte, the higher the performance of the battery as well as it output power, as illustrated by column four of Table 6.13. Research studies undertaken by the author on this electrolyte seem to reveal that a comprehensive knowledge of electrolyte conductivity versus salt concentration is absolutely necessary. The studies further reveal that under high discharge rates, the reduction process of the liquid cathode material is limited by the diffusion phenomenon in a porous structure of the carbon cathode. In this battery, the LiCl is produced at the carbon electrode, which is insoluble in the electrolyte and precipitates in the porous structure of the electrode. In this battery system, SO_2 is very soluble in the flowing electrolyte. The chemical composition of the flowing electrolyte undergoes modification throughout the discharge duration. The absence of sulfur formation during the discharge essentially simplifies the circulation system, thereby leading to higher reliability with better optimization of the overall battery performance in terms of higher power output and energy density over the discharge duration. Optimization of the electrolyte used in the 10 kW, 400 V Volta stack battery design configuration consisting of 24 cells offers about a 10% increase in the battery output power of 7 kW at the low power end and provides for longer discharge at full power of 11 kW, as shown in the last column of Table 6.13. A clear understanding of electrolyte characteristics will allow battery scientists and designers to develop battery systems capable of producing output power close to 600 kW, which is best suited for heavy-weight torpedoes and underwater vehicle propulsion systems. Electrolyte optimization techniques can offer higher power outputs over long-discharge durations exceeding approximately 20 min.

6.6 High-Power Battery Systems Capable of Providing Electrical Energy in Case of Commercial Power Plant Shutdown over a Long Duration

The author has examined the performance capabilities and limitations of various high-power battery systems capable of meeting the electrical power requirements in case of commercial power shutdown. High-power batteries capable of meeting the emergency power requirements include lithium-hexafluoroarsenate (Li-AsF_6), Li-PF_6, dimethyl carbonate (DMC), dimethoxy ethane (DME), and diethyle carbonate (DEC). These batteries have some limitations on the power output capability over very long durations. High-power batteries best suited to provide electrical power during emergency shutdowns include lithium-metal sulfide (Li-M-S) and lithium silicon iron-sulfide (Li_5-Si-FeS_2) batteries. Li silicon is used for cathode electrode, whereas ferrous sulfide is used for the anode electrode. Ceramic or silicon material is used as a separator for stable contact with lithium and lithium-silicon alloy. These high-power batteries provide energy density ranging from 150 to 225 kWh/kg. The typical cost estimate is about \$35/kWh for the LI_5SiFeS_2 cell. Battery cells with 2,500 kWh capability have been produced, which can yield output power in the vicinity of 1,000 MWh. This kind of electrical power is typically generated by a steam turbo-alternator sufficient to power more than 100,000 households.

Material scientists and battery designers have been looking into the design of high-power batteries with capacity in the 15–20 MWh range, which can supply electrical energy for up to 2,000 households. Redox flow batteries are essentially large-scale vanadium-based liquid batteries in which chemical vanadium bonds alternately pick up and emit electrons along the membranes [10]. Pure vanadium is a bright white metal. Some unique properties of this material are briefly summarized for the benefits of readers. This material is soft and ductile. It has high corrosion resistance to alkalis, acids, and salt water. This material has demonstrated high structural strength and is most ideal for high-power batteries operating under harsh environments. Vanadium pentoxide (V_2O_5) is used in ceramic devices. Material scientists claim that small doses of vanadium salts have reversed hardening of arteries.

6.6.1 What Is a Vanadium-Based Redox Battery?

Vanadium-based liquid batteries are capable of generating electrical energy sufficient to power more than 2,000 households or to provide electricity to special forces for conducting covert military missions in a hostile, remote area where electric power is not available. In other words, the vanadium-based redox energy storage system is essentially a flow battery capable of storing electrical energy in the multimegawatt range for a duration of hours or days without any interruption. The unused energy can be returned to commercial grids, if needed.

Vanadium-based flow batteries have been used in Japan, the United Kingdom, and other industrial nations. These batteries use active chemicals in liquid forms that can be stored in portable tanks separate from the battery. Some companies have designed and developed such batteries in the 15 to 25 MW range using sodium bromide and sodium polysulfide and that are based on vanadium or zinc bromide. The liquid electrolyte solution can be added to the battery system when all other batteries are in their respective positions.

6.6.2 Potential Applications of Vanadium-Based Redox Batteries

Vanadium-based redox batteries can be classified in three distinct categories, namely stationary, mobile, and portable, depending of power output level, size, and weight. Output power level and typical applications for these batteries can be briefly defined as follows:

Stationary batteries

■ Output power range: 50 kW to >1 MW
■ Applications: grid reinforcement; integration of renewable energy sources, such as wind or solar

Mobile batteries

■ Output power range: 5 kW to 250 kW
■ Applications: onboard power sources for military vehicles; electric and hybrid drivetrains

Portable batteries

■ Output power range: 1 kW to 2 kW
■ Applications: industrial, military

Studies undertaken by the author seem to indicate that these batteries can be treated as *standalone* systems, which will eliminate the need to build expensive power lines to remote and unknown locations while providing a completely autonomous power generation source.

6.6.3 Structural Details and Operating Principles of Vanadium-Based Redox Batteries

The vanadium charge and discharge occur in the tiny reaction chambers. Several of these chambers known as cells are arranged in stacks to enhance battery power output. The battery designer must ensure that vanadium fluid flows smoothly through the membranes or cells and passes the filter-like carbon electrodes in the cells themselves. According to a German scientist's prediction, high-power vanadium-based liquid batteries will be available within the next five to seven years. The German scientist further predicted that a 20 kW battery system would go into operation at the end of 2012. Research scientists in the United States, Ireland, Austria, and France predict that vanadium-based redox battery systems up to the 100 kW level will be available in the immediate near future [10].

The higher the power output, the greater the size and weight of the battery system. Preliminary estimates indicate that a vanadium-based redox battery system would be about 20 ft. high and would weigh more than 4,500 lbs. When using advanced material technologies, however, such a battery system with a 5 kW rating would be about 4.5 ft. high and would weigh roughly 500 lbs.

Regardless of their weight and size restriction drawbacks, the vanadium-based redox battery system has a potential application as a standalone power source in a remote and unknown location where no power lines exist. In brief, the vanadium-based battery system will be best suited for special forces or covert military missions operating in a remote or hostile location where no commercial power lines exist.

High-power rechargeable battery systems are most suitable for providing electrical energy in the case of a commercial power plant shutdown under an emergency situation. High-power batteries with a power output level exceeding 100 kW may require air cooling during the charge and discharge cycles. Table 6.17 provides estimated values of air-flow rates and power required as a function of battery surface temperature during charge and discharge cycles.

Table 6.17 Cooling Air-Flow Rate and Power Required as a Function of Battery Surface Temperature

Cooling Required	Battery Surface Temperature (°C)	Air-Flow Rate (m³/sec.)	Power (kW)
During charge	100	14	3
	200	23	28
During discharge	100	8	1
	200	14	36

The data presented in Table 6.17 indicate that both the higher air-flow rates and the fan power are required at higher battery surface temperatures during the charge and discharge cycles. Cooling of batteries is necessary to maintain both the higher battery efficiency and reliability.

6.7 Batteries Best Suited for Drones and Unmanned Air Vehicles

This section describes the performance requirements for batteries used for battlefield drones and UAVs. Because both are airborne vehicles, reliability and portability of the batteries are of critical importance. The propulsion power is provided by a compact gasoline engine, whereas low-power electronic devices and sensors would provide the power from rechargeable batteries. Actual battery output requirements for a drone are strictly dependent on the type of mission, the number of sensors and devices deployed, and the amount of electrical energy required to power the electronic sensors and devices. Cruising silently over a hostile battlefield area or a well-protected enemy forward area, the unmanned drone uses its tiny infrared cameras to pinpoint the telltale muzzle flashes from a sniper rifle or machine gun. Knowing these exact locations, the drone can unleash a Hellfire missile under its wing, using the nose-mounted compact laser to accurately guide the missile to the firing target. This is an attack mission, which can be carried out at altitudes less than 5,000 ft.

UAVs can be designed to operate at altitudes ranging from 5,000 to 60,000 ft., depending on the mission requirements. The UAR can be designed for an attack mission, a reconnaissance and surveillance mission, or both missions. In the case of UAV, power requirements could be higher, depending on the number of missions and the duration of loitering.

6.7.1 Battery Power Requirements for Electronic Drones

Electrical energy requirements for a compact laser could as high as 1 kW, which easily could be provided by a compact, sealed Ni-Cd battery pack. The battery pack weighs between 5 and 8 lbs., depending on the power output requirements. This battery is used as a backup power source and offers the best discharge, charge, and storage performance capabilities. This particular battery can be discharged to 100% DOD. The battery requires 3 h for rapid charge and less than 1 h for fast charge. This battery can be stored at 40% SOC and can be stored for five years or more at room temperature with no compromise in battery performance. This battery offers improved reliability, very long life, and safety. If the weight and size are the most critical requirements, one should use zinc-silver oxide (Zn-Ag$_2$) battery instead of the Ni-Cd battery. This particular battery offers a specific power as high as 1.8 kW/kg using conventional cell technology. Using thin electrodes and thin separators, this battery can provide specific power close to 4.4 kW/kg, which can be further increased to 5.5 kW/kg by using bipolar electrodes. The Zn-Ag$_2$ battery offers optimum benefits in terms of weight and size over the Ni-Cd battery pack. A battery pack can consist of several identical cells, ranging from 4 to 16, depending on the output power requirements. The Zn-Ag$_2$ battery suffers from low cycles (50–100 cycles), high procurement costs, and poor performance at low temperature. The latest electronic drones are sophisticated and are designed for attack, intelligence, reconnaissance, and surveillance missions. These drones can fly at altitudes ranging from approximately 5,000 to 50,000 ft. For long-duration missions besides loitering, which could last from 2 h to more than 6 h, batteries must provide emergency power in addition to electronic sensors and devices.

If high cost and excessive weight and size are not acceptable for drone applications, then one should select the Ag-MH battery. These rechargeable batteries are rated for 16 V or 18 V using 18 wafer cells. Each wafer cell comes with a rating equal to 200 mAh/g. At the 16 V rating, each wafer cell will have an energy density of 3.2 Wh/g. The battery designer estimates the weight of each wafer cell to be close to 345 g (0.75 lbs.). Assuming four cells and a weight of 345 g for each wafer cell, the four-cell Ag-MH battery will weigh about 1.485 kg (3 lbs.). This battery offers an energy density of 200 × 16 mWh/g or 3.2 Wh/g, which comes to 3.2 kWh/kg per cell. For a four-cell battery, the energy density will be about 12.8 kWh/kg, which is perhaps more than the drone needs. This clearly demonstrates that the Ag-MH battery will be best suited for electronic drone applications and as well as for certain UAV applications [11].

6.7.2 Battery Requirements for UAVs

Military program managers and field commanders believe that in future military conflicts the UAVs will play a critical role in battlefields and in remote hostile areas. Technical articles published in the June 2009 edition of *Military and Aerospace Electronics* revealed that defense planners are leaning toward the deployment of unmanned vehicles for underwater reconnaissance, mine detection and destruction, and antisubmarine warfare activities. Furthermore, military commanders are considering the deployment of robot vehicles to undertake reconnaissance, surveillance, target acquisition, mine and IED detection, and disposal missions in forward battlefield areas. Some of these vehicles are equipped with deadly Hellfire missiles to destroy hostile targets.

Currently, UAVs are widely deployed in conducting reconnaissance, surveillance, intelligence gathering, target tracking, and attack missions [11]. In case of primary mechanical power failure, the rechargeable batteries aboard these UAVs must be able to provide electrical energy for propulsion. Under these conditions, the battery must provide electric energy for propulsion as well as for the electronic sensors and devices operating aboard the UAV. During missile attack missions, the laser requires high electrical energy from the rechargeable battery so that it can illuminate the target and allow the Hellfire missile to home in on and destroy the target. In attack mode, a sealed Ni-Cd battery pack with output power close to 1 kW will be required, which will be sufficient to power the laser, infrared (IR) camera, IR imaging sensor, compact jam-resistant global positioning system (GPS) receiver, and miniature navigation system, and jam-resistant communications system. This type of battery pack can be discharged to 100% DOD. This battery provide constant current, followed by a trickle charge. The battery requires about 3 h for rapid charge and less than 1 h for rapid charge. These batteries deployed by military aircraft have operating temperatures ranging from –40 to +70°C. The latest Ni-Cd batteries have demonstrated an operating life close to 10 years and a storage life exceeding 15 years. On the basis of these performance capabilities, sealed Ni-Cd batteries are best suited for UCAV applications, for which reliability, portability, longevity, and safety are of paramount importance in conducting military missions over hostile territory and battlefield areas.

In the future, more sophisticated electronic and electro-optic sensors will be available for complex UAV missions. Defense scientists and program managers are seriously considering the deployment of shortwave infrared (SWIR) imaging sensors for precision terminal guidance capability; a wristwatch size, high-security ultra-high-frequency communication system; and a jam-resistant GPS receiver. Scientists working on terminal guidance techniques indicate that a 2D photodiode array, which is sensitive over a 850–1,700 nanometer range, will be most ideal for this particular application. Defense program managers are planning to add a 1-micron target designator to the electro-optic sensor suite. Power requirements for these sensors can be satisfied by a $Zn-Ag_2$ battery, which offers minimum weight and size, high efficiency, and improved reliability. This particular battery has a voltage range of 1.5 to 1.8 V, an operating temperature range from –20 to +60°C, a cycle life of 50 to 100 cycles, a specific energy range of 90 to 100 Wh/kg, and a self-discharge rate of less than 5% per month. The performance characteristics of $Zn-Ag_2$ batteries are most attractive for UAV missions. In summary, this battery is most attractive for UAV applications by virtue of its high specific energy and energy density, proven reliability and safety, and the highest power output per unit weight and volume of all commercially available rechargeable battery systems. Its major drawbacks are high cost and poor performance at very low temperatures.

6.7.3 Batteries for Countering Improvised Explosive Devices

According to a comprehensive technical report published in *IEEE Spectrum* [12], billions of dollars have been spent on defeating IEDs. Most of the operating systems currently deployed to counter IEDs are costly, heavy, and complex. In addition, the batteries used in the current man-pack C-IED versions are not meeting weight, size, and life-cycle requirements. This section identifies the design configuration and performance requirements of the next generation of batteries that will be capable of meeting the weight, size, and performance requirements for anti-IED system applications.

The need for and deployment of anti-IED equipment in the battlefield and hostile territory must not be underestimated, particularly in the presence of terrorist activities and installation of IEDs in worldwide trouble spots. Terrorists have implanted deep-buried and explosively formed penetrators (EFPs) in roadside locations, which are not visible to the eye. The difficulty of countering these IEDs includes the physical, electromagnetic, and chemical environments in which they are deployed. The road-installed IEDs can be detected using a handheld sensor. An IED consists of compact power source, a triggering device, and a detonator. The power source is usually a battery. The function of this battery is to supply enough electrical energy to the detonator (a blasting a cap) to activate the main charge. Once the IED explodes, it can cause tremendous damage to the nearby infrastructure and bodily harm or even death to the person who accidently triggered the IED.

6.7.3.1 History of Property Damage and Bodily Injury to Soldiers

The chronological history of property damage and bodily injury to soldiers operating in hostile territories is extremely important. Research studies conducted by the author indicate that the projected yearly global IED events reached approximately 4,171 during 2006, 4,516 during 2007, and 5,042 during 2008 [13]. These IEDs have claimed the lives of soldiers who were searching for IEDs to diffuse them. Insurgents have used garage door electronic openers and personal cell phones as remote radio-controlled triggering devices. Soldiers have experienced brain injury, amputation, life-threatening mental illness, and even death because of IED explosions. Even the Mine-Resistant Ambush-Protected (MRAP) vehicles and armored Humvees are not safe. IED devices have killed thousands of innocent civilians in Iraq and Afghanistan. MRAP vehicles have been badly damaged by the tremendous explosive power of IEDs. Anti-IED equipment has been designed to minimize property loss and injury to soldiers and currently is being used in hostile areas.

6.7.3.2 Anti-IED Techniques to Minimize Property Damage and Injury to Soldiers

The ability to counter IEDs requires advanced systems capable of neutralizing or destroying the IEDs before they do harm to or kill battlefield personnel. Studies performed by the U.S. Department of Defense (DoD) indicate that fatalities from roadside bomb attacks in Iraq dropped by roughly 90% in 2008 over the previous year (2007). This improvement is due to the deployment of MRAP vehicles and armored Humvees, improved intelligence and surveillance capabilities, use of C-IED detection equipment, and deployment of high-technology explosive-ordinance-disposal (EOD) robots capable of detecting and disabling IEDs [13].

Quasi-portable anti-IED equipment has been designed and developed for soldiers to detect and diffuse roadside IEDs. According to military experts, electronic warfare, of which the C-IED mission must be considered a part, is now viewed as a core competency by the Army. Mobile anti-IED equipment carrying high-power lasers and dismounted or portable anti-IED systems have been deployed on hazardous roads and agricultural alleys to detect and destroy IEDs implanted by the insurgents.

6.7.3.3 Battery Performance Requirements for Dismounted Anti-IED Systems

The batteries deployed by the dismounted anti-IED equipment, which is carried by the soldier on his or her back, must have a minimum weight and size, improved reliability, utmost safety, and excellent portability. The weight and size of the battery must be kept to minimum for the soldier's benefit. Improved and advanced battery technology is critical to meet these weight, size, and power output requirements. In other words, to meet these critical requirements, one has to focus on significant improvements in battery technology. If one can make the battery smaller and lighter, while still providing the same amount of output power or more, great strides will have been made in improving the portable anti-IED systems and dismounted or portable systems. In the case of mounted anti-IED systems, the ability to jam the threat signals without interfering with military communications, civilian radio, or mobile phones is the critical performance requirement. The anti-IED system can be classified into three following categories:

- Dismounted anti-IED system
- Mounted anti-IED system
- Fixed-site anti-IED system

Brief trade-off studies performed by the author on the battery performance requirements indicate that the fixed-site anti-IED system has no strict weight and size requirements for the battery because portability is not a critical requirement for the rechargeable battery. For the mounted anti-IED system, moderate weight and size requirements for the battery are desirable. Dismounted anti-IED equipment, however, has strict weight, size, and power output requirements for the battery. The author has limited this discussion only to the dismounted anti-IED system, because this equipment is widely carried by soldiers on their backs. Therefore, portability is of paramount importance.

Careful review of the battery requirements seems to indicate that a sealed Ni-Cd battery is best suited for dismounted anti-IED equipment. This particular battery offers a nominal voltage of 1.2 V, a high mechanical strength, high charge efficiency, reliable long-term operation, excellent portability, longevity exceeding 10 years, high battery capacity close to 13 Ah, and high electrical efficiency. This battery weighs less than 8 lbs., it can be discharged to 100% DOD with no compromise in performance, and it provides constant current, followed by a trickle charge. The fast-charge duration is less than 1 h, and the rapid charge requires less than 3 h. It has a storage life more than five years at room temperature with no structural or performance degradation. This particular battery is widely used in airborne and space applications, for which weight, size, reliability, structural integrity, and safety are of paramount importance.

If further reduction in weight and size is desired, $Zn-Ag_2$ batteries can be used for anti-IED applications but at a higher battery cost. This battery is attractive for this application because of its high specific energy and energy density, proven track record of reliability, excellent portability, remarkable safety, and highest power output per unit weight and volume of all commercially available batteries. Their major disadvantages include low cycle life between 50 and 100, poor performance at –20°C, and high procurement cost. These batteries have demonstrated battery capacity exceeding 150 Ah at 28 V.

The latest research on battery technology reveals that the lithium-iron-phosphate ($LiFePO_4$) battery could qualify as a candidate for C-IED applications. But reliable information on its safety, reliability, cost, and portability are not fully known. Some battery designers are actively seeking this needed information.

Anti-IED equipment must deploy effective waveforms to optimize the radio-frequency (RF) jammer capability necessary to neutralize or eliminate the threat posed by the roadside-installed IED devices. Ways and means must be improved to optimize and improve the RF jamming technology with a particular emphasis on the jamming-to-signal ratio. Defense research and development activities have focused on development of RF jammers for a variety of jammer platforms, such as ground vehicles, portable systems, UAVs, and UGVs. The RF jammer performance is strictly determined by the IED system's ability to prevent wireless activation from a distance. Geometries between the jammer, IED, and the detonating device are strictly dependent on the wireless technology and remote control–trigging technology. In brief, the RF jammer or the anti-IED system must use a high jamming-to-signal RF-triggering signal to neutralize the activation capability of the remote-controlled RF signal.

6.8 Conclusion

This section briefly summarizes rechargeable battery types and their performance requirements for military applications with an emphasis on reliability, safety, longevity, and portability. Rechargeable batteries and the performance requirements most ideal for C-IEDs, UAVs, UGVs, UUWVs, electric power modules for special forces operating in remote and isolated locations, covert communications systems, and other military equipment applications are discussed with an emphasis on weight, size, reliability, and cost. Performance characteristics and major benefits of

Al-air rechargeable batteries for military applications are identified. Performance capabilities and limitations of saline rechargeable batteries are briefly discussed. Performance parameters and structural aspects of bipolar Ag-MH batteries best suited for various military applications are summarized with an emphasis on reliability and electric efficiency over a wide temperature range. Applications of rechargeable Ag-Zn batteries for specific defense applications are identified, for which high energy density, light weight, and compact packaging are the principal requirements. Disposal requirements for various batteries deployed by various branches of military to comply with EPA laws and guidelines are identified. Performance comparison data on midsize rechargeable batteries, such as Ag-Zn, Ni-Cd, and Ni-MH batteries are provided for the benefit of prospective customers and design engineers. Rechargeable batteries incorporating thin-film technology, microtechnology, and nanotechnology are identified for specific military applications, for which compact packaging, light weight, and consistent performance are the critical requirements. Materials for anodes, cathodes, and electrolytes and their characteristics best suited for high-power thermal batteries are identified with a particular emphasis on longevity, reliability, and structural integrity under harsh operating environments. Performance parameters for thermal batteries best suited for military equipment are summarized with principal considerations of safety, reliability, and consistent power output level over a wide temperature range. Performance requirements for high-power rechargeable batteries for UUWVs are summarized with a major emphasis on safety, survivability, and reliability under various diving conditions. Types of anodes, cathodes, and electrolytes for high-power batteries for possible applications in military electronic equipment are identified with an emphasis on efficiency, energy density, and survivability under harsh battlefield conditions. Performance requirements for high-power rechargeable batteries widely used by armored vehicles and other battlefield vehicles are briefly defined with a particular emphasis on survivability, safety, and reliability under harsh thermal and mechanical environments.

Adverse effects of thermal parameters on the flowing electrolyte or nonstationary electrolyte are discussed in detail with an emphasis on the quality of electrolyte over a wide temperature range. Power output variations as a function of temperature and discharge duration for Volta stack rechargeable batteries with 10 kW and 20 kW ratings are provided for the benefit of the readers. Effects of temperature and DOD on the thermal conductivity and the specific heat of various electrolytes used in thermal batteries are discussed. Performance requirements for high-power battery systems are designed to provide electrical energy in the case of sudden failure of a commercial power plant or under an emergency situation. Design aspects and structural details of high-power vanadium-based redox batteries are given. Battery requirements for electronic drones equipped with Hellfire missiles are defined with an emphasis on weight, size, reliability, and safety under harsh battlefield environments with no compromise of attack mission. Battery requirements for dismounted anti-IED jammers, mounted anti-IED jammers, and fixed anti-IED jammers are defined with an emphasis on battery output power level. The selection of complex waveforms is needed to provide the high jamming-to-signal ratio to effectively jam an RF-triggering signal. A high jamming-to-signal ratio is essential to neutralize the IED's destructive power and to save the lives of military personnel operating in the battlefield.

References

1. E. Howard, "Achieving the information advantage," *Military and Aerospace Electronics*, 21, no. 7 (July 2010), pp. 24–30.
2. Editor-in-Chief, "Power management of integrated circuits," *Power Electronics Technology* (April 2011), p. 24.
3. B. M. L. Rao, R. Cook et al., "Aluminum-air batteries for military applications," *Proceedings of the IEEE* (1993).
4. David Reisner, Martin G. Klein et al., *Bipolar Silver-Metal-Hydride Cell Studies: Preliminary Results*, Danbury, CT: Electro Energy, Inc.
5. Curtis C. Brown, "Long life, low cost, rechargeable Ag Zn battery," IECEC Paper No. EC-57, *American Society of Mechanical Engineers Proceedings* (1995), pp. 245–248.
6. A. R. Jha, *MEMS and Nanotechnology-Based Sensors and Devices for Communications, Medical and Aerospace Applications*, Boca Raton, FL: CRC Press, Taylor and Francis Group (2008), pp. 344–345.
7. Christina D'Airo, "Outlook," *Electronic Products* (January 2011), p. 12.
8. Stephen Kauffman and Guy Chagnon, "Thermal battery for aircraft emergency power," *Proceedings of the IEEE* (1992), p. 22.
9. P. Chenebault and J. E. Planchat, "Lithium-liquid cathode for under water vehicles," *Proceedings of the IEEE* (1992), pp. 81–83.
10. G. Pistogi, *Batteries Used in Both Portable and Industrial/Vehicular Applications*, London: Elsevier Publishing Co. (2005), p. 66.
11. David Schneider, "Drone aircraft," *IEEE Spectrum* (January 2011), pp. 45–52.
12. Glen Zorpette, "Countering IEDs," *IEEE Spectrum* (September 2008), pp. 27–32.
13. Brendan P. Rivers, "Road hazards: Countering the IEDs in Iraq," *Military Microwaves Supplement* (August 2008), pp. 22–32.

Chapter 7

Batteries and Fuel Cells for Aerospace and Satellite System Applications

7.1 Introduction

This chapter is dedicated to rechargeable batteries and fuel cells incorporating advanced electrolyte technology for possible applications in aerospace equipment and space systems. Requirements for secondary or rechargeable batteries widely deployed in aerospace platforms such as commercial aircraft, fighter aircraft, unmanned air vehicles (UAVs), electronic attack drones, missiles, airborne jamming equipment, and helicopters will be defined. Performance requirements for batteries used in communications satellites, surveillance, reconnaissance, and target-tracking satellites, and space-based sensors to monitor upper and lower atmospheric parameters will be established with an emphasis on weight, size, reliability, and conversion efficiency. The performance parameters of rechargeable batteries best suited for a specific airborne platform will be identified with a particular emphasis on reliability and safety under severe vibrations, shocks, and thermal environments.

In the case of batteries specifically deployed for starting an aircraft or helicopter or any other airborne moving platform, the battery power output must be capable of instantly generating the required torque the moment the starting switch is turned to the *on* position.

Studies performed by the author reveal that sealed lead-acid batteries (SLABs), thermal batteries, sealed nickel-cadmium (Ni-Cd) batteries, lithium-ion batteries, and aluminum-air (Al-air) batteries are best suited for aerospace applications for which reliability, high power capability under harsh operating environments, longevity, and high energy density are the principal design requirements. The studies further indicate that Al-air batteries offer high energy density exceeding 500 Wh/kg over alternative lithium and other rechargeable batteries. Using alkaline electrolytes, the Al-air batteries can be deployed in satellite communication applications because of their self-contained and high-portability capabilities. These batteries could provide a cold-start capability within 30 min. from –40°C temperature, a power output from 10 to 400 watts (W), and an energy density as high as 450 Wh/kg. Design modeling and current test data available seem to demonstrate that the sealed Ni-Cd batteries and SLABs are most ideal for satellite communications, unmanned ground and airborne vehicles, and other military applications.

Research studies performed by the author on thermal batteries indicate that these batteries are best suited for providing emergency power for military and commercial aircraft. In the 1980s, comprehensive research and development activities were directed toward the development of high-capacity thermal batteries for aircraft emergency power and aircraft seat ejection system applications. Later, Lucas Aerospace was asked by defense scientists to refine the design of thermal batteries specifically to supply aircraft emergency power with high efficiency, reliability, safety, and portability. Significant research and development efforts were undertaken to develop the lithium-aluminum/ferrous-sulfide ($LiAlFeS_2$) thermal battery and this particular battery came to be regarded as the most ideal option to provide military aircraft emergency power. Thermal batteries were later designed to provide the required power with a fast rise time of less than 1.5 sec. over an environmental temperature range from –40° to +80°C. These batteries were further improved for minimum maintenance and hazardous. The thermal batteries provide direct current (DC) power for the electrohydraulic pump (EHP) drive and the DC emergency bus bar.

To provide ultra-high reliability and independent operational control, two thermal battery types were developed and refined for optimum and reliable performance. One thermal battery was to be used for the EHP application, which requires a square wave current pulse load for its entire operating life. The second thermal battery was designed to provide power to the DC emergency bus bar and was required to meet a constant power output for its entire operating life. The operating life requirement was the same for both thermal batteries. Both of these thermal batteries have met the vibration, shock, and all other applicable military specifications. Specific structural and critical performance parameters will be described in Section 7.8 on thermal battery classification. Ordinance and nonordinance applications will be identified with an emphasis on performance capabilities and limitations. No other battery can outperform the $LiAlFeS_2$ thermal battery in terms of its high reliability and ultra-long shelf life, which is why this battery is highly considered for military aircraft emergency power.

Sealed lead-acid batteries are also attractive for both commercial and military aircraft applications. Comprehensive tests performed by the manufacturers indicate that the cycle life of such batteries is relatively low, as with all other types of secondary or rechargeable batteries. The cycle life is strictly dependent on the depth of discharge (DOD) and ROC. Despite their low cycle life, which is common to all types of rechargeable batteries, these batteries are widely used in commercial and military aircraft applications. Performance characteristics and shortcomings of these batteries will be described later in this chapter. Other rechargeable batteries for commercial and military applications will be discussed. Preliminary cost-effective studies seem to indicate that no other type of rechargeable batteries can outperform lead-acid batteries in terms of superior long-term reliability, ease of recycling, starting torque, and cost.

7.2 Rechargeable or Secondary Batteries for Commercial and Military Aircraft Applications

This section describes the various rechargeable or secondary batteries for commercial and military aircraft applications, including helicopters and UAVs, with an emphasis on electrical, mechanical, and thermal performance levels; reliability; safety; cycle life; and procurement cost. Performance capabilities and limitations of various rechargeable batteries, such as SLABs, sealed Ni-Cd batteries, LiAlFeS$_2$batteries, and Al-air batteries, for commercial and military aircraft applications will be summarized with an emphasis on reliability, safety, cycle life, and power loss per month during storage.

7.2.1 Sealed Lead-Acid Batteries for Commercial and Military Applications

SLABs were first manufactured in 1859, and since then, these batteries have been widely used in automobile and commercial and military aircraft applications. These batteries come in a 6 volt (V) or 12 V voltage rating and are most ideal for low-power device applications. There are two types of lead-acid batteries: high-power SLABs [1] and low-power valve-regulated lead-acid batteries (VRLABs) [2]. The former are widely used by automobiles, trucks, jeeps, aircraft, helicopters, and other automotive and telecommunications systems, whereas the latter are used by computers, cellular phones, and other low-power portable devices. The VRLAB has a capacity of 1.2 Ah and weighs around 300 grams (g). This particular battery uses a limited amount of immobilized electrolyte. The SLAB is generally referred to as a portable energy system, whereas the VRLAB is considered to be a stationary battery, which is recognized as an uninterruptible energy source with greater capacity.

In actual practice, the VRLAB is considered to be a low- to medium-capacity battery and is strictly designed for portable applications such as remotely located telecommunications systems, railroad warning devices, and so on.

The chemistry is the same for both batteries. The positive electrode is made from lead oxide (PbO_2), the negative electrode from lead (Pb), and the electrolyte from a concentrated sulfuric acid (H_2SO_4) aqueous solution. The chemical reaction equations for the lead-acid battery can be written as follows:

$$\text{Negative electrode: } [Pb + HSO_4] = [PbSO_4 + H^+ + 2e] \tag{7.1}$$

$$\text{Positive electrode: } [PbO_2 + 3H^+ + HSO_4^- + 2e] = [2H_2O + PbSO_4] \tag{7.2}$$

$$\text{Overall reaction: } [Pb + PbO_2 + 2H_2SO_4] = [2PbSO_4 + 2H_2O] \tag{7.3}$$

In all secondary batteries reported in this book, the discharge process proceeds from left to right, and the charge process proceeds from right to left.

Comprehensive technical discussions in this chapter will be limited to SLABs. These batteries are widely deployed by automobiles, trucks, automotive systems, and aircraft. In particular, SLABs are extremely reliable in starting engines over a wide temperature range and require minimum maintenance. These batteries are most popular and widely used because of their high affordability, reliability, portability, and safety. Typical characteristics of SLABs are summarized in Table 7.1.

Customer surveys indicate that most customers favor high energy density and compact size over longevity, regardless of applications in some cases. These characteristics are valid for a 6 V SLAB.

Table 7.1 Typical Characteristics of Sealed Lead-Acid Batteries with a 6 V Output Voltage

Characteristics	Typical Values for 6 V
Energy density (Wh/kg)	30–50
Cycle life (to 80% of the initial capacity)	200–320
Fast-charge time (h)	8–12
Self-discharge per month at room temperature (%)	5
Nominal cell voltage (V)	2
Operating temperature (°C)	–20 to +60
Maintenance required (months)	3–6
Typical 6 V battery cost (US$)	25
Cost per cycle (US$)	0.10

In the case of a 12 V SLAB, higher values of energy density, fast charge time, cell voltage, and battery cost will be found, whereas other characteristics will remain unchanged. Preliminary estimates reveal that for a 12 V SLAB, the theoretical energy density will be about 250 Wh/kg. If the weight of sulfuric acid (H_2SO_4) is taken in the calculations, the actual theoretical energy density value will be reduced to 172 Wh/kg, which amounts to a reduction of 25% in the energy density. Modeling studies of energy and power output capability as a function of battery design parameters, which are capable of providing the required battery performance level, must be undertaken to achieve realistic values of energy density and power output.

7.2.1.1 Optimum Charge, Discharge, and Storage Conditions for Lead-Acid Batteries

Optimum charge, discharge, and storage conditions remain the same for both SLABs and VRLABs. The optimum conditions for these lead-acid batteries are briefly summarized as follows:

- *Typical charge conditions*: Constant voltage of 2.4 V is applied followed by float charging at 2.25 V. Float charging can be prolonged, if necessary. A fast-charge method is not possible. A slow charge is required for 14 h to maintain the nominal charge over a long duration. Rapid charge, if required, takes roughly 10 h.
- *Discharge condition*: The limit is about 80% of DOD.
- *Storage condition*: The lead-acid batteries must be stored at full charge level. Storing below the terminal voltage of 2.10 is not recommended, because it produces sulphation, which is known as a neutralizing acidic agent.

7.2.1.2 Pros, Cons, and Major Applications of Lead-Acid Batteries

The following are pros for lead-acid batteries:

- Most ideal for heavy-duty use
- Excellent long-term reliability under harsh operating conditions
- Cost-effective performance
- Easy and least expensive to recycle

The following are cons for lead-acid batteries:

- Relatively low cycle life
- Low energy density
- High self-discharge in flooded batteries

Lead-acid batteries can be used for the following applications:

- Automotive applications, such as scooters, cars, and trucks
- Portable avionic equipment
- Lighting equipment where no commercial power is available
- Telecommunications systems

7.2.1.3 Life Cycle of SLABs for Aircraft Applications

Both energy density and life-cycle parameters are of critical importance for commercial and military aircraft applications. As far as energy density is concerned, higher values are possible with reduced battery weight and higher battery capacity. The improvement in life cycle is not easy, because its value is strictly dependent on the charge rate and DOD. To obtain meaningful life-cycle data, batteries must be tested under various charge and discharge conditions and also at various operating temperatures. To obtain such data, extensive laboratory testing is required, which will be time consuming and expensive. In other words, to visualize the effects of charge rate and DOD on the battery cycle life of SLABs, extensive laboratory test data are essential.

Because these tests could involve several such batteries, the test data obtained can exhibit wide variations in the life-cycle value due to the effect of charge current, effect of discharge current, effect of reconditioning, and overall analysis of cycled batteries. Although this inherent variability can mask the results to a certain degree, definite trends from the data still can be considered reliable.

7.2.1.4 Effect of Depth of Discharge on Life Cycle of the Lead-Acid Battery

The experimental data obtained by the SLAB designer for aircraft applications are summarized in Table 7.2 at a fixed current of 50 amperes (A). These test date indicate that cycle life is significantly higher at the lower DOD and is lower at high

Table 7.2 Impact of Depth of Discharge on Life Cycle of Lead-Acid Batteries at 50 A Charge Current

Depth of Discharge (%)	Life Cycle (cycles)
20	3,858
40	1,824
60	816
80	324
100	15

Table 7.3   Effect of Charge Current Limit on the Life Cycle of the Lead-Acid Battery at 100% Depth of Discharge

Charge Current Limit (A)	Number of Cycles above 15 mA
10	31
20	64
30	87
40	124
50	149

DOD. According to these results, at 20% DOD, the average life cycle is close to 4,000 cycles, and at 100% DOD, the life cycle is only 148 cycles, which is reduced by 27 times. The battery should be operated at lower DOD values, if high cycle life is desired.

These data indicate the trend in life cycle as a function of DOD at the charge current of 50 A. It will be interesting to see how the life cycle is affected as a function of the charge current limit at 100% DOD. The effect of charge current limit on the life cycle at 100% DOD is evident from the data presented in Table 7.3.

7.2.1.4.1 Impact of Charge Current Limit

It is important to understand the effects of charge current limit on the life cycle of the lead-acid battery. To determine the effects of charge current limit on the life cycle requires comprehensive laboratory tests as a function of charge current at 100% DOD. Some battery manufacturers have obtained test results on lead-acid batteries as a function of the charge current limit at 100% DOD. (The data collected on lead-acid batteries are summarized in Table 7.3.)

The values of the life cycle of a battery depend on the DOD expressed in percentage and the charge current limit at 100% DOD. The lead-acid battery selected for the laboratory tests had a capacity of 15 Ah. The values of the parameters shown in Tables 7.2 and 7.3 will not be same if the battery capacity has a different capacity rating other than 15 Ah.

It is evident from the test data presented in Table 7.3 that at low values of charge current (I_c), the cycle life is significantly reduced. It can be estimated from these test results that a life cycle of eight cycles is possible at a charge current of approximately 2.25 A, where a life cycle of 150 cycles is obtained at a charge current of 50 A. If a life-cycle versus charge current limit is obtained at 100% DOD over the range of the charge current values, its slope will be about three cycles. In other words, every ampere increase in charge current limit extends the cycle life by three cycles, which is obtained by dividing 150 cycles by the charge current of 50 A at 100% DOD. This result can be verified by the data summarized in Table 7.3. This linear relationship would be not expected to hold true at higher charge current values, because an asymptotic relationship most likely would be reached as the battery reaches its maximum charge acceptance—that is, its constant potential charging. Furthermore, the correlation between the number of cycles and charge current limit may not be as pronounced at lower DOD levels. Additional laboratory test data may be required to determine the exact relationship between the two critical performance parameters, namely the life-cycle and charge current limit.

7.2.1.4.2 Lead-Acid Battery Capacity Recovery following Recharging or Reconditioning

Battery capacity recovery can be achieved by applying one or two constant current reconditioning charges. In almost all cases, the recovered battery capacity appears to exceed the end-of-life rating, so that a second life can be possible. The presence of second life indicates that the cycling regimes are not causing the battery cells to wear out, but rather, a reversible fading of the battery capacity has occurred. An indication that the battery is below its rated capacity shows that the battery requires immediate maintenance.

For example, a battery cycled using a lower ampere charge current limit of 2 A can achieve a first life of, say, 60 cycles and a second life, say, of nine cycles. Therefore, in this battery test, the second life is found to be much shorter than the first life, and the battery requires maintenance service. Another battery cycled using a higher charge current limit of 50 A can achieve a first life, say, of 85 cycles and a second life exceeding 134 cycles. In this particular battery, the second life is longer than the first life, and the battery does not require further maintenance service. If the end-of-discharge voltage is depressed by approximately 2 V, it indicates the presence of a weak battery cell and the need for a new cell.

7.2.1.4.3 Causes for the Reduction of Battery Capacity

In the case of lead-acid batteries, chemical reactions take place during the charge and discharge processes. Material scientists believe that a minimum degradation of the positive and negative electrodes is possible during these charge and discharge cycles. But a dense layer of lead monoxide (PbO) with a porous layer of lead sulfate ($PbSO_4$) can be seen in the grid interface of the positive electrodes or plates. This verifies that battery capacity loss is caused by the passivating of the positive grids. The capacity loss, however, can be recovered following the cycling regiment described in the previous section. In other words, when the passivating layer is sufficiently reduced by the reconditioning charge, it will allow for a substantial recovery of the positive plate capacity. Both electrodes must use pure grid materials to avoid deposition of impurities on the electrodes, which could further degrade battery performance. Some research scientists believe that passivating is due to a thick porous layer of $PbSO_4$ or due to a thin dense layer of tetragonal PbO. But some material scientists believe that passivating is due to the combination of these two compounds. Regardless of the source of passivating, the passivating layer causes the active material to become electrically insulated or isolated from the grid, thereby limiting the capacity available for the discharge cycle. Essentially, the nature of the passivating layer is strongly affected by the type of grid alloy and by the electrolyte additives. For example, the presence of tin in the grid alloy and the presence of phosphoric acid in the electrolyte solution will minimize the passivation effects.

Chemical scientists feel that capacity fading can be minimized by using higher current charge levels, because these current levels promote the formation of a more porous surface layer on the positive grid. These scientists further feel that even at charge current levels, the passive layer eventually builds up to the point at which the discharge capacity can be severely limited.

If maximum cycle life from the SLAB is the principal performance objective, battery maintenance engineers point out that the DOD must be kept as low as possible, but the charge current limit should remain as high as possible. Increasing the charge current limit requires an increase in the size of the charging source, which could be expensive. Therefore, an important trade-off exists between the size and the cost of the charging source and the cycle life of the lead-acid battery. The optimum system design will be one that produces the best compromise between these two trade-off parameters. Lead-acid designers conclude that deep-cycling conditions must be followed, if optimum battery performance is desired in terms of life cycle, reduced passivation effects, and long-term preservation of battery capacity under a variety of storage conditions. These recommendations will be most useful for SLABs for military aircraft, helicopters, drones, and UAVs for which weight is not a critical requirement.

7.3 Aluminum-Air Batteries for Aerospace Applications

Al-air rechargeable batteries were designed and developed around 1992 for aerospace and other military applications. Portable Al-air rechargeable batteries were developed using both saline and alkaline electrolytes for various military and aerospace applications [3]. These batteries may not be suitable for small aircraft, small UAVs, or compact drone applications, because liquid electrolyte must be added just before use. This is the major disadvantage of these batteries.

7.3.1 Performance Capabilities and Limitations of Al-Air Batteries

Designers of rechargeable batteries claim that the Al-air battery offers a very high energy density exceeding 500 Wh/kg in a dry, unactivated state over alternative lithium and other rechargeable batteries. These batteries can be transported as lightweight collapsible batteries, and the electrolyte can be added at the site of operation before us. Deployment of these batteries in large commercial aircraft is possible, because the battery can be transported or carried in a dry and collapsed state. Saline electrolyte batteries offer a very high energy density of close to 500 Wh/kg and are best suited for applications for which water is readily available. As far as the high energy density is concerned, it is second only to the lithium-air battery. The alkaline electrolyte, however, is most ideal for applications for which moderate power ranging from 50 to 100 W/kg is required and when the alkaline solution can be added just before use. Both battery types can be stored in a dry state and have a long storage life. The battery's inability to operate an aluminum-cathode electrode at its thermodynamic potential is a serious performance limitation. Despite this limitation, the energy density of this battery, which typically varies from 200 to 400 Wh/kg, exceeds that of most other rechargeable battery systems. The key items required in the mass production of these batteries are as follows:

■ A commercially manufactured aluminum alloy that is widely available, which operates at high columbic efficiency over a wide range of current density
■ A low-cost, high-performance air electrode that can be manufactured both in quality and quantity

7.3.2 Impact of Corrosion on Al-Air Battery Performance as a Function of Anode Current Density

The cathode material and its properties play a critical role in the Al-air battery performance. Material scientists have come up with different alloys for the cathode to reduce anode corrosion as a function of anode current density. Material development records kept by these scientists indicate that the ALCAN alloy developed by Alu Power Canada Ltd. in 1990 offers the lowest corrosion current density at less than 2% of the anode current density range of 15 to 1,000 mA/cm^2. For this new alloy, the amount of energy converted into corrosion is significantly less than 2% of the total energy produced over most of the output energy range. This is a key factor because the corrosion not only results in a reduction of battery efficiency but also produces hydrogen during the chemical reaction. This hydrogen must be expelled from the battery system, if higher battery efficiency and lower corrosion current density are desired. The alloy developed with reduced hydrogen content by the same company in 1989 exhibited corrosion current density (mA/cm^2), ranging from 20 to 40% over the same anode current density range. An anode alloy developed in 1960 demonstrated a corrosion current density exceeding 38% of the anode current density range of 200 to 800 mA/cm^2. Significant improvements in the reduction of corrosion density and corresponding improvements in Al-air battery efficiency are evident over the time frame from 1960 to 1992.

7.3.3 Outstanding Characteristics and Potential Applications of Al-Air Rechargeable Battery Systems

The outstanding characteristics of the Al-air battery system can be summarized as follows:

- Its energy density exceeds 450 Wh/kg.
- It has a demonstrated output power greater than 430 W.
- It is highly compatible with a modular design configuration.
- It has demonstrated a cold-start capability at −40°C in less than 30 min. Preliminary studies performed by the author seem to indicate that no rechargeable battery exists that has demonstrated a cold-start capability at −40°C in less than 30 min.
- Its most salient features include quietness of operation, long storage life, excellent portability, minimum maintenance requirement, high energy density, medium to high power output capability, quick recharging, and long-discharge periods.
- The battery system offers a modular design configuration, which is most ideal to provide temporary electrical power for mobile shelters in remote areas with no commercial power, command and covert communication centers, and portable field medical facilities. This battery can be deployed in air commercial and military aircraft, UAVs, and anti-improvised explosive device (IED) systems.

A modular battery system has been designed, developed, and tested for a UAV, which has demonstrated a nominal power of 1.6 kW and a peak power of 4 kW. Another highly modular and extremely portable alkaline Al-air battery system consisting of 10 cells for an aircraft application was designed in 1992 by the same company. This particular 12 V, 12 h modular battery demonstrated an output power for a military aircraft application with a discharge period exceeding 12.5 h, as illustrated in Figure 7.1.

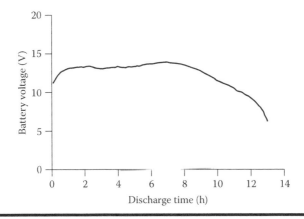

Figure 7.1 **Discharge characteristics of a modular aluminum-air battery.**

This 12 V battery has demonstrated excellent output voltage performance over a discharge time of close to 12 h [3].

7.4 Long-Life, Low-Cost, Rechargeable Silver-Zinc Batteries Best Suited for Aerospace and Aircraft Applications

Studies undertaken by the author reveal that the silver-zinc (Ag-Zn) rechargeable batteries offer the highest output power (watts) and energy density (watt-hours) per unit volume and mass [4]. Because of these two parameters, the Ag-Zn batteries are widely used in aerospace and defense applications for which high reliability, long life, low cost, long calendar life, compact packaging, and minimum weight are of critical importance. The studies further reveal that these batteries have demonstrated cell reliability exceeding 0.9999 probability of success in most military aerospace applications. Because these batteries produce small amounts of gases during normal operation, they must be housed in a vented battery housing.

7.4.1 Vented Secondary Batteries Best Suited for Aircraft and Aerospace Applications

Performance surveys of various rechargeable batteries indicate that the vented Ag-Zn, Ni-Cd, and lead-acid batteries offer unique performance in terms of watts/kg, watt-hour/kg, cycle life, and calendar life. These secondary batteries are best suited for both commercial and military aircraft and other aerospace applications. Their outstanding performance characteristics are summarized in Table 7.4.

Note that the values of the performance parameters summarized in Table 7.4 are accurate within ±5%. The parameters shown here are for batteries developed in

Table 7.4 Unique Performance Characteristics of Secondary Vented Batteries Most Suitable for Commercial and Military Aircraft and Other Aerospace Applications

Characteristics	Rechargeable or Secondary Batteries		
	Silver-Zinc	*Nickel-Cadmium*	*Lead-Acid*
Output power (W/kg)	630	165	110
Energy density (Wh/kg)	110	58	30
Cycle life (cycles)	60–100	1,000–2,000	300–600
Calendar life (years)	2–3	3–6	2–4

the last decade using advanced design concepts and material technologies available in this time frame.

7.4.2 Typical Self-Discharge Characteristics of an Ag-Zn Battery

The author feels that the Ag-Zn battery terminal voltage will fluctuate as a function of ambient temperature and the battery rated capacity. Comprehensive examination of published literatures provides estimated variations in the terminal voltage of this battery as a function temperature and the battery rated capacity, which are summarized in Table 7.5.

It is evident from the data summarized in Table 7.5 that the terminal voltage remains fairly constant at the ambient temperature of 100°F or 40°C. At lower temperatures, however, its terminal voltage drops rapidly when the rated capacity varies from 25 to 75%.

7.4.3 Safety, Reliability, and Disposal Requirements for Ag-Zn Batteries

This section briefly discusses other performance features of the Ag-Zn battery, such as safety, reliability, and disposal requirements. Ag-Zn batteries do produce small amounts of gases during their operation, and these batteries must be housed in the vented enclosures to maintain high operational safety.

This particular battery is best suited for aerospace and defense applications because it offers the highest power (watts) and energy (watt-hours) per unit volume and mass. This battery has demonstrated its suitability for aerospace, defense, and manned suborbital flight applications, for which severe operating conditions exist. The weight of an Ag-Zn battery is roughly 50% of an Ni-Cd battery and close to 25% of a SLAB. This battery has demonstrated high reliability in space environments. The battery cell has demonstrated a reliability of 99% or better in aerospace and defense applications.

Table 7.5   Terminal Voltage of an Ag-Zn Battery as a Function of Temperature and Rated Battery Capacity (V)

Battery Rated Capacity (%)	Ambient Temperature (°F)		
	−20	+40	+100
25	1.28	1.47	1.51
50	1.30	1.46	1.50
75	1.22	1.45	1.51
100	∞	1.43	1.49

Some secondary batteries such as Ag-Zn batteries are not disposed of but are claimed at the end of their useful life. The battery manufacturers offer a service for reclamation and precious metal can be recovered. A trace amount of mercury is embedded with the zinc oxide if the AG-Zn cell is used to suppress outgassing during normal battery operation. There are disposal requirements for certain secondary batteries deployed in aerospace and military programs, and readers or battery designers must be aware of such requirements.

7.4.4 Typical Battery Voltage Level and Cycle Life

Rechargeable batteries can be charged over a very short time or a very long time, depending on the charge rate. According to battery designers, slow-charge rates offer a technique that is capable of maintaining battery performance over relatively longer durations. Typical battery voltage levels as a function of charging time and at a slow-charging rate for 1.8 V rechargeable batteries are summarized in Table 7.6

The rated capacity of a rechargeable battery decreases with the increase in cycle life. In other words, the rated battery capacity decreases as the rechargeable battery gets older. Table 7.7 provides the estimated rated capacity values of a rechargeable battery as a function of cycle life and the rated capacity values.

Table 7.6   Typical Charge Voltage as a Function of Charge Time

Charge Time (h)	Battery Voltage Level (V)
0	1.50
1	1.50
2	1.50
4	1.52
6	1.58
8	1.82
10	1.84
12	1.85
14	1.86
16	1.86
18	1.88
20	1.90

Table 7.7 Typical Cycle Life of a Rechargeable
Battery versus Rated Capacity of the Battery

Cycle Number	Percentage of Battery Rated Capacity
0	100
5	95
10	87
20	83
30	79
40	72
50	63
60	61
70	58
80	54
90	51
100	50

Normally, a battery's rated capacity is 100% when it is brand new, but its rated charge goes down as the battery gets older.

7.5 SLABs for Commercial and Military Aircraft Applications

Despite their excessive weight, SLABs have been used by the commercial and military aircraft applications for a long time. These batteries have demonstrated improved cycle life as a function of ROC, DOD, and operating temperatures. Although the cycle life of SLABs is relatively low, their service life of three to six years can be reached before replacement is necessary. These batteries consist of 6 V and 12 V modules with capacity ratings exceeding 18 Ah at the 1 h rate. Each cell is generally composed of 10 negative plates and 9 positive plates, with a porous fiber glass separators surrounding each plate. The grids used by the battery are made of pure lead and the electrolyte contained in the pasted active material, with only a small amount contained in the separator. The modules are sealed by means of a one-way vent valve. The typical module's overall dimensions are 25 cm in length, 15.2 cm in height, and 9.7 cm in width. The maximum battery weight is about 19 lbs.

7.5.1 Performance Aspects of SLABs

This section briefly describes outstanding performance aspects and characteristics of SLABs. These rechargeable batteries are being manufactured by EaglePicher Company, Hawker Energy, and others. The first two battery suppliers are currently undertaking research and development efforts for the next generation of SLABs with significant improvements in charge and discharge characteristics, cycle life, and DOD. In addition to SLABs, nickel-hydrogen and lithium-ion batteries are getting serious consideration for deployment in energy lighting, military telecommunications system, military aircraft, aerospace, and satellite applications.

7.5.1.1 Performance of the EaglePicher Battery Ultralife UB1-2590

EaglePicher's Ultralife UB1-2590 is widely deployed in various military system applications. This rechargeable battery is a SLAB [5] and is widely used for its unique performance capabilities, including ultra-high reliability and safety while operating under severe thermal, mechanical, and aerodynamic environments. Typical applications and performance parameters can be summarized as follows:

- *Applications*: Aircraft, military communications, robotics, unmanned underwater vehicles (UUVs).
- *Class*: 9HAZMAT using absorbent glass material (AGM) technology.
- *Maintenance requirements*: Batteries using AGM technology are carefree.
- *Price*: 2010 estimated price of $494.
- *Design characteristics*: Highly reliable, light weight, high-energy density, no memory loss effect, wide operating temperature range.
- *Electrical performance parameters*
  - *Voltage characteristics*: Two sections with nominal voltage of 14.4 VDC.
  - *Operating modes*: Series mode with voltage range of 20–33 VDC and parallel mode with voltage range of 10–16.5 VDC and maximum voltage of 16.5 VDC.
- *Battery capacity*: 6 Ah (series mode) and 13.6 Ah (parallel mode).
- *Maximum discharge current magnitude*: 6 A (series mode) and 12 A (parallel mode).
- *Maximum pulse discharge*: 18 A for 5 sec. (series mode) and 36 A for 5 sec. (parallel mode).
- *Typical operating mode capacity*: 2.06 Ah.
- *Typical energy density*: 143 Wh/kg.
- *Cycle life*: Greater than 300 cycles.
- *Operating temperature range*: –32° to +60°C.
- *Storage temperature range*: –32° to +60°C.

- *Self-discharge*: Less than 5% per month.
- *Charging requirements*: 6.6 VDC with charging current level not exceeding 3 A over a charging duration not exceeding 3 h.
- *Shock and vibration*: Meets all such military specifications.
- *Special material used*: Wolverine advanced material is used for brake shims and solution dampers to minimize vehicle vibrations.

7.5.1.2 SLAB from EaglePicher for Commercial Applications

The CF-12V-100 SLAB rechargeable battery is widely used by UPS vehicles. UPS has found this battery to be most cost-effective and highly reliable. Furthermore, most UPS vehicles use this particular battery for its safety and reliability. Major highlights of this SLAB [5] are summarized as follows:

- *Terminal voltage*: 12 VDC
- *Battery capacity*: 100 A h
- *Energy density*: 37.5 Wh/kg
- *Physical dimensions (length, width, and height in inches)*: 6.73 × 6.73 × 8.43
- *Estimated retail price*: $299

7.5.2 Test Procedures and Conditions for SLABs

SLABs are widely used in several commercial and military applications and require frequent charging and maintenance procedures. The following recommendations must be followed if cost-effective performance and maximum cycle life are critical performance requirements:

- DOD should be kept as low as possible, but the charge current limit should be as high as possible to achieve optimum battery life.
- The high-charge current technique requires charging equipment with increased weight, size, and cost.
- Trade-off studies must be performed between the size and weight of the charging equipment and the cycle life of the SLAB.
- At higher charge currents, capacity fading is minimized because of the formation of a more porous surface layer on the positive grid.
- When the passive layer builds up, the discharge rate is severely limited.

7.5.3 Impact of Charge Rate and Depth of Discharge on the Cycle Life of SLABs

Conventional lead-acid batteries are suited for various aerospace and other military applications, excluding compact aircraft applications, for which safety and reliability

Table 7.8   Impact of Charge Current on
the Cycle Life of the SLAB

Charge Current (A)	Cycle Life (cycles)
10	32
20	63
30	85
40	124
50	153

Table 7.9   Impact of Depth of Discharge on
Cycle Life of the SLAB

Depth of Discharge (%)	Cycle Life (cycles)
20	3,980
40	1,825
60	745
80	338
100	169

Source: From Vutetakis, D. G., and H. Wu, "The
effect of change rate and depth of discharge on the cycle life of sealed lead-
acid aircraft batteries," *Proceedings of
the IEEE,* © 1992 IEEE. With permission.

are prime considerations. For a fighter or bomber or close air-support applications,
however, SLABs are best suited because these rechargeable batteries offer high electrical performance, impressive safety, and ultra-high reliability under severe thermal
and mechanical environments. Because charge rate and DOD are the critical performance parameters of the rechargeable SLAB, their impact on cycle life requires a better understanding in terms of these parameters. Cycle life data as a function of charge
current and as a function of DOD are shown in Tables 7.8 and 7.9, respectively [5].

7.5.4 Life-Cycle Test Conditions

If reliable test data as a function of charge current and duration, discharge current, and terminal voltage are of prime importance, then life-cycle test conditions
must be strictly followed. Discharge-cycle test conditions are defined in terms of

Table 7.10 Recommendations for Charge-Cycle and Discharge-Cycle Test Conditions at 100% Depth of Discharge

Discharge-Cycle Test Conditions	Charge-Cycle Test Conditions	Rest Duration (h)
15 A at 9.2 V	2.25 A for 12 h	1
15 A at 9.2 V	5.0 A for 12 h	1
15 A at 9.2 V	10 A for 12 h	1
15 A over 1 h	15 A for 3 h	0
15 A over 1 h	25 A for 3 h	0
15 A over 1 h	50 A for 3 h	0
15 A over 1 h	50 A for 2 h	1
22.5 A over 40 min.	50 A for 2 h	1

Source: Vutetakis, D. G., and H. Wu, "The effect of change rate and depth of discharge on the cycle life of sealed lead-acid aircraft batteries," *Proceedings of the IEEE*, © 1992 IEEE. With permission.

ampere-hours at a specified battery terminal voltage. Discharge current at 100% DOD is of critical importance, and therefore, discharge tests are performed at 100% DOD. Charge-cycle tests involve current and test duration, and discharge-cycle test conditions involve current, duration, and battery terminal voltage. Test parameter recommendations for current-cycle and discharge-cycle test conditions for SLABs are identified in Table 7.10.

For a 20% DOD cycle test condition, a current of 20 A over a 9 min. duration is involved. For the charge-cycle test, a current of 50 A over 51 min. is involved with no rest duration. When these test conditions are satisfied, the SLABs are fully qualified for their specific application and are ready for deployment in a specified electrical system.

7.6 Thermal Battery for Aircraft Emergency Power and Low-Earth-Orbiting Spacecraft

Studies performed by the author indicate that thermal batteries using lithium-aluminum/ferrous-sulfide (LiAl/FeS$_2$) are most ideal for supplying emergency electrical power for electronic systems and hydraulic power in fighter aircrafts. A compact version of this advanced thermal battery has been deployed in low-earth-orbiting (LEO) spacecraft. Critical components of this advanced thermal battery are shown in Figure 7.2. These thermal batteries can be deployed in nonordinance applications for which high power capability and long life are the principal requirements.

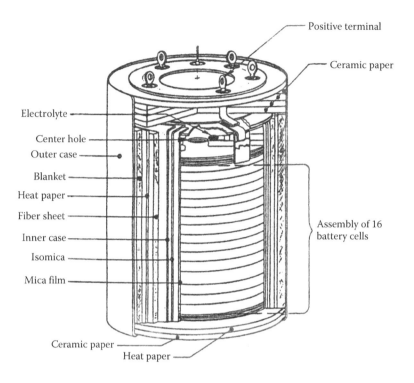

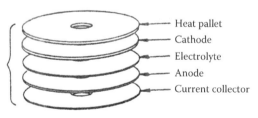

Cell configuration in a mini-assembly housing

Figure 7.2 Elements of the thermal battery designed for space applications. (From Embrel, J., M. Williams et al., "Design studies for advanced thermal batteries," *Proceedings of the IEEE*, © 1992 IEEE. With permission.)

These batteries have been designed particularly for military weapon system applications such as military aircraft, missiles, and space communications systems operating over temperatures ranging from –40 to +80°C. The company has developed two high-power thermal batteries using LiAl/FeS₂ systems. This battery is intended to supply emergency electronic and hydraulic power in a high-speed fighter aircraft [6].

7.6.1 Performance Capabilities of LiAl/FeS₂ Thermal Batteries

The first thermal battery (TB1) was designed to provide DC power for an EHP drive application, and the second thermal battery (TB2) was specifically developed to provide DC power to the DC emergency bus bar that supplies electrical power to aircraft electronic devices and sensors. These thermal batteries have the following unique performance capabilities [7]:

- *Applications*: Ordinance and nonordinance (aircraft seat ejection mechanism system).
- *Fast rise time*: Less than 1.5 sec. over the temperature range from –40 to +80°C.
- *Current pulse*: TB1 provides square wave current pulse.
- *Minimum and maximum voltage*: 26 VDC/38 VDC (for TB1); 23 VDC/31.5 VDC (for TB2).
- *Operating life*: 9 min. for both batteries.
- *Capacity*: 25 Ah (TB1) and 11.7 Ah (TB2).
- *Average or CW power*: 5.1 kW (TB1) and 2.0 kW (TB2).
- *Nonoperation vibration*: 10.75 g root mean square for 1.5 h per axis × three axes (for both batteries).
- *Operation vibration*: 10.75 g RMS for 10 min. per axis × three axes (for both batteries).
- *Bump shock*: 1,000 shocks × 3 axes, 40 g, 6 m.sec, 1/2 sine (for both batteries).
- *Standard shock*: 30 g, 3 axes, 2.5 m.sec, 1/2 sine/7.5g RMS, 3 axes, 40 m.sec, 1/2 sine.

7.7 Rechargeable Batteries for Naval Weapon System Applications

The lithium-thionyl chloride (Li-SOCL₂) battery is specifically designed for multiple defense system applications. This battery uses a metallic lithium anode and a liquid cathode comprising a porous carbon current collector filled with thionyl chloride (SOCL₂). The battery cell is rated at 3.6 V and has a cylindrical shape with spiral electrodes to meet specific power levels and a bobbin construction for prolonged discharge. This battery uses a flowing electrolyte and has a conductivity of 13 ms/cm, which is considered ideal for a flowing electrolyte. The electrolyte composition is modified through the discharge cycle [7]. The Li-SOCL₂ cell possess built-in high-energy capability because of its high nominal voltage of 3.6 V. The bobbin construction feature can deliver energy density in excess of 760 Wh/kg with a current capacity of 18.5 Ah at 3.6 V in D format. This particular battery has been developed for an underwater propulsion system, and the battery offers high rate capabilities and a long-duration discharge associated with a high theoretical energy density. Underwater systems like minisubmarines, torpedoes, and antitorpedo weapons require batteries with a long-duration discharge in the range of 20 to 30 min. [7].

7.7.1 Performance Characteristics of Li-SOCL$_2$ Batteries

The performance characteristics and unique capabilities of this particular battery can be summarized as follows:

- *Output power*: Close to 600 kW
- *Open-circuit voltage*: 3.95 VDC (maximum)
- *Maximum discharge rates*: 200 mA/cm^2
- *Typical battery stack*: 40 cells, but maximum cells can be inserted in the stack
- *Output power level*: 20 kW (maximum) at the end of 20 min. discharge
- *Battery output power requirement*: 600 kW (typical) for a heavy-weight torpedo

7.8 Thermal Battery Design Configurations and Requirements for Launch Vehicle Applications

Preliminary studies undertaken by the author reveal that thermal batteries using advanced titanium alloys could meet stringent structural, performance, and reliability requirements. Batteries for launch vehicle applications require stringent electrical, thermal, and mechanical specification requirements. Such a battery must use advanced materials for its critical elements. Thermal batteries incorporating titanium alloy as early as 1999 were designed and developed for space applications, and these batteries can operate reliably in a high-acceleration environment [8].

7.8.1 Design Aspects and Performance Capabilities of Advanced Thermal Batteries

Two designs for the advanced thermal battery were evaluated by SAFT engineers and were found to be satisfactory for space applications. One battery was designed for a lightweight pulse operation with a lifetime operation of more than 300 sec., and the second battery was a thermal battery specifically designed for a high-acceleration environment. This battery was designed to meet a 200,000 g launch-force acceleration [8]. Critical components of this advanced thermal battery are illustrated in Figure 7.2.

The use of titanium alloy provided reduced weight, greater mechanical strength, and significantly improved thermal properties, which yielded extended thermal lifetimes, leading to remarkable improvements in the battery's specific energy. The baseline chemistry used was LiAL/LiCL-KCL/FeS$_2$, which provided the cell terminal voltage of 1.6 VDC at a nominal current density of 200 mA/cm^2. Advanced concepts were used in the enclosure design for both batteries to reduce stress, and the use of titanium in the case material provided reduced weight and high mechanical strength. The use of titanium for the case material improved the discharge life at low temperatures, reduced battery weight, and enhanced the specific energy for the pulse-power application. Titanium was also used as a case material for the high-acceleration thermal battery. Construction details of the thermal battery involving several critical elements are shown in Figure 7.2. The cell stack is contained and crimped under a specified pressure in an inner steel can. As illustrated in the figure, the inner can assembly is wrapped in a heat source equipped with thermal insulation, which is placed in an outer steel can.

The battery voltage leads are spot welded to a cover with glass-sealed terminal feed-throughs (see Figure 7.2). The inner case, outer case, and cover are made from 1010 high-strength steel material. The battery cover is hermetically sealed and welded to the outer can. This thermal battery design meets all space mission requirements for acceleration, shock, vibration, and launch-force magnitude. Important elements of this rechargeable battery along with construction details of the 1.6 V cell are shown in Figure 7.2. The outer battery case or housing typically contains 16 such 1.6 V cells, which provides an overall battery voltage of 22.5 V if the cells are connected in a series. The nominal current density of the cell is about 200 mAcm^2.

7.8.2 Unique Performance Capabilities of Thermal Batteries

The most notable electrical and mechanical aspects and benefits of using titanium alloy for the battery case can be summarized as follows:

- The use of titanium alloy offers a significant reduction in the weight of the battery.
- Improvement in both mechanical strength and integrity.
- Unique thermal properties of advanced titanium alloy provide extended thermal lifetimes.
- Substantial increase in specific energy of the battery.
- Most ideal for a high-acceleration environment.
- Offers a stress-handling capability to meet the 200,000 g launch force with no structural damage to the battery. This particular capability of the thermal battery makes it most attractive for underwater missile and satellite launch applications.
- Reliability of this battery approaches better than 95% even under severe mechanical and thermal environments.

- Heat paper made from zirconium-based material ($Zr/BaCrO_4$) yields the most efficient thermal insulation.
- This particular battery consists of 16 cells to meet operating voltage requirements.
- Cylindrical battery design configuration offers compact packaging in launch-vehicle applications.
- ISOMICA materials are used when thermal insulation is required, particularly when the thermal battery is located in an ultra-high temperature region.

7.9 High-Temperature Lithium Rechargeable Battery Cells

These high-temperature lithium rechargeable battery cells were developed under the sponsorship of the Electronics and Power Sources Directorate of the U.S. Army Research Laboratory, Fort Monmouth, New Jersey. The high-temperature rechargeable lithium cells use a solid solution of lithium germanium oxide (Li_4GeO_4) and lithium vanadium oxide (Li_3VO_4) as the lithium-ion-conducting solid electrolyte. This solid electrolyte complex formula can be written as $Li_{3.6}Ge_{0.6}V_{0.4}O_4$. This solid electrolyte is reported to have the lowest electrical conductivity of about 0.08 S/cm at a temperature of 300°C. This solid-state electrolyte has an electrochemical stability window of more than 4 VDC. The solid-state rechargeable battery cell includes an Li-Al alloy anode and a chemical vaporization deposition (CVD) thin film of titanium disulfide as the cathode. This cell has demonstrated a good discharge rate capability up to a current density of 20 mA/cm² and a cycle life exceeding 200 cycles. Further improvements are needed in the rate capability of the solid-state cells to develop these cells as viable power sources for high-rate pulse-power applications.

7.9.1 Unique Performance Parameters and Design Aspects of Solid Electrolyte Cells

Outstanding performance capabilities and design aspects of solid electrolyte cells are briefly summarized as follows:

- These cells employ a 50–50 weight percentage of Li_4GeO_4 and Li_3VO_4 pellets as the lithium-ion-conducting solid electrolyte.
- The incorporation of lithium ion in a solid electrolyte yields higher pellet conductivity and significantly improved high-rate performance of the solid-state rechargeable battery cells.
- The cells can be made of either an Li-Al or lithium-silicon alloy as the anode and a CVD thin film of titanium disulfide as the cathode.

- These cells have demonstrated an open-circuit voltage of better than 2.1 V at a temperature of 300°C.
- The cells can be discharged at current densities up to 0.5 A/cm².
- These solid-state cells are best suited for high-rate pulse-power applications.
- Although no reliable data are available for these cells, the author can predict reasonably high reliability, especially at operating temperatures as high as 300°C.

7.10 Solid Electrolyte Technology for Lithium-Based Rechargeable Batteries

Technical reports and papers on solid electrolytes indicate that the use of plasticizers in polyether-based solid polymer electrolytes (SPEs) has become the most promising approach to increase room-temperature ionic conductivity, leading to significant improvement in the performance of rechargeable batteries [9]. Some plasticizers have detrimental effects on the interfacial impedance and cycle life of SPE-based rechargeable lithium batteries. This section identifies the development of some rechargeable lithium batteries using this SPE technology that has minimized or eliminated these detrimental effects.

7.10.1 Critical Role of Solid Electrolytes

Rechargeable battery designers initially used linear-homo-polymers of ethylene oxide (EO). The initial design of polymer-ethylene-oxide (PEO) batteries demonstrated good performance over the temperature range of 60° to 125°C. At these temperatures, an ionic conductivity of 0.001 S/cm was observed for fully amorphous solid polymer electrolytes. The use of these plasticized networks could yield ionic conductivities better than 0.001 S/cm even at room temperature (27°C).

7.10.2 Improvement in Performance Parameters of Lithium Rechargeable Batteries

This section focuses on the improvement in performance parameters of lithium-based rechargeable batteries due to the use of solid electrolytes. According to the battery scientists, these performance improvements are strictly due to the development of a reversible nonmetallic lithium source anode, namely the lithium-carbon (Li-C) intercalation compound; the implementation of high-voltage cathode materials, such as lithium cobalt oxide ($LiCoO_2$) and lithium nickel oxide ($LiNiO_2$); and the development of liquid electrolytes based on solvent mixtures capable of yielding high rates and significantly improved electrochemical stability in oxidation. The use of Li-C-based anodes also improves the safety of the rechargeable lithium-based batteries. Some important attributes associated with the use of solid polymer electrodes are summarized in Table 7.10, and these attributes would signifi-cantly improve the battery's performance characteristics compared with a bat-tery using a liquid electrolyte. Both the energy density and cycle life are strictly associated with the gain in the chemical process and the electrochemical sta-bility that can achievable by using an immobile solvent matrix. Improvements in power density can be achieved with SPEs of high ionic conductivity, and even more improvements can be achieved with a high cationic transport num-ber combined with low interfacial impedance levels. Research and develop-ment studies performed by material scientists indicate that, at present, neither high- nor low-temperature SPE technologies can provide the desired attributes. The studies further indicate that the high-temperature SPEs are based on EO homopolymers and copolymers. The ionic conductivity of these SPEs is close to 0.001 S/cm when they are fully amorphous. The material scientists believe that the amorphous state or condition is dependent on the cationic number, which varies between 0.4 and 0.6.

The material scientists further believe that most polyethers have a low static dielectric constant as well as low donor and acceptor numbers. These factors promote the formation of ion aggregation in pairs, eventually leading to larger aggregates. The low-temperature SPE formulations generally include a large number of low-molecular-weight plasticizer, which is considered to be a good solvent for the lithium salt and is highly mobile in the reduced viscosity SPE. It is possible to obtain ionic conductivity, cationic transport number, and elec-trochemical stability similar to liquid electrolytes for the low-temperature plas-ticized SPEs.

Performance enhancement in lithium-based batteries using liquid electrolytes is of significant importance and must be investigated. On the basis of research studies performed by the author, the enhancement in performance parameters due to various attribute categories have been identified and are briefly summarized in Table 7.11.

Table 7.11 Enhancement in Lithium-Based Batteries Using Liquid Electrolytes

Attribute Category	Requirements for Parameter Enhancement
Power density	High surface-area-to-volume ratio
	High cationic transport number (typical number is 0.5)
	Monolithic low-impedance cell
Electrical energy density	High-voltage composite bipolar cathode
Cell cycle life	Immobilized electrolyte
Manufacturing capability	High-speed continuous process
Safety and reliability	No liquid electrolyte, but solid-state matrix essential

Table 7.12   Impact of Lithium-Chloride-Oxide Concentration on Ionic Conductivity

LiCLO$_4$ Concentration Weight (%)	Ionic Conductivity (S/cm)
0.5	0.00080
1.0	0.00300
5.0	0.00308
10.0	0.00315
15.0	0.00312

7.10.3 Impact of Lithium Chloride Oxide Salt Concentration in the Solution of Liquid Plasticizer on Room-Temperature Ionic Conductivity

The effect of lithium-chloride-oxide (LiCLO$_4$) concentration in the solution of plasticizer solution on ionic conductivity deserves serious consideration. Data presented in Table 7.12 reveal that when the concentration weight exceeds by 1%, higher variation in the ionic conductivity can be expected.

Comprehensive examination of the tabulated data seems to indicate that the lowest ionic conductivity is possible when the salt concentration is less than 1% by weight. The data further indicate that the increase in ionic conductivity is negligible even when the salt concentration increases to 15% by weight [9].

7.11 Rechargeable Batteries for Electronic Drones and Various UAVs

It is desirable to describe the principal functions of UAVs and the types of sensors and devices that are deployed by the vehicles before identifying the battery requirements for each vehicle. Performance requirements and design aspects for rechargeable batteries deployed by various military vehicles will be defined with an emphasis on reliability, safety, weight and size, and life cycle. A preliminary survey made by the author on the procurement of rechargeable batteries indicate that there are at least three distinct U.S. suppliers of high-power batteries for military applications: EaglePicher, Energex, and SAFT American.

7.11.1 Performance Requirements for Batteries Best Suited for Electronic Drone Applications

A drone is a pilotless micro air vehicle (MAV), which is fully remote-controllable and reusable as a combat aircraft. Essentially, the drone provides almost all of the functions and capabilities similar to those available from a fighter or surveillance aircraft. Images collected by the drone are transmitted back to the operator on the ground via a video link. In addition, the drone is proliferated in a miniaturized form and can operate at very high altitude in the battlefield or hostile area without being detected by the enemy radars. The drone is generally used for reconnaissance or sustained surveillance functions. If someone wants to use it for an attack mode, it can be equipped with miniaturized air-to-ground missiles called Hellfire missiles.

When a drone is used for surveillance or reconnaissance missions, microelectrical mechanical system (MEMS) or nanotechnology-based sensors or electromechanical devices are given serious consideration. The drones are generally equipped with tiny gyroscopes, accelerometers, compact and reliable global positioning system (GPS) receivers, and airspeed sensors. The flying power is provided by a compact gasoline engine, and the electrical power for the miniaturized sensors and devices is provided by Ni-Cd batteries. Next-generation drones may use lithium-ion batteries using carbon nanotube (CNT) technology. Research and developments activities undertaken by MIT scientists indicate the CNT-based cathodes in lithium-ion batteries would yield an output power 10 times that of the conventional lithium-ion batteries, because the lithium storage reaction on the surface of the CNTs is much faster than the conventional lithium intercalation reactions. The MIT laboratory-tested CNT electrodes have demonstrated remarkable stability over time, because no detectable change in the CNT material was noticed after 1,000 cycles of charging and discharging. The battery will have to provide additional power if a laser illumination source is required to guide air-to-air missiles to their targets. Electrical power required for such a source could be as high as 1,500 W over a period not exceeding a couple of minutes, because the target may have been detected and tracked.

7.11.2 Rechargeable Battery Requirements for UAVs, Unmanned Combat Air Vehicles, and MAVs

This section describes the primary functions of UAVs, unmanned combat air vehicle (UCAVs), and MAVs. The UAV is widely deployed by the ground operator to collect surveillance and reconnaissance data on hostile territory or battlefield regions. Sensors and electronic devices carried by a UAV were discussed in the last section. UCAVs use a synthetic aperture radar (SAR) to provide surveillance, reconnaissance, and high-resolution images of moving targets of combat importance in a hostile region or battlefield area. The data can be transmitted to ground control center via a video data link. This particular vehicle is deployed for attack missions and can carry a high-capacity, lightweight battery pack consisting of two sealed Ni-Cd batteries each weighing less than 8 lbs. Rechargeable batteries for UAVs and MAVs must be selected with a particular emphasis on weight, size, and reliability.

The UCAV's maneuvering and control functions are handled by the ground operator. The UCAV uses a compact GPS navigation system, infrared (IR) camera, SAR (only in a large UAV that requires high-resolution images of moving targets), video camera, forward-looking infrared sensor to provide precision navigation, electro-optical colored camera, and multispectral targeting system best suited for precision attack missions using laser illuminator, and air-to-air missiles. Most of the electro-optical cameras have a limited amount of tilt and zoom capability. Therefore, a digital stabilization system must be deployed to allow UAV or UCAV to keep the target in the operator's sight. Because the UCAS system carries lots of electro-optical sensors, an illuminating laser, three different cameras, and other electronic devices, a high-capacity and lightweight Ni-Cd or nickel-metal-hydride (Ni-MH) battery will be best suited for this vehicle. Both the UAV and UCAV carry the most common sensors to detect wind speed, wind direction, and other vital targeting information for the benefit of the ground operator. Because the overall weight of next-generation MAVs will be less than 16 lbs., the rechargeable batteries must be extremely compact with a weight not exceeding 6 lbs. Specifically designed Ni-Cd batteries incorporating nanotechnology-based materials are most ideal for MAV applications.

The MAV was especially designed, developed, and evaluated to save lives from the explosion of the improvised explosive devices (IEDs) [10]. The initial version of MAV weighs only 16 lbs. dry and 18.5 lbs. when filled with fuel. Three distinct fuels can be used, namely regular gasoline, jet engine fuel (JP 8), and diesel, depending on the vehicle's endurance for hover-and-stare capability for a period close to 1 h. The MAV's takeoff and landing function with hover-and-stare capability make it the best and most compact airborne UAV to detect and destroy roadside vehicles equipped with explosives and buried IEDs [10].

Several rechargeable batteries are available for UAVs, UCAVs, and MAVs whose performance characteristics are summarized in Table 7.13. The user can select the most suitable battery for the application once he or she performs a brief trade-off study in terms of battery performance, cost, and reliability. Because weight and output power of the battery are the most critical parameters, energy density (Wh/kg) must be given serious consideration in the selection of a rechargeable battery for UAV, UCAV, or MAV application. When a specific battery meets all other performance parameters, battery cost becomes a secondary selection parameter.

Ni-Cd rechargeable batteries are best suited for high-altitude and long-endurance UAV or MAV applications. Ni-Cd batteries offer compact size, lower weight, and high reliability. These batteries are widely used where procurement cost and life cycle are the most critical factors. These batteries allow the UAVs, MAVs, and electronic drones to accomplish their missions over extended durations. Ni-MH batteries are heavier than Ni-Cd batteries and therefore must be avoided for UAV applications, if weight is a critical parameter. In general, both of these batteries are used in portable system applications. If cost is the most critical factor, Ni-MH and lithium-ion batteries are not suitable for certain applications.

Table 7.13   Rechargeable Battery Types for Various UAVs and Their Performance Specifications

	Battery Type			
Performance Parameters	*Nickel-Cadmium*	*Nickel-Metal-Hydride*	*Lithium-Ion*	*Nickel-Hydrogen*
Energy density (Wh/kg)	50–80	60–120	110–160	45–55
Cell voltage (V)	1.25	1.25	3	1.5
Fast-charging time (h)	1	2–4	2–4	2–4
Self-discharge/month (%)	20	30	10	60
Maintenance requirement	30–60 days	60–90 days	Not required	30–50 days
Operating temperature (°C)	–40 to +60	–20 to +60	–20 to +60	–10 to +30
Cycle life (to 80% capacity)	1,500	300–500	500–1,000	>2,000
Estimated battery cost (US$)	50	60	100	65
Estimated cost/cycle (cents)	4	12	14	6
Best suited for:	UAV/MAV	UCAV	UCAC	MAV

7.11.3 Rechargeable Batteries for Glider Applications

According to naval experts, fleets of gliders [11] are deployed to collect data on ocean currents and on acoustic properties that could affect the performance of military sonar systems. Gliders have been used to study ocean circulation, climate events, and marine life, and to track sediment flow off coastal regions during heavy storms. A glider contains a flooded chamber, rudder mechanism, bladder, pump, oil reservoir, pressurized chamber, and batteries for pitch-and-roll control functions. A glider rises by inflating a bladder with oil and sinks by deflating the bladder. The fixed wings of rudder converts the vertical motion of the glider, while the battery-operated mechanisms control the pitch-and-roll motions. According to glider designers, these vehicles are best suited for specific naval applications, for which covert naval operations are desired. The glider has a torpedo-shaped aluminum hull with a length of less than 2 m (6 ft.) and filled with sensors, electronics, and batteries. To move the glider, a pump is required to inject or remove oil from a bladder located in the flooded part of the hull. To ascend, a robot expands the bladder, displacing the water and increasing its buoyancy. To descend, the bladder must be empty. The rudder's fixed wings convert a part of the vertical displacement into a horizontal motion, forcing the glider to travel in a saw-tooth trajectory. To change pitch while climbing or sinking or to turn left or right, the gliders must change their center of mass by shifting the positions of the battery packs. Studies performed by the author indicate that sealed Ni-MH batteries or lithium-ion batteries are best suited for glider applications. Advanced versions of gliders deploy integrated robot technology, which yields a longer operating range close to 4,000 ft. and a battery longevity exceeding 6 months, which are ideal for long-range and long-duration naval missions.

7.12 Rechargeable Batteries for Space-Based Military Systems and Satellite Communications

Battery requirements for military-based space sensors are strictly dependent on mission requirements and mission duration. For example, rechargeable batteries are required to power all of the sensors capable of providing electrical energy during surveillance, reconnaissance, and target tracking missions. Battery requirements for satellite communications equipment in general will be moderate compared with military airborne sensors for specific missions and their durations. Following are the seven distinct characteristics of rechargeable batteries required for their deployment in space system and satellite communications applications:

- Energy performance
- Power performance
- Operating life or longevity
- Safety
- Reliability under severe mechanical and thermal environments
- Uninterrupted power delivery
- Maintenance-free performance

7.12.1 Rechargeable Battery Requirements for Military Space-Based Sensors Requiring Moderate Power Levels

Military space-based sensors have been deployed for more than 4 decades to provide surveillance, reconnaissance, and tracking of space-borne targets such as infrared airborne missiles, rocket plume signatures during launch, midcourse correction, and terminal phases. Military communications satellites are generally designed to provide covert communications between the various assets, such as fighter aircraft, support aircraft, naval vessels, aircraft carriers, and army divisions.

Rechargeable battery power performance is strictly dependent on ambient temperatures. For moderate power outputs, sealed Ni-Cd, Ni-MH, and other batteries

Table 7.14   Impact of Ambient Temperature on Discharge Capacity of a Battery

Temperature (°C)	Battery Discharge Capacity (%)
0	70
10	96
20	100
30	100
40	95
50	82
60	74

can be deployed. Space temperatures would affect the discharged capacity and storage time even of sealed batteries. Table 7.14 illustrates the impact of temperature on the discharge capacity of a sealed Ni-Cd battery as a function of ambient temperature.

The data presented in Table 7.14 reveal that Ni-Cd batteries exhibit slow self-discharge as a function of ambient temperature. It is possible that these batteries can retain 62 to 70% of their capacity after one month. These rechargeable battery stacks consisting of several cells will be well suited for space applications involving passive sensors, such as IR cameras and low-power electro-optical sensors. Sealed Ni-Cd or Ni-MH or lithium-ion battery packs are ideal for moderate-power applications. Performance capabilities of these batteries are summarized in Table 7.15.

Table 7.15   Performance Parameters of Three Rechargeable Batteries with Moderate Power Outputs

Performance Parameters	Ni-Cd	Ni-MH	Li-Ion (BP)
Power density	Excellent	Good	Excellent
Specific energy (Wh/kg)	45–65	60–95	160–200
Cycle life (to 80 % capacity)	1,100–1,500	650–1,100	700–1,250
Self-discharge at 20°C/month (%)	<20	<25	<3.5
Fast-charging time (h)	1	1	2–3
Nominal cell voltage (V)	1.20	1.20	3.6
Operating temperature range (°C)	−20 to 60	−20 to 60	−20 to 65

Note: BP: battery pack.

Table 7.16   Reduction in Battery Residual Capacity as a Function of Storage Time at 45°C

Storage Time (days)	Residual Battery Capacity (%)
0	100
10	76
20	55
30	38
40	29

In addition to these performance parameters, discharge characteristics and residual battery capacity as a function of temperature and storage duration are of paramount importance and must be given serious consideration, regardless of the type of batteries selected for space applications.

Fundamental problems are associated with rechargeable batteries. One dominant problem is long-term storage, which can produce permanent damage to the battery's critical elements, such as seals and separators, particularly at high temperatures. In space applications, the storage periods are typically close to 30 days or more; therefore, residual battery capacity would tend to decrease rapidly as a function of storage time, particularly at higher temperatures. Laboratory tests performed by the suppliers on space-based rechargeable batteries indicate that continuous exposure to 50°C reduces cycle life by 52%. In addition, the residual battery capacity undergoes rapid reduction as a function of storage time at a temperature of 45°C, even when the batteries are designed for space applications. Table 7.16 shows the storage characteristics for a sealed Ni-Cd battery as a function of storage period.

Our preliminary review of sealed Ni-MH rechargeable batteries with moderate capacity indicates that charge is a critical step in determining battery performance and overall life of Ni-MH batteries because of their sensitivity to charging conditions. Performance highlights of Ni-MH batteries are summarized as follows:

■ Faster self-discharge in Ni-MH batteries has been observed compared with Ni-Cd rechargeable batteries.
■ Charge is the most critical step in determining a battery's overall life as well as the performance of this battery, regardless of battery capacity.
■ Ni-MH batteries can deliver between 2 and 40% of the battery's capacity after one month, whereas Ni-Cd batteries can retain approximately 60 to 70% of the capacity.
■ Ni-MH batteries are charged at constant current. Furthermore, the current levels have to be limited to avoid overheating and incomplete oxygen recombination.

- Battery voltage raises sharply around 80% charge because of more significant oxygen evolution.
- The charge process is exothermic in Ni-MH batteries, whereas it is endothermic in Ni-Cd batteries.
- The voltage drop and temperature rise are the indicators of the end of the charge.
- Ni-MH rechargeable batteries will suffer in longevity and performance if repeatedly overcharged.
- In the case of rechargeable batteries, service life is one of the most critical performance parameter. Published reports claim that Ni-Cd rechargeable batteries are considered maintenance-free power sources. It is believed that the Ni-Cd rechargeable batteries, which are used in electric vehicles, have a service life in the range of 10 to 15 years. Even some lithium-based rechargeable batteries have claimed a service life close to 25 years.
- Batteries with long service life are most ideal for UAV, MAV, and UWUAV applications.

7.13 High-Power Fuel Cells for Satellites with Specific Missions

When satellites are required to provide surveillance, reconnaissance, and tracking of airborne targets, advanced thermal batteries using the latest battery technology such as lithium-ion battery packs is required. Fuel cells are available to power the host of microwave, millimeter wave, and electro-optical systems and devices aboard the satellite. Typical drawbacks associated with rechargeable batteries such as discharge rate, safety concern, and disposal issue have compelled system engineers to locate an alternate power source, such as fuel cells. Power system engineers are increasingly leaning toward the deployment of fuel cells for high-power sources [12]. Because of ultra-high-power capability and enhanced reliability, fuel cells were used to provide electrical power for manned spacecraft as early as 1960. Preliminary studies undertaken by the author indicate that zinc-air fuel cells [13] yield electrical energy density in excess of 4 kWh/kg, which is roughly 1,000 times the energy available from lead-acid batteries and three times the energy available from gasoline.

Fuel cells are classified by the types of electrolytes used in the fuel cell module. Four distinct types of fuel cells have demonstrated remarkable performance capabilities as high-power sources. The most prominent fuel cell types are as follows:

- Low-temperature phosphoric acid fuel cell (PAFC)
- High-temperature molten carbonate fuel cell (MCFC)
- Proton exchange membrane (PEM) fuel cell (PEM-FC), which is illustrated in Figure 7.3
- Solid oxide fuel cell (SOFC)

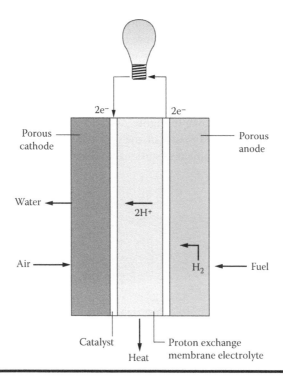

Figure 7.3   Proton exchange membrane-based fuel cell showing the critical elements.

Research and development activities pursued by various fuel cell companies and research organizations indicate that PEM-based fuel cells offer simple design, improved reliability, low operating cost, lower procurement cost, and a small footprint. Dow Chemical Company and Ballard Power Systems played a key role in commercializing the PEM-FC as alternate power-generating sources [13]. The fuel cells have demonstrated the following unique capabilities:

■ The fuel cell technology has demonstrated a proven "stack" technique to accommodate different output requirements with the same cell design.
■ Fuel cells generally have high efficiencies irrespective of cell size and performance characteristics.
■ Fuel cells are easy to site because of extremely low environmental intrusion.
■ Some fuel cells can use a variety of fuels with quick-change-out provision.
■ Specific details on the performance capabilities of various fuel cell designs are briefly discussed in Chapter 1.
■ Fuel cells offer operational advantages, such as electrical energy control, quick ramp rate (which is most desirable for some space-based systems), remote and unattended operations, and high reliability due to an inherent redundancy feature.

Because of this high-reliability capability, these PEM-FCs have been used consistently to provide onboard electrical power for the manned spacecraft as early as 1960. Interestingly, the exhaust product produced is safe drinking water for the astronauts. Within the last few years, the U.S. military has provided significant research and development support for the development of practical fuel cells with high-power capabilities. Fuel cells design configurations using semisolid electrolyte and molten electrolyte are shown in Figure 7.4a and b, respectively. Both fuel cells operate at high temperatures because of the types of electrolytes used. These fuel cells are best suited for deployment as stationary power sources. Such fuel cells [14]

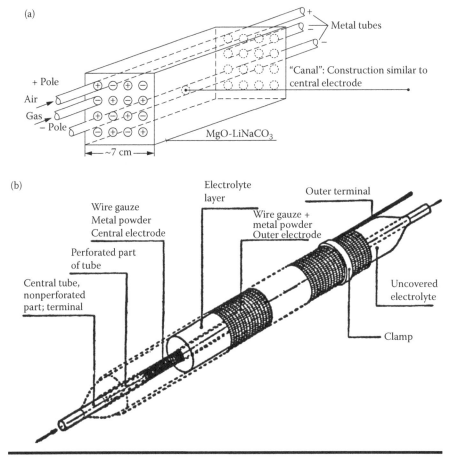

Figure 7.4 (a) High-power fuel cell consisting of MgO-$LiNaCO_3$ semisolid electrolyte containing metal tubes as electrodes as electrodes for air and gas. (b) Critical elements of a high-capacity fuel cell using an electrolyte paste consisting of fine grain solid MgO and molten electrolyte for optimum performance. (From Pettie, C. G., "A summary of practical fuel cell technology," *Proceedings of the IEEE,* © 1963 IEEE. With permission.)

are not suitable for aircraft or space-based systems because of high temperature, excessive weight and size, and frequent maintenance requirements.

Based on these facts, it is reasonable to believe that PEM-based fuel cells may be deployed to power surveillance, reconnaissance, and tracking sensors aboard satellite or spacecraft. High-power sealed Ni-Cd and lithium-based batteries can be just as appropriate for deployment in microsatellites to power the various electronic systems, electro-optical sensors, and high-resolution infrared cameras.

A hydrogen fuel cell known as a mono-skeleton catalyst (MSK) has demonstrated optimum cell performance. The cell was designed and developed by Brown, Bovary and Company of Switzerland around 1959. Swiss scientists have selected the best materials and technology for electrodes. The metal electrodes were selected to obtain an adequate storage capacity in case the cells were subjected to overloading for a long time. The materials selected for the electrodes offered optimum electrochemical properties.

7.13.1 Performance of the MSK Hydrogen-Oxygen Fuel Cell for Communications Satellite Applications

The MSK hydrogen-oxygen (H_2-O_2) fuel cell was first designed, developed, and evaluated by Swiss Scientists. The anode and cathode electrodes were made from silver-based alloys. The alkaline electrolyte was used at 70°C. The scientists found that the electrodes made from alloyed silver provided better cell performance over the electrodes made from pure silver. The fuel cell exhibited hardly any change with time, even when similar cells were operating at full-load conditions for several months. This illustrates that these fuel cells have high reliability. During tests, the anode and cathode spaces must be separated by a polyethylene diaphragm. The Faraday efficiency of this cell is remarkable. Typical cell characteristics of the H_2-O_2 fuel cell with alkaline electrolyte are summarized in Table 7.17.

Table 7.17 Typical Characteristics of an H_2-O_2 Fuel Cell Using Alkaline Electrolyte at a Temperature of 70°C

Current Density (mA/cm²)	Terminal Voltage (V)	Cell Power Output (W)
0	1.13	0
100	0.91	0.817
200	0.78	1.512
300	0.65	1.921
400	0.58	2.212
500	0.42	2.207
600	0.37	2.000

The terminal voltage and power output are from a single H_2-O_2 cell. The cells must be connected in a series to obtain higher power output. To collect data for a meaningful endurance test, the test must involve a large number of H_2-O_2 fuel cells. During the endurance test, the electrode potential can be continuously monitored and the data are available to plot Faraday efficiency.

7.14 Classification of Fuel Cells Based on the Electrolytes

Preliminary analysis by the author on fuel cells indicates that a fuel cell is a device that combines an oxidation reaction and reduction reaction in such a way that generates electricity. Primary and the secondary (rechargeable) batteries do the same thing. But in a fuel cell, both the fuel and the oxidant are added from an external source to react at two separate invariant electrodes. In brief, the fuel cell is an energy conversion device in which chemical energy is isothermally converted into DC electricity.

7.14.1 Performance Parameters of Fuel Cells Using Various Fuels and Their Typical Applications

Fuel cells have been classified into two major categories: high-temperature fuel cells and low-temperature fuel cells. High-temperature fuel cell technology is best suited for stationary plant applications, where they can meet the emergency power requirements in case the main power plant generators are shut down for repairs or schedule maintenance. High-temperature fuels cells also can be deployed to power submarines.

Low-temperature fuel cells are generally used to provide power for public buses, heavy-duty trucks, and electric and hybrid electric vehicles. Regardless of the fuel cell type, the three most critical elements of the fuel cell are the anode, cathode, and electrolyte. Three distinct types of electrolytes are used to design fuel cells, which include the following:

■ *Aqueous electrolyte or fluid-based electrolyte fuel cells*: In such devices, the electrons are liberated during oxidation of the fuel, which passes through an external circuit and provides the electrical energy. Most aqueous electrolyte fuel cells use alkaline electrolyte, and the water is formed at the fuel electrode. Another popular aqueous fuel cell is known as the ion-exchange membrane (IEM) cell, in which the water is formed at the oxygen (air) electrode. The fuel electrode works best when pure hydrogen is used. Current densities in excess of 400 mA/cm^2 have been observed. This IEM fuel cell and its performance capabilities will be described later on under appropriate title. This

IEM fuel cell was used as an electrical power source for the NASA Gemini project. Since then, IEMs have been deployed by spacecraft and surveillance and reconnaissance satellites.

■ *Molten electrolyte fuel cells*: In these devices, molten carbonate electrolytes are used at very high temperatures, ranging from 500° to 750°C. These fuel cells operate on impure hydrogen and do not require an expensive catalyst agent, thereby yielding the relatively cheapest device. These cells are best suited for industrial and commercial applications.

■ *Solid electrolyte fuel cells*: These cells were first designed and developed at the research laboratories of General Electric and Westinghouse companies. These devices use solid fuels and operate at a cell temperature close to 1,100°C. The cell is designed to use natural gas as a fuel for self-starting operation. The cell is enclosed in a heating jacket and the initial heat is supplied by burning the natural gas in the heating jacket. When the cell temperature reaches an appropriate value, the natural gas is fed directly into the cell structure.

The electrolyte used by the fuel cell is a solid gas–impermeable zirconia known as zirconium oxide (ZrO_2). This ZrO_2 is doped with calcium oxide (CaO) to supply enough oxide ions to carry the cell current. The oxidant air or oxygen is bubbled through the molten silver cathode, which is held inside the zirconia cup. At the fuel electrode or the carbon-based anode electrode, the oxide ions are combined with carbon monoxide (CO) and give up their electrons to an external circuit. The cell by-products CO and hydrogen, which are formed in the initial fuel decomposition, are burned outside the cell to keep the fuel cell at operating temperature. The hydrogen is not involved in the electrochemical cell reaction.

Fuel cells using solid fuels initially demonstrated current densities in excess of 170 A/ft² at a DC voltage of approximately 0.7 V. These fuel cells have operated more than 3,500 h without any cell deterioration. The maximum efficiency of such cells with a carbon anode was 35%. Devices designed later using improved materials demonstrated greater efficiency close to 50%.

The fuel cells designed and developed using solid fuel at Kaiser Aluminum and Chemical Corporation in Louisiana have demonstrated the ability to meet the power plant's requirements at 6,000 A/ft.² The estimated current densities exceed 3,000 W/ft.²

7.14.2 Comparing Fuel Cell Parameters

Table 17.8 compares fuel cells in terms of fuels, oxidants, electrolytes, operating temperatures, efficiencies, current power output levels, and specific applications. The performance parameters, such as efficiency, temperature, and current estimates of power output levels, shown in Table 7.18 are accurate within ±5% [13]. More research and development efforts are required in the areas of reliability and safety of the device.

Table 7.18   Comparison of Various Fuel Cells

Performance	Types of Fuel Cells				
	Proton Exchange Member	*Phosphoric Acid*	*Alkaline*	*Molten Carbonate*	*Solid Oxide*
Electrolyte	Polymer	H_3PO_4	KOH/H_2O	Molten salt	Ceramic
Oxidant	Oxygen, air	Oxygen, air	Oxygen	CO_2, air	Oxygen
Fuel	Hydrogen	Hydrogen	Hydrogen	Hydrogen	H_2, CO
Temperature range (°C)	80	195	80–200	650–750	1,000
Efficiency (%)	45–55	40–50	40–50	>60	>60
Power output	150 W/ 10 MW	200 kW/ 10 MW	100 W/ 10 W	>150 MW	>100 MW
Application	EV/HEV, small utility	Small utility	Aerospace	Utility	Utility
Technology maturity (%)	100	85[*]	100	75[*]	85[*]

Source: Gilcrist, T., "Fuel cell to the fore," *IEEE Spectrum*, © 1998 IEEE. With permission.

[*] More research, development, and testing efforts are needed to bring performance to 100%, particularly in the field of reliability and safety.

7.15   Battery Sources for Spacecraft Applications

Studies performed by the author on rechargeable batteries specifically designed for spacecraft operations must meet not only the electrical performance requirements but also the safety and reliability requirements of the devices while operating under severe space environments. The studies further reveal that interphase conversion provides a reliable method for aging cell predictions. Sodium sulfur, sealed Ni-Cd, and nickel-hydrogen (Ni-H$_2$) rechargeable batteries have been approved by NASA administration for deployment in spacecraft.

7.15.1   Application of the First Principle Model to Spacecraft Operations for Aging

The decision of deploy these batteries comes after conducting rigorous laboratory tests for aging using the First Principle Model to spacecraft operations. The principal objective of the application of the First Principle Model is to review the biphasic nickel electrode performance, the effects of aging on hydrogen pressures, the effects of aging on the nickel electrode, and the adverse effects of aging on the cadmium electrode under space environments. These tests revealed the following:

- Single-phase reaction-cycled cell behavior cannot be predicted with a single phase.
- Interphase conversion provides reliable information on film aging.
- Aging cell predictions display typical behaviors, such as voltage fading upon cycling, second voltage plateau on discharge of cycled cells, negative limited behavior for the Ni-Cd cells, and pressure variations in Ni-H$_2$ cells.
- Studies of negative limited Ni-Cd cells have identified causes and effects. The causes include increased positive electrode capacity, fading of cadmium electrode capacity, and changes in the precharge level. The effects observed include higher cell potentials, reduced charge currents, reduced charge efficiency, reduced state of charge (SOC), and divergence of SOC and the voltage in the battery packs.
- A sealed Ni-Cd battery with 9 Ah capacity was first launched in August 1996 and subsequently deployed in several LEO spacecraft. According NASA scientists, the sealed Ni-Cd cells were activated for about 63 months before launch and were maintained on trickle charge at 5°C, whenever possible.

7.15.2 Typical Performance Characteristics of the 40 Ah Sodium-Sulfur Battery Cell

The 40 Ah sodium-sulfur (Na$_2$S) battery cell was developed for space applications by EaglePicher. Following are the typical performance characteristics:

- *Capacity*: 40 Ah
- *Physical dimensions*: length, 9.44 in.; diameter, 1.39 in.; weight, 1.28 lbs. (0.58 kg)
- *Specific energy*: 138 Wh/kg
- *Electrolyte used*: alumina (Al$_2$O$_3$)
- *Cell voltage*: 1.40 V at the start; 1.34 V at 19 Ah capacity; 1.30 V at 20 Ah capacity; 1.25 V at 30 Ah capacity; and1.15 V at 40 Ah capacity

7.16 Conclusion

This chapter identifies various secondary or rechargeable batteries and their potential applications. Lead-acid batteries are considered the oldest working horse among the rechargeable batteries. Design improvements in these lead-acid batteries have been implemented from time to time and these improvements have been briefly mentioned throughout the chapter. The SLAB has demonstrated remarkable electrical performance as well as significant improvements in terms of reliability, safety, and residual discharge rate. Optimum charge, discharge, and storage conditions of SLABs are identified with particular emphasis on device longevity. Ways and means to improve the life cycle of SLABs for aircraft applications are briefly discussed. The impact of DOD on the life cycle of SLABs has been thoroughly evaluated. The author feels that the DOD must be kept as low as possible, if maximum cycle life of the rechargeable battery is the principal goal. Capacity fading is the classical problem for almost all rechargeable batteries as they age. Studies undertaken by the author on capacity fading indicate that this can be minimized or almost eliminated by using higher charge current levels. Performance capabilities and limitations of Al-air batteries most ideal for aerospace applications are summarized with an emphasis on energy density, reliability and safety. Key performance parameters of Al-air batteries, such as columbic efficiency and lost cost, are discussed in great detail. The impact of corrosion on the electrical performance of these batteries as a function of anode current density is identified. Outstanding electrical and structural characteristics of Al-air batteries are summarized. The potential application of this rechargeable battery in anti-IED systems is identified for personnel looking for roadside-buried IEDs because of reduced weight, size, and modular construction. Performance capabilities of Ag-Zn batteries best suited for aerospace and aircraft applications are identified with an emphasis on weight, energy density, reliability, and safety. The author has investigated the benefits of vented or sealed design aspects of rechargeable batteries. Sealed battery designs prevent the loss of electrolyte, whereas vented battery designs allow the safe escape of vapors generated in the battery enclosure or housing. The author has identified the significant performance improvement in vented Ag-Zn and sealed Ni-CD, Ni-MH, and sealed lead-acid rechargeable batteries. Performance improvements in power output level (W), energy density (Wh/kg), life cycle (cycle), and calendar life (years) in Ag-Zn and Ni-Cd batteries and SLABs are observed.

Safety, reliability, and disposal requirements of Ag-Zn batteries are discussed in great detail. The typical charge voltage for Ag-Zn batteries as a function of charge duration is provided. The typical life cycle of rechargeable battery as a function of rated battery capacity is summarized to visualize any indications of the aging effect. Performance characteristics and physical parameters of rechargeable batteries developed by EaglePicher for commercial and military applications are summarized. Memory effect and voltage depression phenomenon associated with lithium-based rechargeable batteries are identified as well as their impact on battery performance. The impact of charge rate and DOD on the life cycle of SLABs has been discussed as well as their effects on the battery's electrical performance. Recommendations for charge-cycle and discharge-cycle test conditions are summarized. Performance characteristics of $LiAl/FeS_2$ thermal batteries best suited for aircraft energy and power applications are summarized with particular emphasis on energy density and uninterrupted battery power. Performance capabilities and the limitations of Li-SOCL$_2$ rechargeable batteries most ideal for naval weapon system applications are described. These batteries are specifically designed for underwater propulsion systems, torpedoes, and antitorpedo applications. Performance requirements and design aspects of the latest thermal batteries for space applications are summarized with an emphasis on reliability, safety, and extended longevity. These batteries use advanced titanium alloys to meet stringent structural, thermal, and reliability requirements while operating under 200,000 g environments.

Performance capabilities and major benefits of high-temperature lithium batteries with solid electrolyte consisting of Li$_4$GeO$_4$ and Li$_3$VO$_4$ are summarized, highlighting reliability at elevated temperatures as high as 300°C. Solid electrolyte technology offers several performance benefits, which are most suited for military applications. The use of plasticizers in polyether-based SPE has demonstrated room-temperature ionic conductivity better than 0.003 S/m, leading to a significant improvement in the electrical properties of these rechargeable batteries. Batteries using SPE technology yield improved high-rate pulse power and enhanced reliability under high ambient temperatures. Battery designers have observed performance improvements in lithium-based batteries using liquid electrolytes. Improvements in power density, energy density, cycle life, reliability, and safety have been observed over long operating durations. Chemical scientists have observed the benefit of LiCLO$_4$ salt concentration in the solution of liquid plasticizers on room-temperature ionic conductivity, which is considered a primary requirement for improved performance of such batteries. Battery performance requirements for electronic drones, UAVs, UCAVs, and MAVs are identified with a particular emphasis on weight, size, reliability, and longevity. Rechargeable battery requirements for anti-IED equipment are summarized with a major emphasis on weight and reliability with zero maintenance. Light weight, compact size, and reliable rechargeable batteries, namely sealed NI-Cd, Ni-MH, and Ni-H$_2$, are most suitable for unmanned vehicle applications. Rechargeable batteries best suited for glider applications are identified.

Battery requirements for spaced-based military systems for surveillance, reconnaissance, and target tracking missions are identified with a major emphasis on longevity, uninterrupted power capability, and maintenance-free operation. Effects of space temperature and radiation on discharge capacity and residual battery capacity are briefly mentioned. Reduction in the residual capacity of an Ni-MH battery as a function of storage duration at a given ambient temperature is specified to determine the aging effects of the battery. Reduction in residual capacity is critical for all batteries recommended for space applications involving long-duration surveillance and reconnaissance missions. Various types of fuel cells for stationary and space-borne applications are discussed with an emphasis on cost, complexity, weight, size, and reliability. Low-temperature PAFCs, high-temperature MCFCs, PEM-FCs, and SOFCs are the most suitable power sources for space-based systems for specific missions and satellite communication equipment.

Performance parameters of low-temperature H_2-O_2 fuel cells with alkaline electrolytes are provided for communications satellite applications. Estimated terminal voltage and output power level of H_2-O_2 fuel cells as a function of current density are provided for space system applications [14]. Performance characteristics of various fuel cells are summarized in terms of fuel, oxidant, electrolyte, temperature, efficiency, power output, and suitability for specific applications. High-capacity batteries and fuel cells are suggested for which high output power is the principal requirement. The critical electrical performance parameters of a 40 Ah Na_3S battery best suited for military applications are summarized.

References

1. Kerry Langa, "What is the best type of battery?" *Electronic Products* (March 2011), pp. 40–44.
2. G. Pistoiria, *Batteries, Operating Devices, and Systems*, London: Elsevier Publishing Co. (2009), pp. 77–78.
3. G. D. Deuchars, *Aluminum-Air Batteries for Military Applications*, Kingston, Ontario, Canada: Alu Power Ltd. (nd), pp. 34–35.
4. Curtis C. Brown, *Long-Life, Low-Cost, Rechargeable Ag-Zn Battery*, Joplin, MO: EaglePicher Technologies, LLC, pp. 245–246.
5. D. G. Vutetakis and H. Wu, "The effect of charge rate and depth of discharge on the cycle life of sealed lead-acid aircraft batteries," *Proceedings of the IEEE* (March 1992), pp. 103–105.
6. Stephen Kauffman and Guy Clagnon, "Thermal battery for aircraft emergency power," *Proceedings of the IEEE* (March 1992), pp. 227–230.
7. P. Chenebault and J. P. Planchast, "Lithium/liquid cathode for underwater vehicles," *Proceedings of the IEEE* (March 1992), pp. 81–83.
8. Janet Embrel, Mark Williams et al., "Design studies for advanced thermal batteries," *Proceedings of the IEEE* (March 1992), pp. 231–232.
9. D. Fauteux and B. Banmet, *Recent Progress in Solid Polymer Electrolyte Technologies*, Cambridge, MA: Arthur D. Little Inc.
10. David A. Fulghum, *Aviation Week and Space Technology* (August 6, 2007).
11. Erico Guizzo, "Gliders to gather data on ocean currents," *IEEE Spectrum* (September 2008), p. 56.
12. Michael J. Riezenman, "Metal fuel cells for transportation," *IEEE Spectrum* (June 2001), pp. 55–59.
13. Tom Gilcrist, "Fuel cell to the fore," *IEEE Spectrum* (November 1998), pp. 35–40.
14. C. Gordan Pettie, "A summary of practical fuel cell technology," *Proceedings of the IEEE* (May 1963), pp. 795–804.

Chapter 8

Low-Power Batteries and Their Applications

8.1 Introduction

This chapter is dedicated to low-power batteries widely used for various commercial applications, such as cameras, smoke detectors, security sensors, cell phones, home phones, medical devices, minicomputers, and a host of a other electronic components. This chapter discusses the battery requirements for low-power consumer electronics and devices. It is important to distinguish between the batteries for low-power applications for which the battery rating varies from nanowatt or nW (10^{-9} W) to milliwatt or mW (10^{-3} W). Power ratings for low-power electronics for which the battery rating vary from 10 to 25 W. In more explicit terms, it can be stated that the low-power applications can vary to cover the range of a few microwatts for watches and cardiac pacemakers and a range of 10 to 25 W for minicomputers or subnotebook computers. Some nanotechnology-based devices or circuits may require electrical power of 100 to 200 nW. Low-power batteries are available in both categories, namely primary batteries and secondary batteries (rechargeable). The primary batteries or cells are often rated at a current that is one-one hundredth of the cell capacity in ampere-hours or C/100, where C stands for capacity, whereas secondary cells are rated at C/20.

This chapter discusses the existing and next generation of low-power batteries in terms of energy density, shelf and cycle life, and other critical performance parameters to give a performance prospective of the various low-power batteries. Energy density and cycle life, in particular, are application dependent and will vary with discharge rate, operating voltage of the device, and duty cycle. This energy variance could lead to confusing conclusions for which a wide range of performance characteristics exist.

Advances in battery technology and materials used have been linked to electronic device applications. Incorporation of microelectronics, microelectrical mechanical systems (MEMS), and nanotechnologies in the design of these batteries could lead to significant improvements in energy density, shelf life, cycle life, and residual leakage. Ultimate size, weight, reliability, and energy density are limited by the chemistry used in the design of batteries. A review of technical papers on low-power batteries reveals that advancements in materials and packaging have resulted in significant changes in older batteries, including alkaline-manganese batteries, nickel-cadmium (Ni-Cd) batteries, and lead-acid batteries. Existing batteries such as primary and secondary (rechargeable) lithium-based batteries, zinc-air (Zn-air) batteries, and Ni-Cd batteries are fully commercialized to meet ever-increasing demands for energy density without affecting weight and size. This commercialization represents the greatest challenge for the battery designers and material scientists.

Performance characteristics of batteries for low-power applications will be somewhat different from the performance characteristics of batteries required for low-power electronics applications. This chapter identifies the major difference between the performance characteristics of batteries deployed for these two different applications. Progress in both battery development and technology is strictly tied to progress in electronics. Low-power electronics applications are referred as microelectronics, whereas low-power electronics are called low-power electronic devices, such as disc players, subnotebook computers, and minicomputers.

8.2 Performance Capabilities of Lithium-Based Batteries for Low-Power Applications

Lithium metal is considered to be the most suitable material for realizing high-energy density batteries, because of its low equivalent weight and built-in constant voltage characteristic. Because of lithium's high reactivity, the material comes with safety concerns, which can be eliminated by using carbonaceous materials. Furthermore, the disposal of lithium rechargeable batteries, irrespective of their capacity, must be followed within industry guidelines and procedures.

The anode, cathode, and electrolyte play a critical role in maintaining the performance of lithium-ion (Li-ion) batteries for low-power applications [1]. The anode uses the spinel (lithium-titanium-oxide [$Li_4Ti_5O_{12}$]) material for its fabrication because it offers the optimum electrochemical performance for the battery. The spinel is considered to be a complex chemical compound. This anode can deliver a very high retention capability in excess of 150 mAh/g or 150 mAh/kg. This kind of energy density and retention capacity is remarkable and can significantly reduce the weight and size of the battery. As a matter of fact, spinel-based anode is widely used in solid-state, plastic, Li-ion batteries for low-power applications [2]. High-potential materials, such as lithium manganese oxide ($LiMn_2O_4$) and lithium cobalt oxide ($LiCoO_2$), are used for the fabrication of the cathode to provide a cell with an operating voltage of 2.5 V. Solid electrolyte is used for the Li-ion-based batteries best suited for low-power applications. Comprehensive laboratory tests undertaken by the battery designers on the electrochemical performance of $LiMn_2O_4$ indicate that cathodes made from this material have demonstrated exceptional stability over a wide temperature range (from $-60°$ to $80°C$). Commercial $LiMn_2O_4$ material can be used to develop polymer electrodes on aluminum foil substrate by deploying a screen-printing deposition technique. An alternating current (AC) impedance method can be used to evaluate the transport properties of symmetrical polymer-electrode and polymer-electrolyte cells. In brief, solid-state, plastic, Li-ion batteries can be fabricated with a cell configuration consisting of a cubic spinel ($Li_4Ti_5O_{12}$) anode electrode, polymer electrolyte, and an $LiMn_2O_4$ cathode electrode.

Material scientists working on the lithium-based rechargeable batteries believe that overcharging these batteries will lead to the deposition of a reactive metallic lithium that could generate flammable gaseous mixers by reacting with the solvent. In rare situations, the overlithiated carbonaceous anode electrode could cause an explosion with flames, leading to a serious safety concern. This safety concern could be reduced through the following measures:

- By introduction of anode electrode materials characterized with higher voltage ratings. New anodic materials that can intercalate lithium at higher voltages include cobalt nitrides (CoN_3), manganese nitrides (MnN_3), and several transition metal oxides characterized by a low intercalation potential.
- By deployment of solid electrolytes with higher stability.

8.2.1 Benefits of Solid Electrolytes in Lithium-Based Rechargeable Batteries

Deployment of solid electrolytes generally requires no added liquid solvents. Therefore, the absence of vapor pressure makes the rechargeable battery most ideal for unique applications, such as space aircraft and satellites.

Scientists working on solid electrolytes believe that polyethylenoxide (PEO), a lithium-salt compound that has been recognized as a liquid-free solid polymer electrolyte (SPE), offers appreciable electrical conductivity at room temperature ($27°C$) because of its unique structure. Low-power battery designers believe that a new approach involving the use of two different polymers in the formulation of the cathode electrode, and of the separator, can significantly enhance the performance of the lithium-based batteries. Essentially, a solid low-molecular-weight polyethylene glycol (PEG) was used as a fast ionic-transport matrix in the composite cathode electrode, and a high-molecular-weight PEO blended with PEG was used in the separator. This particular design configuration achieves high-power, solid-state, lithium-based batteries operating at temperatures as low as 64°C. The electrical performance of such a device at lower temperatures is still satisfactory, particularly for low-power applications.

Assembly of the cells is straightforward. The battery cells can be assembled by sandwiching a polymer electrode disc between the composite cathode and the composite anode. The diameter of the composite electrodes is less than 8 mm. The electrochemical device can be contained in Teflon cells. The cells are thermostated and galvanostatically cycled between a fixed terminal voltage ranging from 0 to 3.0 V. The battery package could be made compact, which is best suited for applications for which size and weight are the critical design requirements.

Studies performed by the author on SPEs demonstrate that polymer electrolyte-based PEO has shown exceptional stability of the $Li_4Ti_5O_{12}$ anode electrode, particularly over temperatures ranging from −60° to 85°C. This anode material offers uniform phase purity over this temperature range. The studies further reveal that all solid-state lithium-based batteries using the solid polymer material for the spinel $Li_4Ti_5O_{12}$-based anode, $LiMn_2O_4$ for the cathode electrode, and appropriate material blended with low-molecular-weight PEG for the separator material would provide excellent electrochemical performance in the battery. The studies further reveal that this particular design configuration is best suited for high-power, solid-state lithium-based batteries operating at temperatures as low as −60°C. Using glass fiber for the separator, stainless steel as current collectors, and polymer for electrolyte, the solid-state composite batteries have demonstrated overall satisfactory performance even at temperatures as low as −55°C.

Cell assembly is accomplished by sandwiching the polymer electrolyte between the composite cathode and the composite anode electrodes. The diameter of the electrodes can be made less than 3/16 of an inch or 0.08 mm. The electrochemical phenomenon is completely contained in the cells made from Teflon plastic material. The battery is assembled in the most compact package with minimum weight and size.

The spectral purity of spinel $Li_4Ti_5O_{12}$ material is of paramount importance. Material scientists have confirmed the presence of two broad-spectrum bands resulting from the symmetric and asymmetric stretching vibrations of the octahedral group lattice. The material scientists also confirmed the presence of the normal spinel. Furthermore, the lack of the absorption bands, which is typical of titanium oxide (TiO_2), is a clear indication of the phase purity in the spinel ($Li_4Ti_5O_{12}$) material.

8.2.2 Total Conductivity of the Battery Material

Lithium-based battery performance is dependent on the total conductivity of the composite material used by the Li-ion battery. The total conductivity of the composite material can be obtained by the impedance measurements [1]. Material scientists believe that the conductivity mechanism strictly obeys the Arrhenius Law, which can be written as follows:

$$Ln\ (\sigma\ T) = [Ln\ (\sigma_0) - E_a/kT] \tag{8.1}$$

where σ is the conductivity (S cm^{-1}), T is the absolute temperature (K), σ_0 is the at-zero degree (K), k is the Boltzmann constant (eV/K), and E_a is the activation energy, which is equal to 0.51 eV/mol. The extrapolated value of the conductivity at 25°C is approximately 10^{-9} S cm^{-1}.

The author has used the parametric information from the logarithmic plot of the total conductivity of the Li$_4$Ti$_5$O$_{12}$ material as a function of inverse temperature. The values of total conductivity as a function of inverse temperature (1,000/T), absolute temperature (T), and standard ambient temperature (t°C) are summarized in Table 8.1.

Comprehensive review of these tabulated data indicate that total conductivity is approximately equal to $10^{-10} \times 3.35$ S cm^{-1}at ambient temperature of 25°C, which is slightly better than 10^{-9} S cm^{-1}. Furthermore, the total conductivity at 60°C can be estimated better than 10^{-8} S cm^{-1}. It is evident from these data that total conductivity improves at lower operating temperatures.

The authors of *Solid State Ionics* [1] eloquently describe the performance of the all-solid-state, plastic, Li-ion batteries and the cycling behavior for the Li$_4$Ti$_5$O$_{12}$ cell and for the LiMn$_2$O$_4$ cell as a function cycler number. The battery composite structure best suited for low power applications consists of the Li$_4$Ti$_5$O$_{12}$ anode

Table 8.1 Total Conductivity as a Function of Absolute Temperature and Standard Temperature

Total Conductivity (S cm^{-1}K)	1,000/T	T (K)	t (°C)	Conductivity Value (S cm^{-1})
10^{-1}	1.18	847	847	$10^{-4} \times 1.18$
10^{-2}	1.58	633	360	$10^{-5} \times 1.58$
10^{-3}	1.95	513	240	$10^{-6} \times 1.95$
10^{-4}	2.33	429	156	$10^{-7} \times 2.33$
10^{-5}	2.73	366	93	$10^{-8} \times 2.73$
10^{-6}	3.15	317	44	$10^{-9} \times 3.15$
10^{-7}	3.35	298	25	$10^{-10} \times 3.35$
10^{-8}	3.60	278	5	$10^{-11} \times 3.60$

electrode, polymer electrolyte, and $LiMn_2O_4$ cathode electrode. An SPE can be used as separator in these batteries. The performance of all-solid-state, plastic, Li-ion battery and the cyclic behavior of the cells at 20°C can be briefly summarized as follows:

- The $Li_4Ti_5O_{12}$ cell demonstrated a specific capacity better than 150 mA/g over the entire cycle, ranging from 0 to 100 cycles. The $LiMn_2O_4$ shows specific capacity of about 100 to 110 Ah/g over the entire cycle range.
- The cells show very low capacity, fading even after 100 cycles, which means the cells have longer storage durations and longer operating lives.
- The electrodes made from $Li_4Ti_5O_{12}$ are most ideal for microbattery applications for which compact size and minimum weight are the principal requirements.
- The excellent cyclability of $Li_4Ti_5O_{12}$ material can be attributed to the stability of the spinel framework and to the minimum dilation upon cycling.
- The cell's performance could be adversely affected if it is stored at elevated temperatures.
- The ionic resistance of the polymer electrolyte appears relatively low compared with charge transfer resistance (R_{ct}). The values of both resistances decrease as the temperature increases. But the value of the R_{ct} remains high because of the poor electrical conductivity of lithium titanate, which is about 10^{-9} S cm^{-1} at the room temperature of 25°C as shown in Table 8.1.
- The high value of cell total resistance indicates that the device can be discharged only at very low currents, thereby limiting its use only for low-power applications or microbatteries.
- Cycling test data on polymer microbatteries cycled at 20°C indicate that the cell capacity was reduced to 55 mAh/cm^2. During the discharge cycle at C/rate, however, the cell exhibited a specific capacity of the cell around 70 mAh/g.
- A $Li_4Ti_5O_{12}$-based electrode cycled at C/25 rate showed a constant specific capacity better than 150 mAh/g.
- Preliminary test data collected by the these battery designers indicate that cell performance at room temperature was not very impressive because of the transport limitation in the polymer electrolyte. Nevertheless, the feasibility to fabricate thin electrodes will enable the design and development of microbatteries for various applications.
- Next-generation batteries using $Li_4Ti_5O_{12}$-based anode, $LiMn_2O_4$-based cathode, and polymer as a separator must reduce the charge transfer resistance to improve battery performance, particularly at room temperature.
- Next-generation batteries must make serious efforts to fabricate thin electrodes for such batteries to achieve specific capacity better than 160 mAh/g for a host of commercial applications.

8.3 Batteries for Low-Power Electronic Devices

Advancement and progress in low-power battery technology is closely tied to that in electronics [2]. In other words, battery development is strictly dependent on advancements in electronic circuit technology. Major improvements in batteries, including energy density, reliability, shelf life, and longevity, have occurred ranging from vacuum-tube technology to transistor technology to microcircuit technology. Progress in battery technology is rather slow compared with advancements in electronic circuits and devices. This is true, in particular, for lithium-based cells operating in the 2.5 to 3.0 V range. Older electrochemical devices such as carbon zinc (C-Zn), Zn-air, alkaline, nickel cadmium (Ni-Cd), and lead acid continue to get better over the time, and sustain the market for the electronic devices that can deploy them. A comprehensive review of primary lithium batteries shows that they are growing at a rapid rate as new electronic circuits are designed and developed for higher operating voltage, improved reliability, and superior shelf life. For example, $LiMnO_2$ batteries currently dominate the commercial market because of lower cost and improved safety compared with lithium-based batteries.

Market demand and survey for secondary or rechargeable batteries seem to indicate that two battery systems, namely Li-ion and nickel-metal-hydride (Ni-MH), are growing at a very fast rate in response to the environmental concerns and the demand for higher energy density requirements. In addition, other rechargeable batteries such as Zn-air and lithium with SPE devices are currently in commercial production. Such devices are available for specific applications for which demand for higher energy density, longer service life, and improved reliability is of critical importance.

Strict environmental regulations and guidelines have a great impact on battery use and its disposal, leading to greater interest in secondary batteries that can be recharged several times before their disposal. Mercury and other toxic materials have been barred from use in batteries because of health reasons and strict disposal guidelines.

This section has discussed the current and emerging battery devices with an emphasis on specific energy density, cycle life, shelf life, and other critical battery characteristics. Specific energy density and cycle life are particularly dependent on applications and will vary with discharge rate, duty cycle, and operating voltage of the device.

8.3.1 Impact of Materials and Packaging Technology on Battery Performance

Performance capability, battery capacity, and energy density are limited by battery system chemistry, use of the latest fabrication materials, and packaging technology. Advances in battery materials and packaging technology have resulted in significant progress and changes in older types of batteries, such as alkaline manganese, Leclanché, Ni-Cd, and lead acid. New battery systems, namely primary and secondary lithium, Zn-air, and Ni-MH batteries, have been commercialized to meet demands for higher energy density. Low-power applications are generally considered to cover the power range from a few microwatts for watches, electronic toys, electric toothbrushes, electric savers, smoke detectors, infrared cameras, and cell phones to a few tens of watts for heart pacer devices, subnotebook computers, and other electrical devices requiring power levels in a specified range.

Battery characteristics for low-power applications include energy and output power per unit weight to meet general portability requirements. Energy and power output per unit volume is more critical for many portable devices. Battery designers claim that as the output current is increased, the amount of energy available from the battery is decreased. The energy delivered by a specific device is strictly dependent on the rate at which power is withdrawn. In addition, battery energy output and power output are also affected by the materials used in the fabrication of the device, the cell size, and the highest duty cycle employed. Battery manufacturers rate the battery capacity for a given cell or battery as ampere-hours or watt-hours when discharged at a specific rate to a specific voltage cutoff value. Battery engineers often select the discharge rate, frequency, and cutoff voltage to simulate a specific device application, such as a camera, a smoke detector, or a security-warning device.

As far as the rating base is concerned, in general, primary cells or batteries are rated at a current level that is one-one hundredth of the capacity (C/100) expressed in ampere-hours, whereas secondary or rechargeable cells or batteries are rated strictly at one-twentieth of the capacity (C/20).

8.3.2 Glossary of Terms Used to Specify Battery Performance Parameters

This section defines glossary terms used to specify battery performance parameters. Prospective battery designers who want to be involved in the design and development of next-generation batteries for electric vehicles (EVs) and hybrid electric vehicles (HEVs) should become familiar with these terms.

- *Current rating or C rate*: The discharge or charge current expressed in ampere-hour capacity of the battery. Multiples or fractions of the C rate are used to specify higher or lower current such as 2C or C/10, where C stands for charge.
- *Energy density*: Nominal energy content of a battery expressed in watt-hours per liter or watt-hours per kilogram. This is sometimes also known as a specific energy parameter.
- *Cycle life*: Cycle life is the most critical performance parameter and is defined as the number of cycles delivered by a rechargeable or secondary battery before the battery's rated capacity falls to 80% of the initial capacity in a particular application.

■ *Power density*: The power delivered by a battery; expressed in watts per liter.

■ *Rated capacity*: The nominal capacity delivered by a battery as a specific rate of discharge in secondary cells and rated at a C/10 current and at C/100 in the case of primary cells. This is particularly used to calculate the energy density.

■ *Specific energy*: The nominal energy of a battery expressed in watt-hours per kilogram.

■ *Self-discharge*: The loss of capacity of a rechargeable or secondary battery; expressed in percent of the rated capacity lost per month at a given temperature.

■ *Intercalation electrode*: A solid electrode that operates by storing and releasing ions and electrons during its operation. The ions are *intercalated* or placed between the lattice planes of the host material—for example, a titanium-sulfide (TiS_2) cathode accepts lithium ions and electrons during the discharge and gives them up during the charge. A carbon anode, however, stores lithium ions and electrons during the charge cycle and gives them up during the discharge cycle.

■ *Metal hydride*: A result of incorporating hydrogen ions and electrons or hydrogen into a metal. It is used to describe the anode of a metal-hydride cell.

■ *Solid polymer electrolyte (SEP)*: A combination of a polymer (such as polyethylene oxide) with a salt to form a solid material that conducts ions but no electrons. This is also used to describe chemical mixtures that contain organic solvents.

■ *Solvent*: A solution or liquid substance that is capable of dissolving or dispersing one or more other chemical substances.

■ *Solid electrolyte*: The most efficient and highly reliable electrolyte best suited for batteries and fuel cells in spacecraft and satellite applications.

■ *Polyethylenoxide (PEO)*: A lithium salt complex that has been recognized as a liquid-free SPE. Because of its unique structure, the SPE electrical conductivity becomes appreciable only above room temperatures. According to the material scientists, this material is best suited for a positive electrode in a polymer electrolyte based on PEO. It yields exceptional stability of the electrode at operating temperatures from −60° to 80°C.

8.3.3 Fabrication Aspects of Batteries for Low-Power Electronic Device Applications

Fabrication of a battery or a cell determines how the device will function and how much it will cost to construct it. Several primary cells with aqueous electrolytes seem to use a single electrodes arranged in parallel or concentric configuration. Specific construction of this type of cell includes *cylindrical* or *bobbin, button*, and *coin*. Small secondary cells use a *wound* or *jelly roll* construction feature, in which long thin electrodes are wound into a cylinder and placed in a metallic housing or container. This particular cell construction yields higher power density, but it does so with reduced energy density and a significantly higher construction cost. Because of the use of electrolytes with low electrical conductivity, lithium primary cells deploy the wound construction configuration to provide higher discharge rates. An increasing number of both the primary and secondary cells, however, are being made in prismatic and thin, flat construction to keep battery size to a minimum. Studies performed by the author on potential form-factors seem to indicate that prismatic and thin form-factors allow better utilization of the volume available in batteries but that such form-factors generally yield lower energy density. The author recommends undertaking trade-off studies for potential form-factors in terms of energy density, construction cost, and device volume. A form-factor optimized for one specific form-factor may not be optimum for another form-factor.

8.3.4 Performance Capabilities and Limitations of Various Primary and Secondary Batteries for Low-Power Applications

8.3.4.1 Carbon-Zinc Primary Batteries

According to a market survey, C-Zn primary batteries continue to dominate the market on a worldwide basis. They are widely used for most of the electrical and electronic devices requiring power ranging from approximately 50 mW to 15 W. C-Zn primary batteries are widely used because of their exceptionally reliable performance, improved and enhanced shelf life, and significantly reduced leakage.

These batteries use zinc for the anode, manganese dioxide for the cathode, and ammonium chloride for the electrolyte. High rate capacity batteries use a complex electrolyte consisting of a manganese dioxide electrolyte and zinc chloride electrolyte, which offers high rate capability and significantly improved reliability. The majority of the standard size batteries come in D, C, and AA configurations. C-Zn batteries with these configurations are widely used worldwide because of their good electrical performance, lower cost, and reliability. Such batteries are manufactured in the United States, Europe, Japan, China, and other countries. China alone produces more than 7 billion C-Zn batteries because of high demand in the country. Both the Leclanché and zinc chloride are most popular worldwide, because they are capable of meeting the power requirements for several electronic devices for which cost and rapid availability are the most demanding requirements. Older C-Zn batteries may stop or reduce the production rate of these batteries as new types with better performance are available for new applications.

There is environmental pressure to remove not only the mercury but also cadmium and lead from the zinc cans. A couple of decades ago, California announced a prohibition on the sale of C-Zn and alkaline cells containing mercury. The restriction on the use of mercury in batteries eliminates the need for recycling, thereby eliminating storage and transportation problems. In other words, the restriction on the use of zinc plus the need to remove mercury would favor the sale of premium cells. Removal of mercury from cells also will lower performance at high current levels. An alternative approach to overcome these problems is to use alkaline cells in such applications. C-Zn batteries are not considered suitable for some electronic devices such as disc players, automatic cameras, flash units, and certain toys because they do not provide the power required for satisfactory operation of these devices.

8.3.4.2 Alkaline-Manganese Batteries

The growth of alkaline batteries is strictly due to a reduction in power requirements. C-Zn batteries had not been able to meet the energy requirements for satisfactory performance of disc players, automatic cameras, camera flash units, toys, and newly developed motors, displays, and electronic devices. The power requirements for these devices, however, can be met with alkaline-manganese rechargeable batteries.

Major improvements in the performance of alkaline batteries were introduced around 1988 with the introduction of the plastic label construction feature instead of a cupboard insulation tube over the cell body and a steel outer jacket. Battery designers claim that deployment of this insulating plastic label provided a 15 to 20% increase in the internal volume, which allowed for additional space for the active materials needed for performance enhancement. Material scientists believe that the use of advanced materials and improvements in the module's construction have demonstrated significant improvement in the battery's capacity.

8.4 Performance Capabilities of Primary Lithium Batteries

This section briefly describes primary lithium batteries with an emphasis on their performance capabilities and limitations. Most lithium-based batteries are designed for 3 V ratings, but they can be designed for other voltage ratings. Lithium-based batteries have gained popularity, particularly in the past decade. This is because lithium is the lightest metal with a density equal to 5.34 g/cc. In addition, significant improvements in lithium-based batteries have been observed in terms of capacity, reliability over wide temperature range, longevity, and long-term storage exceeding 15 years. Li-ion batteries have become a popular choice for battery designers because they are best suited for jobs demanding high power and energy density with minimum weight and size. Because of these performance capabilities, lithium cells are widely used in consumer electronics for which weight and size are of critical importance. Larger lithium cells are widely used in commercial applications for which high power and energy density are the principal requirements.

Mass production of lithium-based batteries started in Japan about 3 decades ago. Most low-capacity lithium-based coin cells were manufactured about 20 years ago in Japan, Germany, France, and the United States, but high-capacity lithium cells with cylindrical configurations are emerging as their applications grow. Most lithium-based batteries operate at 3 V. Rechargeable lithium-based batteries with higher voltage ratings are available in the market for specific applications requiring higher energy capacity. Lithium-based batteries possess several distinct characteristics, such as higher energy density, longer operating live, excellent electrical performance over a wide temperature range, and ultra-long shelf life. An application survey of lithium-based batteries indicated that these batteries, except $LiMn_2O_4$ batteries, are most suited for medium- to high-power applications. The next sections describe the performance capabilities and limitations of various lithium-based batteries.

8.4.1 Lithium-Iodine Batteries

Lithium-iodine ($Li-I_2$) batteries have been in use in cardiac pacemakers for more than 25 years. These batteries offer very high energy density, but they are low in power. The designers of $Li-I_2$ batteries have used low electrical conductivity solid-state electrodes, which limit the current levels to a few microamperes and keep the output power level to a minimum. Material scientists claim that with advances in both the battery and pacer technologies, reliable operation between 10 and 15 years can be expected. Higher power implanted cells using lithium-silver vanadium oxide ($Li-AgVO_2$) technology are now being deployed to power the pacers that can also supply the electrical energy for automatic, portable defibrillation applications. This automatic, portable defibrillation equipment provides the most reliable and rapid medical services under medical emergency conditions.

8.4.2 LiMnO₂ Battery

The $LiMnO_2$ battery is the most popular lithium battery, and it is produced worldwide by more than 14 manufacturers. The battery cells are available in a wide range of sizes, shapes, and capacities. Types of construction features for this battery include coin shape, cylindrical bobbin, cylindrical wound, cylindrical D cell configuration, and prismatic feature. Performance characteristics of commercially available $LiMn_2O_4$ rechargeable batteries are summarized in Table 8.2.

Highlights of $LiMn_2O_4$ batteries can be summarized as follows [3]:

- Operating temperature range of –20° to 60°C.
- Energy density of 300 to 430 Wh/kg.
- Major advantages are low cost and safe operation.
- Major drawback is low gravimetric energy density.
- Typical applications include consumer electronics devices, military communications equipment, transportation, automatic meter reading, medical defibrillators, and memory backup.

Table 8.2 Performance Capabilities of Lithium-Manganese-Dioxide Batteries with Various Construction Features and Design Configurations

Performance Parameters	*Lithium-Manganese-Dioxide Battery Construction Features*				
	Coin	*Cylindrical Wound*	*Cylindrical Bobbin*	*High-Rate D Cell*	*Flat*
Voltage rating (V)	3	3	3	3	3, 6
Capacity (mAh/cell)	30–1,000	160–1,300	65–5,000	10,000	150–1,400
Rated current level (mA)	0.5–7	20–1,200	4–10	2,500	20–125
Pulse current level (mA)	5–20	80–5,000	60–200	N/A	N/A
Energy density (Wh/L)	500	500	620	575	290

Source: Powers, R., "Batteries for low power electronics," *Proceedings of the IEEE,* © 1995 IEEE. With permission.

8.4.3 *Lithium-Carbon Fluoride Battery*

The lithium-carbon fluoride (Li-CF$_x$) battery offers impressive capabilities in terms of energy density of 500 to 700 Wh/kg or 700 to 1,000 Wh/L and a wide operating temperature range from –60° to 160°C. Its major performance capabilities and potential applications can be summarized as follows:

- High energy and power density.
- Long shelf life and operating life.
- Very wide operating temperature range.
- Low to high discharge rates in portable electronics devices.
- Potential applications include military search and rescue communications equipment, some industrial systems and transportation.
- LiCF$_x$ long-term storage capability exceeding 15 years, high reliability, and wide operating temperature range from –60° to 160°C.
- Real-time storage of 15 years at room temperature (27°C) has shown an impressive capacity loss less than 5%.
- This battery has a voltage rating of 2.5 V.
- A patent indicates that a version of this battery with modifications to the electrolyte, separator, and seal of the CF$_x$ coin cell has demonstrated satisfactory operation of the coin cell close to 125°C, which is remarkable.
- Next-generation LiCF$_x$ batteries deploy composite cathode materials involving a special additive conductive agent and a chemical binder. The battery

designers claim a significant reduction in voltage drop during initial discharge, improved energy density, higher running voltage, and increased discharge rate ceiling. Potential applications of this newly designed battery include medical devices, electronic products, aerospace sensors, and military missiles. The battery designers claim that these newly designed batteries using composite cathodes are expected to cost less than the existing $LiCF_x$ batteries.

8.4.4 Lithium-Sulfur-Dioxide Battery

The lithium-sulfur-dioxide ($LiSO_2$) battery offers moderate energy density, ranging from 240 to 315 Wh/kg, but an ultrawide operating temperature range from –55° to 70°C. Because of its ultra-low-operating-temperature capability, this battery is best suited for military and aerospace communications systems. Its major advantages are low cost, high pulse-power capability, and low temperature, but it suffers from passivation. Safety concerns are visible during high sustained discharge, which can cause overheating and pressure buildup. This battery also generates toxic waste that must be removed occasionally. Maintenance problem associated with this battery can prevent its deployment for long-term aerospace missions. This particular battery with higher capacity cells is best suited for military communications equipment. According to existing published reports, there are no commercial applications for $LiSO_2$ batteries. The outstanding features of this battery can be summarized as follows:

■ Lithium cells with liquid cathodes, such as sulfur dioxide, offer higher energy density, improved rate capability, and enhanced low-temperature performance compared with those with solid electrolytes.
■ Because the cathode material is also the electrolyte, the packaging efficiency will be much higher than for solid cathodes.
■ The liquid cathode has demonstrated improved electrochemical kinetics.
■ Because of some inherent reliability problems with large-area cells, the commercial market for such batteries is limited to small-area $LiSO_2$ cells.
■ Large-area cells are used by military systems.
■ The U.S. Department of Transportation limits cells to less than 1 g of lithium for ordinary shipment to avoid safety concerns. One gram of lithium is equal to a 3.86 Ah capacity rating.
■ Disposal of large quantities of lithium batteries is a problem.

8.4.5 Lithium-Thionyl-Chloride Battery

The lithium-thionyl-chloride ($Li-SOCl_2$) battery is best suited for military and transportation applications because of its high energy density and very wide

Table 8.3   Performance Characteristics of Primary Lithium-Based Batteries

Performance Characteristics	Lithium-Based Batteries			
	$Li\text{-}MnO_2$	$Li\text{-}SO_2$	$Li\text{-}SOCl_2$	$Li\text{-}CF_x$
Gravimetric energy density (Wh/kg)	300–430	240–315	500–700	500–700
Volumetric energy density (Wh/L)	500–650	350–450	600–900	700–900
Operating temperature range (°C)	−20 to 60	−55 to 70	−55 to 150	−60 to 160
Shelf life (years)	6–12	14	16–20	15–18
Environmental impact	Moderate	High	High	Moderate
Relative price/performance	Fair	Good	Fair	Good

operating temperature range from –55° to 150°C. Its major advantages and disadvantages are as follows:

■ High energy density of 500 to 700 Wh/kg.
■ Wide operating temperature capability from –55° to 150°C.
■ High volumetric energy density from 600 to 900 Wh/L.
■ High pulse-power capability.
■ Long service and shelf life.
■ Reliability can be affected by the generation of toxic waste.
■ It suffers from passivation.
■ Safety concerns are visible, particularly during high sustained discharge, which causes pressure buildup and generates toxic waste.
■ Its typical applications include commercial and consumer electronics devices and components, military communications equipment, transportation, memory backup, and automatic meter reading.
■ The weight and size of a battery of this type are critical, especially when multiple batteries are needed to deliver adequate power for a particular application.

Important performance characteristics of primary lithium batteries are briefly summarized in Table 8.3.

8.4.6 Lithium-Ferrous Sulfide (Li-FeS₂) Battery

Lithium-ferrous sulfide ($Li\text{-}FeS_2$) AA cells were first introduced in the market around 1992. The cells were very popular in Canada, Japan, and European countries. The 1.5 V cell, which has a wound construction feature, was directed at high-rate, high-voltage-cutoff photoelectronic applications, such as disc players, cell phones, flash units, and minicomputers. In such applications, it offers up to three times the alkaline service to high-voltage endpoints, but it is not suitable for low-cutoff-voltage devices.

8.4.7 Conclusions on Lithium-Based Batteries

This section provides conclusive remarks on lithium-based primary batteries, in particular on energy density capability, power density limit, the impact of temperature on battery performance, shelf life, safety and reliability, the environmental impact of batteries, and price-to-performance ratio [3]. Each item is summarized as follows:

- *Energy density*: Both the weight and size of the battery are critical, particularly in aerospace, military, and spacecraft applications when multiple battery modules are required to meet the output power level for the desired service life. For both the gravimetric and volumetric energy densities, $LiCF_x$ batteries enjoy a distinct advantage.
- *Power density*: The total capacity of a battery is characterized by its power density, which varies as a function of current level required for a specific application. In some applications, power density variation is based on a form-factor of the battery. Performance comparison must be made under specific circumstances involved in an application. Customization of a battery takes place during the manufacturing phase by altering how fluorine is introduced into the carbon structure at the atomic level. Customization of a battery is a complex and cumbersome process.
- *Operating temperature*: The operating temperature range is generally not a significant factor in typical applications because all four lithium-based batteries (see Table 8.3) operate over a fairly wide temperature range, with $Li-SOCl_2$ and Li-CFx batteries exhibiting the best performance characteristics. Only one of the batteries, namely $LiMn_2O_4$, fails to operate reliably below −20°C. This means that this particular battery is not suitable to operate in frigid environments.
- *Safety*: The use of liquid or gaseous material for either the cathode or electrolyte can raise safety and reliability concerns during battery deployment. The worst battery candidates are $LiSO_2$ and $Li-SOCl_2$, which both require the use of pressurized canisters or housings, which can rupture and vent or leak corrosive and toxic gases. The deployment of pressurized canisters provides safety to the equipment and to the people working nearby.
- *Shelf life*: Shelf life is critical in applications for which a device is required to provide adequate safety during storage to protect its structural elements. Shelf life is particularly important for public safety, search-and-rescue operations, and certain medical equipment. The four distinct lithium-based batteries

summarized in Table 8.3 provide adequate shelf lives, with the Li-SOCl$_2$ battery demonstrating the longest shelf life.

- *Environmental impact*: On the basis of a review of manufacturing processes, it can be stated that all lithium-based batteries are manufactured with some fairly toxic and dangerous chemicals, which can have adverse impacts on the environment during their storage or manufacturing or improper disposal. According to material scientists, the worst offenders include LiSO$_2$ and Li-SOCl$_2$, because both contain toxic chemicals that can poison groundwater, which then cannot be used for drinking. Toxic chemicals are generally found near battery manufacturing companies.

- *Price-performance comparison*: It is difficult to have a meaningful comparison between the price and performance of a battery, because so many factors in the performance of a battery are involved. Li-CF$_x$ battery systems enjoy some significant advantages over all other types of primary battery systems. Its major drawback, however, is its higher cost. But when demand increases for this battery system, its cost will certainly go down. The reduction of cost is dependent on the deployment of the battery for various commercial and industrial applications. The battery's increased performance is strictly based on its high energy and power density capabilities. According to the battery designers, the Li-CF$_x$ battery offers the best price-to-performance ratio over all alternative systems. Battery designers hope that with a reduction in manufacturing cost, advancements in battery chemistry and technology, and deployment increases in quantity, the price-to-performance ratio of this particular battery system can be expected to improve significantly. Only time and demand for this battery in large numbers will determine the price-to-performance ratio of this battery system.

8.5 Applications of Small Rechargeable or Secondary Cells

Thus far, lithium-based primary batteries have been described with an emphasis on their performance characteristics and their various applications. This section focuses on the performance capabilities and applications of small secondary or rechargeable cells. Published reports indicate that the small rechargeable battery market has been experiencing more than 25% growth over the past 2 decades because of an explosion in the use of cell phones, camcorders, minicomputers, and electronics entertainment devices. These devices require more power consumption than can be provided by the primary cells. These devices are experiencing rapid growth, which can be satisfied by Li ion, Ni-MH, sealed-nickel-cadmium (S-Ni-Cd), and lithium-polymer-electrolyte rechargeable batteries. These batteries come in different sizes and with various power ratings. The following sections briefly describe the performance capabilities and applications of such batteries.

8.5.1 Sealed Lead-Acid Batteries

Lead-acid batteries have been in use for various applications over the past two centuries. These batteries are most cost-effective, can be recharged multiple times, and have been widely deployed in automobiles and trucks and as emergency power sources. These batteries are relatively heavy and take longer times for recharging.

In some applications, it is necessary to use sealed lead-acid batteries (SLABs), for which safety, reliability, and long service life are important design considerations. These cells are produced in various sizes and shapes and have capacities in the range of 1 to 2 Ah. These SLABs offer more than 200 cycles of life at discharge times as short as an hour. Such batteries with larger sizes are available for notebook computers or minicomputers. SLABs offer the following benefits:

- Initial low cost
- Low self-discharge
- Reliability and safety
- Excellent rate capability
- Small cells with cylindrical configurations

8.5.2 Small Li-Ion Rechargeable Batteries

Small rechargeable Li-ion batteries are available that are best suited to power laptops, cell phones, and certain electronics devices with power requirements that are compatible with such batteries. These devices require more power than can be provided by the primary batteries or cells described in Section 8.4. Over the past 3 decades or so, a number of Li-ion rechargeable batteries using solid lithium anodes, liquid organic electrolyte, and cathodes consisting of multiple oxides were designed and developed. Some versions of Li-ion rechargeable batteries were discontinued because of safety and cycle-life problems.

The solid lithium anode was replaced with a carbon material that merely stores electrons and lithium ions on charge and gives them up during the charge. The cathode in such a battery design employs a number of oxides, which also store electrons and Li-ion on discharge and give them up on charge. This particular battery design has demonstrated a cycle life exceeding 1,200 cycles and is free from safety and reliability problems. Because this particular battery design contains no lithium metal, it is exempt from stringent shipping regulations.

In 1995, Japan has installed a production facility with an 80-million cell capacity for mass-scale production of practical rechargeable batteries using $LiCoO_2$ as the cathodes. These rechargeable batteries are rated at 3.6 V and have demonstrated a density exceeding 350 Wh/liter. Performance improvements have been observed with improvements in the carbon anode, electrolyte, and cathode materials. Battery scientists may deploy lithiated oxides of nickel and manganese on the basis of cost reductions and reduced environmental impact.

8.5.3 S-Ni-Cd Rechargeable Batteries

S-Ni-Cd rechargeable batteries were first designed and developed in Europe in the early 1950s. Worldwide production of S-Ni-Cd rechargeable batteries was accelerated to produce more than 1 billion units in 1994. Production of such batteries has continued. Major improvements have been made in past 2 decades in terms of energy density capability, improved electrodes, and packaging technology. Typical capacity for an AA cell is now well above 900 mAh. Improvements in such cells are due to the use of high-porosity nickel foam to replace the traditional sintered nickel as a carrier for the active materials, charge retention, and the elimination of memory effect. The S-Ni-Cd cell offers high discharge rates of 10 min. or less and is tolerant of overcharging and undercharging conditions. All chemical reactions occurring in a S-Ni-Cd cell are shown in Figure 8.1. Chemical reactions at the positive electrode and at the negative electrode are displayed at the right and left sides of Figure 8.1, respectively.

This battery design suffers from the toxic effects of cadmium and, thus, the manufacturers and users must comply with the shipping and disposal guidelines established by their respective governments. Proposed changes in regulations and guides are established by the respective countries for the disposal and shipping of the S-Ni-Cd rechargeable batteries. Highlights of this battery are as follows:

- Chemical reactions determine battery capacity.
- The reversal reduction of NiOOH to Ni(OH)$_2$ involves intercalation of a hydrogen ion into the layered structure of the former, giving rise to a solid solution. This process favors the long-term reversibility of an electrode.
- In an S-Ni-Cd batter, there is recombination of the oxygen generated at the positive electrode during the overcharge process.

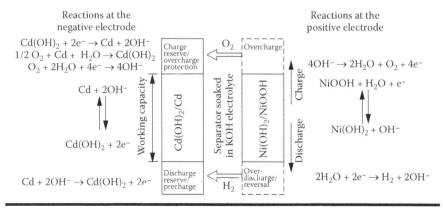

Figure 8.1   All chemical reactions occurring in a sealed nickel-cadmium battery cell.

■ In this particular battery, the negative-to-positive capacity ratio is typically between 1.5 and 2.0.

■ The overcharge protection of this battery is around 30% of the positive capacity, and the discharge reserve is between 15 and 20%.

■ The construction of the positive electrode is of critical importance because it determines battery performance.

8.5.4 Nickel-Metal-Hydride Rechargeable Batteries

The Ni-MH battery was first developed in 1993, and since then, it has been growing at a remarkable rate of 20% per year. The cadmium electrode of the Ni-Cd battery was replaced with a complex alloy that absorbs hydrogen on charge and releases it on discharge. The operating voltage of the hydride anode is similar to that of the Ni-Cd battery and therefore, complete interchange between these batteries is possible. In addition, the environmental concern of cadmium is completely eliminated in the Ni-MH battery. In this battery design configuration, the hydride electrode provides more capacity in the same size AA cells with more than 1,400 mAh ratings, which have been boosted to exceed 1,700 mAh using the latest materials for the anode and cathode along with advanced battery technology. The self-discharge rate for this battery is slightly higher than that of the Ni-Cd battery. In addition, the rate capability and low-temperature performance are inferior. Proper charging appears to be more critical. Smart charger chips are integrated directly into the battery package to control the charge and monitor the self-discharge. Currently, high-power Ni-MH batteries are manufactured for deployment in electric and hybrid electric cars, such as the Toyota Prius, because they have demonstrated the greatest suitability for this application.

There are two types of hydride alloys, namely AB_5 type and AB_2 type. The former design is based on combinations of rare earth materials with nickel and the alloy is widely used in small sealed cells. The AB_2 is based on complex alloys of titanium, vanadium, and nickel, and this particular alloy is best suited for the batteries deployed in large EVs and trucks. These rechargeable batteries currently cost between $15,000 and $20,000, depending on the size of the EV. The batteries have demonstrated a service life exceeding 15 years. Such batteries or cells do not fall under the category "small secondary or rechargeable cells," but the cell size AA, with a rating close to 1,400 mAh, does fall under the category "small secondary cells."

8.5.5 Lithium-Polymer-Electrolyte Cells

Decades ago, battery designers predicted that rechargeable lithium batteries using solid polymer as the electrolyte would yield better electrical performance and would do so at reasonably lower manufacturing costs. But the battery designers later realized that it is not easy to develop a rechargeable battery using an SPE, because important characteristics of potential SPE material were not fully known [2]. The material scientists perceived that advantages of this approach included high-speed production using web equipment, flexible thin cells, form-factor, shelf life, and energy density. Despite comprehensive research and development activities for about 15 years, no commercial battery using solid electrolyte has been developed successfully. One of the major problems is that the electrical conductivity of most solid electrolytes at room temperature (27°C) is very low (10^{-6} S cm). This conductivity problem can be partially overcome by using thin electrolyte layers; however, operating temperatures in excess of 60°C are required for any reasonable current level. Another approach to overcome this problem is to add an organic solvent to the polymer. This approach reduces the electrical conductivity of the solid electrolyte with solvent to an acceptable level (10^{-3} or 0.001 S cm). But this method reduces cycle life, and in addition, it introduces the safety problems associated with liquid electrolytes.

Material scientists have devised a new technique to overcome this problem. The proposed technique combines the electrodes of the Li-ion cell with highly conductive polymers. Some material scientists have solved the outstanding problem by using carbon anodes and lithiated manganese oxide with proprietary electrolytes. Battery designers using this technique were successful in achieving an initial energy density slightly higher than 200 Wh/L. Material scientists later obtained an energy density close to 400 Wh/L. Batteries using this technology were manufactured for various applications. Around 2005, design and development efforts were undertaken using pure polymer and solid lithium. The tests conducted on these batteries were so remarkable that these devices were recommended for EVs and HEVs. Cycle life and manufacturing cost data are not readily available. Such cells or batteries do not fall under the category of "small rechargeable batteries." But this polymer electrolyte technology could be used to manufacture AA or coin-type small rechargeable batteries. Typical performance characteristics of AA small rechargeable or secondary batteries such as Li-ion (CoO_2), Li-polymer (E), and $LiMnO_2$ devices are summarized in Table 8.4.

$LiMn_2O_4$ batteries are considered to be primary batteries. They have demonstrated high energy density and power density over alkaline batteries, excellent storage life, self-discharge rate less than 0.5% per year, and flat discharge characteristics over operating temperatures ranging from −40° to +60°C. They are best suited for cardiac pacemakers and consumer products such as complementary metal-oxide semiconductor (CMOS) memory storage devices, digital cameras, portable power tools, heavy-duty flashlights, watches, and other commercial products requiring minimum weight and size. These batteries come in button and cylindrical configurations.

Examination of the data presented in Table 8.4 shows that storage loss for Ni-MH batteries is very high (20%), which means that these batteries would require more frequent charging. Among the AA-size rechargeable batteries, Li-polymer (E) and $LiMnO_2$ devices seem to retain a charge for much longer periods because of their low storage loss or self-discharge loss per month. Both the

Table 8.4 Typical Performance Characteristics of AA-Size Small Rechargeable Batteries

Performance Characteristics	Small AA Rechargeable Battery Types				
	Ni-Cd	*Ni-MH*	*Li-Ion (CoO$_2$)*	*Li-Polymer (E)*	*Li MnO$_2$*
Voltage rating (V)	1.2	1.2	3.6	2.5	3.0
Capacity rating (mAh)	1,000	1,200	500	450	800
Charging rate (C)	10 C	2 C	C	C/2	C/2
Energy density (Wh/L)	150	175	225	200	280
Energy density (Wh/kg)	60	65	90	110	130
Cycle life (cycle)	1,000	1,200	500	200	280
Storage loss per month (%)	15	20	8	1	1

Ni-Cd and Ni-MH batteries do suffer memory loss. Battery users and designers should consider the following:

- Li-ion devices are intrinsically safe because they do not contain metallic lithium.
- Lithium-polymer batteries do contain metallic lithium and require compliance with authorities for disposal and storage to preserve safety.
- Ni-Cd and Ni-MH rechargeable batteries do suffer from memory effects.
- Self-discharge is the lowest (less than 1% per month) for rechargeable lithium-polymer batteries.
- Solid polymer enhances the cell's specific energy capability by acting as both an electrolyte and separator. Most suitable solid polymers have a low ionic conductivity ranging from 10^{-8} to 10^{-6} S cm at room temperature. Potential lithium salts, organic solvents, and other dopants must be investigated to reduce the ionic conductivity to a reasonable value at room temperature.

For reliable and comprehensive information on rechargeable small batteries, readers are advised to refer to the *Handbook of Batteries* (1995) and Volume 3 of the *Encyclopedia of Chemical Technology* (4th ed., 1992).

8.6 Thin-Film Batteries, Microbatteries, and Nanobatteries

This section describes the performance characteristics and applications of thin-film batteries, microbatteries, and nanobatteries. The typical power output of battery sources is generally given in milliwatts, microwatts, and nanowatts, respectively.

8.6.1 Structural Aspects and Performance Capabilities of Thin-Film Batteries

The fabrication of thin-film battery involves thin-film technology, appropriate materials, and various film deposition techniques and guidelines, as well as photo-resistant patterns, etching techniques, and fabrication technology [4]. Specific details on these topics are well described in Section 8.2. Low-loss metallic film with a typical thickness ranging from 0.004 to 0.012 in. is used. The battery assembly size, which includes the anode, cathode, and housing assembly, can have dimensions ranging from 0.800 in. (20 mm) to 1.625 in. (42 mm). In general, thin-film technology is best suited for three-dimensional (3D) microbatteries and its major advantages are briefly summarized in *Performance Parameters of a 3-D Thin-Film Micro-Battery* (2008) [5].

The thin-film battery weighs less than approximately 50 g. Its circuit voltage (OCV) is around 3.8 V, and the battery can be designed for a capacity ranging from 0.1 to 5 mAh, depending on the battery size and thickness restrictions. The electrical connections are made from metal foils with typical dimensions of 10 mm (0.394 in) in length, 2 mm (0.0788 in) in width, and 0.1 mm (0.00394 in) in thickness. Battery size is strictly a function of its capacity. Technical information and performance characteristics of microbatteries can be summarized as follows:

- The battery chemistry is composed of solid-state films and uses lithium-phosphorus-oxynitride (LiPON) solid-state electrolyte developed by Oak Ridge National Laboratory. The cathode is made from $LiCoO_2$ and the anode uses a very small amount of lithium metal.
- This microbattery contains no toxic liquid electrolyte and is free from outgassing and explosion problems.
- The thin-film battery package contains five critical components, namely the anode, electrolyte, cathode, current collector, and ultra-thin substrate.
- The world's thinnest battery, developed by Front Edge Technology, Inc., Baldwin Park, California, has a thickness of 0.05 mm (0.002 in.). This battery can be charged to 70% of its capacity within 2 min. It has a cycle life of 1,000 cycles and has a self-discharge rate of less than 5% per year. This thinnest battery can be twisted with no deterioration in performance or damage to the battery housing.
- The thin-film battery is perfectly safe and is hermetically sealed to eliminate gas leakage.
- Battery charging is easy, foolproof, and requires 4.2 V. Continuous charging at 4.2 V does not degrade battery performance and does not overheat the device.
- This microbattery can be charged at very high rates without any impact on its performance.

- The 0.25 mAh capacity battery can be charged to 70% of the rated capacity in less than 2 min. and to full capacity within 4 min. A 0.9 mAh capacity battery can be charged at 4.2 V; the battery achieves 70% of the capacity within 6 min. and 100% capacity in 20 min. The use of cobalt in the cathode takes more charging time.
- Batteries using thin-film technology are best suited for cell phones, smoke detectors, hearing aids, and other medical devices, for which available space is of critical importance.
- Typical charging characteristics of a 0.9 mAh thin-film battery showing the charging current and charging capacity as a function of charging time are illustrated in Figure 8.2a.
- The discharge characteristics of a 0.9 mAh thin-film battery showing battery capacity as a function of OCV or various discharge rates are illustrated in Figure 8.2b.
- Battery capacity loss in mAh over 1,000 charge-discharge cycles is illustrated in Figure 8.2c.
- Discharge curves at the 1st, 500th, and 1,000th cycle are illustrated in Figure 8.2d.
- Charge current and battery capacity as a function of charge time for various cycles are illustrated in Figure 8.2e.
- Discharge curves of a 0.1 mAh thin-film battery at 25, 60, and 100°C operating temperatures are shown in Figure 8.2f.
- These batteries can be stored over the temperature range from –40° to +80°C with no deterioration in performance. The battery can be operated at temperatures as high as 170°C, but there will be capacity drop much faster during the cycling.
- Thin-film battery capacity ratings as a function of physical dimensions are summarized in Table 8.5.

It is evident from the data presented in Table 8.5 that battery capacity rating is strictly dependent on the physical dimensions (length, width, and thickness) of the device. Note that the weight of the battery will increase with its physical dimensions.

8.6.2 Thin-Film Metal-Oxide Electrodes for Lithium-Based Microbatteries

In the opinion of material scientists, thin-film technology is essential in the development of rechargeable lithium-based microbatteries for potential applications, such as smart cards, nonvolatile memory backup devices, MEMS sensors and actuators, and miniaturized implantable medical devices. Battery designers predict that for such applications film thickness should not exceed a few tens of micrometers or microns (10^{-4} cm). This means that the film thickness must be at least ten micrometers or 0.001 cm (0.0025 in.), which may be suitable for minimum battery power. Higher battery output levels will require film thicknesses of 20 to 50 mm. Regardless of the film thickness, the quality of the film, the material for the film, and surface conditions must meet stringent requirements to achieve the optimum performance of the metal-oxide electrodes, anodes, electrolytes, and current collectors shown in Figure 8.3.

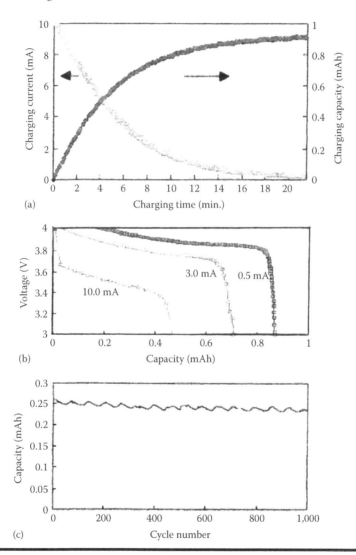

(a)

(b)

(c)

Figure 8.2 Electrical performance of thin-film microbatteries showing (a) charging current versus time, (b) terminal voltage versus capacity, and (c) battery capacity versus cycle number. Performance parameters of thin-film microbatteries showing (d) charging current versus time, (e) terminal voltage versus capacity at various temperatures, and (f) capacity versus cycle.

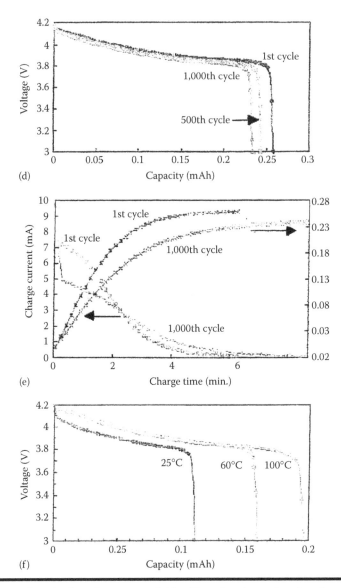

Figure 8.2 Electrical performance of thin-film microbatteries showing (a) charging current versus time, (b) terminal voltage versus capacity, and (c) battery capacity versus cycle number. Performance parameters of thin-film microbatteries showing (d) charging current versus time, (e) terminal voltage versus capacity at various temperatures, and (f) capacity versus cycle. (Continued)

Improvements in the microelectronics circuits and the miniaturization of electronic devices have reduced the current and power requirements of some electronic devices to an extremely low level, which can be satisfied only by thin-film batteries

Table 8.5   Thin-Film Battery Capacity as a Function of Physical Dimensions

Physical Dimension (mm/in.) (Length × Width × Thickness)	Battery Capacity (mAh)
20 × 25 × 0.1 (mm) 0.7900 × 0.9875 × 0.0040 (in.)	0.1
20 × 25 × 0.3 (mm) 0.7900 × 0.9875 × 0.0119 (in.)	1.0
42 × 25 × 0.1 (mm) 1.6590 × 0.9875 × 0.0040 (in.)	0.5
42 × 25 × 0.4 (mm) 1.6590 × 0.9875 × 0.0158 (in.)	5.0

using transition metal-oxide technology for the battery's critical elements, such as anodes, cathodes, current collectors, and electrodes. Thin-film battery technology offers minimum weight, small size, and thin form-factor, which are most desirable for batteries to power micro devices.

Research studies performed by various material scientists reveal that titanium (Ti), tungsten (W), and molybdenum (Mo) oxysulfide thin films and high-performance solid electrolyte are best suited for current and next-generation rechargeable microbatteries. Typical voltage ratings for a 3D lithium-Mo-oxysulfide battery as a function of current densities are shown in Figure 8.4. Studies performed by the author on various thin-film oxides reveal that on the basis of their electrochemical properties. The most promising oxide materials for microbatteries are $LiCoO_2$, lithium nickel oxide ($LiNiO_2$), and $LiMn_2O_4$.

The studies further reveal that $LiCoO_2$ thin films have received significant attention in the past decade, because this particular material offers optimum performance for the positive electrode in the thin-film battery. The reversible capacity per unit area

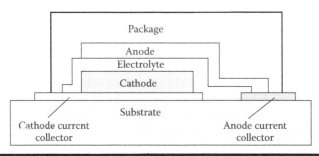

Figure 8.3   Schematic layout of a thin-film battery showing the critical elements of the metal-oxide electronics.

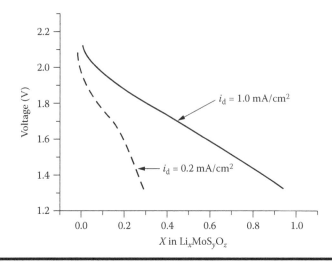

Figure 8.4 **Typical voltage rating for the 3D lithium-Mo-oxysulfide battery as a function of two current densities.**

and thickness for this material is about 69 mAh/cm² mm for an exchange of 0.5L/mol. of oxide (137 mAh/g) in a fully dense film. This means that the maximum specific capacity can vary from 55 to 69 mAh/cm² mm. The unit of measurement used for the film area is cm² and the unit of measurement for film thickness is the micron. The optimization of the microbattery capacity can be achieved by using thicker positive oxide films and then depositing the electrode material on a flexible polymer substrate. By using such a cathode film of 6.2 mm thickness, it is possible to achieve a high discharge capacity battery of about 250 mAh/cm² with good cyclability.

8.6.3 Performance Capabilities and Applications of Microbatteries

Thin-film batteries with a power output less than 100 microwatts can be designed using MEMS technology [5]. The 3D thin-film microbatteries are most ideal for satellite, spacecraft, and space sensors for which size, weight, and ultra-low power consumption are the design requirements. The critical elements of the 3D thin-film microbattery using MEMS technology are the graphite anode, cathode, nickel-current collector, and hybrid polymer electrolyte (HPE), which are shown in Figure 8.5a. The multichannel plate (MCP) with perforated holes plays a key role in achieving significant cathode volume (see Figure 8.5b). An increase in geometrical area, cathode thickness, and cathode volume provides the maximum energy density and the battery capacity. Studies performed by the author indicate that cathode volume can be achieved with a perforated substrate instead of full substrate.

Area gain (AG) is strictly a function of substrate thickness, the aspect ratio (height/hole diameter or thickness/hole diameter), and the number of holes or

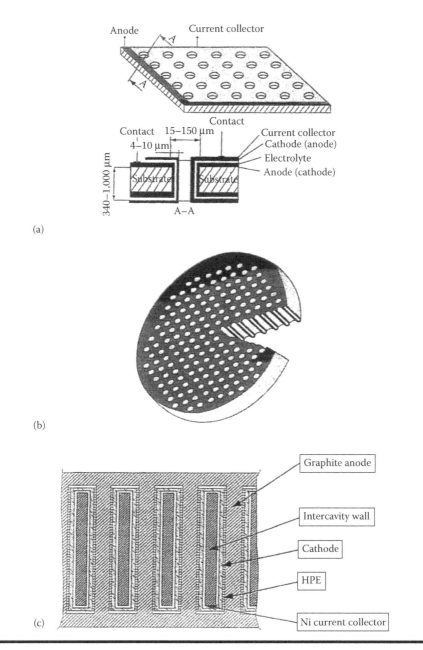

Figure 8.5 Three-dimensional microbattery design using MEMS technology showing (a) critical elements, (b) MCP substrate with perforated holes, and (c) cross-sectional view of the 3D microbattery.

microchannels in the MCP substrate. Computed values of AG as a function of hole diameter for given values of interchannel spacing (s) and substrate thickness (t) are shown in Figure 8.6a. AG values as a function of substrate thickness (t) for a given hole diameter (d) and interchannel spacing (s) are displayed in Figure 8.6b. A review of mathematical models indicates that a slight tilting of the footprint area

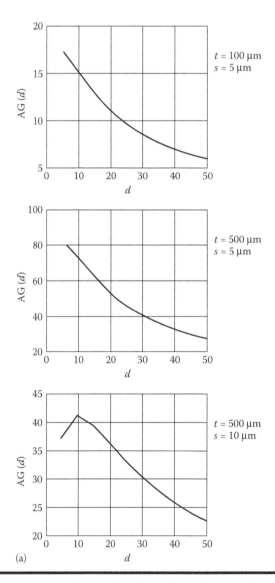

(a)

Figure 8.6 (a) Area gain (AG) as a function of hole diameter for assumed valued of spacing (s) and thickness (t). (b) AG as a function of substrate thickness (t) for assumed values of hole diameter (d) and hole spacing (s).

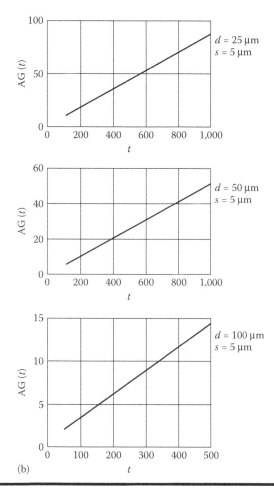

Figure 8.6 **(a) Area gain (AG) as a function of hole diameter for assumed valued of spacing (s) and thickness (t). (b) AG as a function of substrate thickness (t) for assumed values of hole diameter (d) and hole spacing (s). (Continued)**

with various hole geometries could yield larger AG by 10 to 15%. In other words, the geometry with a hexagonal hole configuration and separated with a constant thickness provides optimum AG in a perforated substrate, which will yield both the optimum capacity and energy density needed for MEMS-based microbatteries.

Trade-off studies performed by the author seem to favor the deployment of conformal thin-film structures using a combination of thin-film technology and MEMS technology to achieve miniaturization of the geometrical dimensions. The studies further indicate that the planar two-dimensional (2D) thin-film battery using MEMS technology would require large footprints of few square centimeters to achieve a battery with reasonable capacity [5]. The maximum energy available from a thin-film battery using MEMS technology is about 2 J/cm³, where J is the electrical energy of the battery.

Commercial thin-film batteries with a footprint of 3 cm³ will have a capacity of 0.4 mAh, which comes to 0.133 mAh/cm². Thin-film lithium-based microbatteries can be expected to meet the power requirements for the miniaturized power sources for MEMS-based sensors.

Research studies performed by various material scientists reveal that that the 3D microbatteries using MCP substrate are best suited to power the MEMS-based sensors and devices because of minimum size, minimum power consumption, and higher volumetric efficiency. The microbattery illustrated in Figure 8.5 has a configuration using a nickel cathode current collector; a low-cost, low-toxic Mo-sulfide cathode structure; lathinated graphite anode electrode; and an HPE and MCP substrate. The MCP substrate shown in Figure 8.5b has a diameter of 12 mm and a thickness of 0.5 mm with several 50-mm diameter hexagonal holes or microchannels. This substrate is the most critical element of a 3D microbattery best suited to power MEMS-based sensors and devices [5].

8.6.4 Electrical Performance Parameters of Nanobatteries

Limited research and development activities have been pursued on nanotechnology-based batteries. Preliminary laboratory tests conducted by nanomaterial scientists have demonstrated a cutoff voltage for a 3D nanobattery ranging from 1.3 to 2.2 V at room temperature with current densities ranging from 100 to 1,000 mA/cm². The slope characteristics of charge-discharge curves for a nanotechnology-based battery are similar to those of planer cells. Limited test data seem to indicate that 3D nanobatteries remain practically unchanged over extended cycling periods and at current densities ranging from 100 to 500 mA/cm². The nanobattery designers predicted that a 3D nanobattery with a cathode thickness of 1 mm would demonstrate 20 to 30 times higher capacity over 2D devices. Computer simulations indicate that with a 3-mm-thick cathode, the energy density of a 3D battery can be increased to greater than 100 mWh/cm² with the same footprint.

Toshiba of Japan was the first company to use nanotechnology in the design and development of Li-ion polymer-based nanobatteries. These nanobatteries have demonstrated significant performance improvements, such as a faster charge time, greater capacity in ampere-hours, and a longer cycle life. The use of nanotechnology in the development of batteries will bring significant improvements in performance irrespective of their capacities. Toshiba's principal objective is to develop batteries for EVs and HEVs, which will provide a reduction in their size, weight, and cost, while achieving significant enhancement in energy density (Wh/L) and specific energy (Wh/kg), which are essential for EV and HEV applications. A reduction in weight and size is of paramount importance when manufacturing EVs and HEVs. With lower weight and size, the automobiles have more room for storage and can accommodate more passengers. Studies conducted by the author on the use of rare earth materials and nanomaterials in manufacturing EV and HEV could affect the sale price of these vehicles by ~15 to 20%. The author did not perform cost analyses as part of these studies.

8.6.4.1 Applications of Nanomaterials, Carbon-Nanotubes, and Carbon-Nanotube Arrays in Development Batteries

Comprehensive knowledge of nanomaterials and carbon-nanotudes (CNTs) is essential to understand the operational and design aspects of nanobatteries. The anode, cathode, and current collector elements of the nanobattery must deploy appropriate nanotechnology-based materials. Studies performed by the author on potential nanomaterials indicate that CNT arrays represent a new class of material with multifunctional capabilities [6]. The studies further indicate that CNT-polymer arrays are most ideal for high-efficiency photovoltaic cells (solar cells), microbatteries, nanobatteries, and electrodes for micro- and nanobatteries. CNT can be grown in the form of powders or loose nanotubes and the CNT arrays can be grown on silicon substrate. Thermally driven chemical vapor deposition (CVD) technology uses hydrocarbon molecules as the carbon source, and iron, nickel, and cobalt as the catalysts to grow high-density nanotubes or CNT arrays on silicon substrate material. A CNT array can be fabricated using either single-wall CNT (SWCNT) or multiwall CNT (MWCNT) or both. CNTs are considered to be smart materials because of their high mechanical strength, improved electrical conductivity, multifunctional capabilities, piezoresistive and electrochemical sensing, and actuation properties. More research and development activities are required toward the development of synthesis, material characterization, processing, and nanotechnology-based device fabrication technology. The CNT growth mechanism is complicated, which affects the length of the nanotubes, and is strictly dependent on the strength of the interaction between the nanoparticles and the support. The growth kinetics of a CNT is a complicated phenomenon, which is based on its molecular dynamics. A nanotube array is typically 4 mm in length and has an area of 1 mm × 1 mm. Knowing the fundamentals of CNTs and nanomaterials, a battery engineer can fabricate the nanobattery with desired electrical parameters. The terminal voltage of this particular battery will be between 1.2 and 2.2 V, and its current density can be expected to be somewhere between 100 mA and 500 mA/cm^2. Its weight can be estimated to be less than 10 g on the basis of good engineering judgment.

8.7 Batteries for Health-Related Applications

Preliminary studies undertaken by the author on the batteries required for medical applications and other medical diagnostic procedures reveal that these batteries must meet stringent performance specifications, such as accuracy, dependability, portability, reliability, and long service life. This section describes the performance capabilities of batteries best suited for cardiac rhythm applications. These batteries will have unique characteristics with an emphasis on safety, reliability, and near-zero voltage fluctuations. This section summarizes the battery requirements for other medical applications, such microgastrointestinal investigation, self-regulating therapeutic treatments, hearing aids, high-intensity illuminators for eye-related diagnosis, and medical diagnostic medical procedures.

8.7.1 Battery Requirements for Cardiac Rhythm–Detection Applications

According to heart specialists, four distinct types of medical devices can be used to treat cardiac diseases, namely pacemakers, cardioverters, defibrillators, and left-ventricular assist devices. In addition, the total artificial or mechanical heart needs to be powered by batteries with an emphasis on reliability, safety, and OCV with no electrical surges and voltage fluctuations. Generally, cardiac pacemakers are prescribed by heart specialists when the cardiac rhythm is too slow or when the patient has an abnormal heart beat. This device is implanted in the patient's chest. The implanted pacemaker detects the slow heart rate and sends electrical impulses to stimulate the heart muscle. An electrical signal from the muscle is fed back to the device to make appropriate corrections in the stimulation parameters, which will normalize the heart rate within a hundred microseconds or so.

When pacemaker devices were first being installed, heart surgeons had to open the chest cavity of the patient under general anesthesia. The device and terminal of the device were attached to the heart surface to sense the electrical signals coming from the heart. Later on, some changes were made to the leads reaching the heart through a vein. The life of early pacemakers, continuously pacing the heart at a fixed heart beat, was less than two years because the life of the zinc-mercury-oxide batteries used in the device was less than two years, which meant this particular battery had to be changed every two years.

New pacemaker devices have been introduced that act only when a preset heart rate is not detected by the pacemaker. After 1975, $Li-I_2$ batteries with an operating life of up to 10 years were available for pacemaker applications. Critical elements of an $Li-I_2$ battery are shown in Figure 8.7. Continuous design improvements have improved the battery's service life to nearly 14 years. Other performance characteristics and design aspects of $Li-I_2$ batteries will be discussed in Section 8.7.2.1.

With remarkable progress in materials technologies, the use of titanium casing was introduced to replace the plastic material used previously. Use of titanium materials and special microfilters in the device enclosure housing shielded the pacemaker from electromagnetic fields emitting from nearby power transmission lines; radio frequency (RF) signals emitting from cellular phones, other personal communication devices, and microwave ovens; and other electrical appliances emitting RF signals. The external RF signals are in the range of 10 to 100 hertz to avoid interference with the pacemaker's

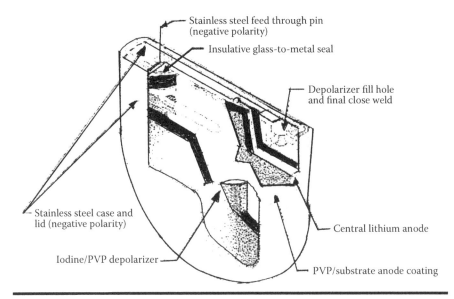

Figure 8.7   Isometric of a Li-I₂ battery for implantable pacemaker.

oscillator function, which operates at 167 hertz. The critical specification requirements for implantable pacemakers using Li-I₂ batteries, as noted in the October 2004 edition of *Indian Pacing and Electrophysiology Journal* [7], can be summarized as follows:

- *OCV*: 2.8 V.
- *Control-circuit minimal voltage*: 2.2 V.
- *Control-circuit current drain*: 10 mA.
- *Battery resistance*: 10,000 Ω.
- *Holding capacitor rating*: 10 mF.
- *Device oscillator frequency*: 167 Hz.
- *Duty cycle*: 16.7%.
- *Typical ampere-hour rating*: 2 Ah.
- *Failure rate*: 0.005% failures per month, which is equal to 0.5 failures per month.
- *Reliability*: 99.6% probability of survival beyond eight years of continuous service.
- *Space occupied by the battery in the pulse generator unit*: 5–8 cm³.
- *Battery weight*: 12.5–15.5 g depending on the pacemaker manufacturer; in addition, variation in battery weight is strictly dependent on the battery's longevity requirements and current drain capabilities.
- *Shape*: Most cardiac pacemakers make the devices with circular or elliptical shapes to avoid sharp corners to avoid any damage to skin or surrounding tissues. The pacemaker is shaped to conform to overall device geometry optimized by the device designer with a semicircle radius of about 3 cm and a depth of 6 to 8 cm.

8.7.2 Various Batteries Used to Treat Cardiac Diseases

Various batteries are used by implantable devices for the treatment of cardiac diseases. Some of these devices have been found to be ideal in terms of reliability and longevity, some have been found to be marginal in performance, and only few have been found to be suitable for pacemaker applications. The batteries selected for the treatment of various cardiac diseases include $Li-I_2$, $LiAgVO_{12}$, $LiMnO_2$, Li-ion, and $LiCF_x$. The performance of the batteries used in pacemakers is somewhat affected by such factors as body temperature, dissolved blood oxygen, physical exertion, and body movements.

8.7.2.1 Li-Ion Batteries Best Suited Primarily for Medical Devices Used to Treat Cardiac Diseases and to Detect Unknown Ailments

Li-ion batteries have been recommended to treat various ailments ranging from cardiac disease to hearing loss because of their excellent service lives exceeding 10 years. Their performance capabilities and unique features can be summarized as follows:

- Li-ion battery's self-discharge rates are less than 2% per year.
- The battery's operating life ranges from 8 to 12 years.
- The current requirements are minimum and vary from 10 to 20 mA.
- Typical battery capacity is in the range from 70 to 600 mAh.
- Its charge-discharge cycles exceed 500.
- A battery pack involving three cells could generate electrical power close to 500 W for a short-duration application.
- This battery comes in both primary and secondary categories.
- The battery is available in AA, AAA, C, and D formats with lower voltage ratings, but its CR123A battery has an OCV of 9 V, which is useful for applications requiring 9 V.

This battery is suitable for multiple applications requiring various operating voltage levels, including the following:

- This particular battery is used to activate miniature transmitters to track birds, wild animals, and large living objects in the ocean.
- One unique application of this battery involves swallowing a microbattery-powered pill with a built-in microcamera to view the digestive tract of a patient.
- Its consumer applications include cellular phones, iPods, electrical toothbrushes, low-cost cordless drills, electronic toys, and other miniature electronic devices.

8.7.2.2 Li-I₂ Batteries for Treating Cardiac Diseases

Critical design aspects and benefits of Li-I$_2$ batteries were briefly described in the previous section. Its critical features are as follows:

■ The battery construction requires a molten cathode material, which is a mixture of iodine and poly-2-vinylpyridine (PVP); it is poured into the cell, and a layer of lithium forms at the anode electrode, producing in situ a separator layer.
■ The chemical equation for this battery can be written as follows:

$$Li + (1/2) I_2 = [Li\ I] \tag{8.2}$$

■ The energy density of the Li-ion-PVP system is a result of the high energy density of the iodine (I$_2$). Discharge voltage variations as a function of capacity for the Li-I$_2$ battery at an ambient temperature of 37°C are displayed in Figure 8.8.
■ During the discharge, the Li-I$_2$ layer grows in thickness, which increases the cell's internal impedance.
■ Because of the solid electrolyte being used in the battery, the Li-I$_2$ cells are capable of supplying low current levels in the microampere range (of 10 to 20 mA).
■ Batteries that can generate currents in this microampere range are considered ideal for implantable pacemakers that are capable of operating for more than 10 years and for which high safety and reliability are of paramount importance.
■ The normal cell designs have demonstrated an operating life close to 10 years. But the larger implantable Li-I$_2$ batteries have demonstrated longevity exceeding 15 years or so.

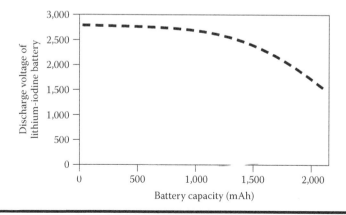

Figure 8.8   Discharge characteristic of Li-I₂ battery at a temperature of 37°C.

8.7.2.3 Li-AgVO₂ Batteries for Treatment of Cardiac Diseases

According to cardiologists, the Li-AgVO$_2$ battery is best suited for implantable pacemaker devices. The anode of this battery is made from lithium metal and the cathode is made from silver vanadium oxide (Ag$_2$V$_4$O$_{11}$). The cell chemical reaction is given by the following equation:

$$7 \text{ Li} + \text{Ag}_2\text{V}_4\text{O}_{11} = [\text{Li}_7\text{Ag}_2\text{V}_4\text{O}_{11}] \tag{8.3}$$

where chemical symbol V$_4$O$_{11}$ stands for vanadium oxide. Critical design aspects and performance characteristics are as follows:

- Lithium arsenic fluorine (LiAsF$_6$) has been selected as the substrate material for this battery.
- Electrolyte additives, such as an organic carbonate like dibenzyl carbonate (DBC), have a significant impact on the long-term performance of Li-AgVO$_2$ batteries. In addition, this electrolyte additive reduces the internal resistance by modifying the surface layer at the lithium anode.
- Self-discharge for recently designed batteries is less than 2% per year.
- The addition of seven equivalents of Li to AgVO$_2$ corresponds to a theoretical specific capacity of about 315 mAh, which results in a specific energy of 270 Wh/kg and a volumetric energy density of Wh/L for a practical Li-AgVO$_2$ cell.
- The reduction of AgVO$_2$ over the range of $o < x < 2.4$ for the chemical compound Li$_x$Ag$_2$V$_4$O$_{11}$ gives rise to the formation of metallic silver, which greatly increases the electrical conductivity of the cathode material. This essentially contributes to the high current capability of the Li-AgVO$_2$ battery. As the value of the parameter exceeds 2.4 the reduction of V^{5+} to V^{4+} and V^{3+} can be identified. Furthermore, when x exceeds 3.8, mixed-valent materials are found containing V^{3+}, V^{4+}, and V^{5+} in the same sample, but in reverse order.
- The presence of several different oxidation states of vanadium (V), as well as Ag$^+$ reduction, will result in stepped discharge characteristics as illustrated in Figure 8.9. This stepwise change in battery voltage provides a state-of-charge indication for the implanted battery. Figure 8.9 also shows the effect of superimposing a current pulses to the background current.
- The use of high rate pulses of 2 A each with a pulse duration of 10 sec. and applied in sets of four, with 15 sec. apart, can significantly enhance the long life characteristics of this kind of battery.
- Two types of Li-AgVO$_2$ implantable batteries are available, namely a high-rate battery capable handling ampere-level current pulses, which is most ideal for cardioverter defibrillators, and a medium-rate battery operating in the milliampere range, which is best suited for atrial defibrillators.
- Cardioverter defibrillators are capable of delivering one or two electric shocks within 5 to 10 sec. to start a heart that has started beating too fast. The power level of these shocks may be as high as 8 W.

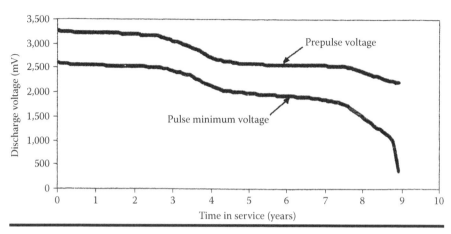

Figure 8.9 Discharge voltage characteristics of a Li-AgVO battery during the pulse condition test at 37°C.

- The battery is unable to do that but can charge with current levels of 1.5 to 3.0A in 5 to 10 sec. But the use of capacitors can deliver the shocks.
- This battery must be capable of delivering a continuous current in the micro-ampere range and it should last for 6 to 10 years yielding the specified electrical performance.
- The high current charging the capacitor can be achieved only with large-area electrodes.
- Battery suppliers claim that the latest Li-AgVO$_2$ battery has a volume of less than 6.5 cm^3 and a weight close to 19 g or 1.2 oz.
- These batteries are best suited as implantable medical devices.

8.7.2.4 Batteries for Critical Diagnostic Procedures

This section describes the types of batteries and their performance capabilities for critical medical diagnostic applications. Three distinct diagnostic systems will be described: (1) microgastrointestinal systems, (2) implantable gastric stimulation system, and (3) a self-regulating response to a therapeutic drug delivery system.

1. *Microgastrointestinal system (MGS)*: The MGS system deploys a zinc-silver-oxide (ZnAg$_2$O) battery-powered diagnostic capsule and is widely used to map the small intestine with a mini video camera. The ZnAg$_2$O battery is a primary or nonrechargeable device and is best suited for this particular application. Design features of the capsule and performance capabilities of the ZnAg$_2$O battery and other system elements are summarized as follows:
   - The endoscopy capsule battery provides excellent photos and streaming videos during its 8 h journey through the digestive system of the patient under investigation.

- High-quality images are transmitted at the rate of two per second to a data recorder, powered by a second battery, carried by the patient. The microgastrointestinal diagnostic system essentially consists of a compact light source, mini video camera, processing microelectronics, data radio minitransmitter, and the $ZnAg_2O$ battery.
- The principal object is to scan the small intestine, which is not accessible using other noninvasive diagnostic techniques.
- This capsule provides colored images suitable to identify diseases, such as ulcers, cancer, internal bleeding source, and intestinal obstruction.
- The capsule deploys CMOS semiconductor technology, which offers miniaturization of electronic circuits and substantial reduction in energy consumption.

2. *Implantable gastric stimulation (IGS) system*: The second system is implanted under the patient's abdomen and is designed to stimulate the stomach for the treatment of obesity.
   - This system includes an external programmer and a gastric stimulation lead that send electrical pulses to the smooth muscle of the stomach.
   - The programmer communicates with the pulse generator and is capable of modifying the electrical parameters of the pulses, such as pulse width and rise and fall times of the stimulating pulses.
   - The stimulator is about the size of a matchbox and also includes a lithium sulfur oxide chloride ($LiSOCl_2$) primary battery with an operating life ranging from three to eight years depending on the stimulating pulse parameters.
   - High reliability, utmost safety, and power output with minimum variations.

3. *Self-regulating responsive therapeutic drug delivery system*: This particular system essentially is a biosensor-drug delivery system that eliminates the need for telemetry and human in the loop. The system is enclosed in a microcapsule.
   - The critical elements of the capsule include an optical dome with high visibility, compact lens holder, small lens, illuminating light-emitting diodes, images generated by CMOS technology, microbattery, application-specific integrated circuit (ASIC), and a miniaturized antenna.
   - The cylindrical-capsule has a length of 26 mm and diameter of 11 mm.
   - The capsule delivers the drug as per the prescribed instructions to the patient.
   - This particular drug-delivery system is foolproof.
   - The system offers a highly reliable, error-free technology.
   - The system is very complex and expensive.

8.8 Batteries for the Total Artificial Heart

The total artificial heart (TAH) is a mechanical pump capable of replacing the heart of a patient suffering from heart-related problems. Several experimental

TAHs have been designed and implanted in patients. Only a few TAH designs are considered to be reliable and are still operating satisfactorily in a limited number of patients.

This section provides specific details on the critical elements of the pump and the types of batteries most suitable for the artificial heart pump application. The design of this replacement heart pump consists of two chambers each capable of pumping more than 7 liters of blood per minute. The artificial pump deploys an implantable Li-ion battery pack that is recharged through the patient's skin, if and when necessary. All implantable Li-ion battery packs must meet the following stringent requirements:

- Safety against all electrical and mechanical abuses
- Reliability with a particular emphasis on critical battery pack performance parameters
- Predictability on electrical performance parameters, such as voltage, current, and time relationships
- Ultra-low self-discharge (reduced parasitic reactions in the battery)
- High energy density
- Minimum battery pack weight and size
- End-of-life indication with audible sound
- High cycle life
- Safe charging capability within a specified period
- High cell voltage ranging from 3.6 to 4.2 V, which means that the battery pack for a TAH can be designed with a minimum number of cells

8.8.1 Major Benefits of Li-Ion Batteries Used for Various Medical Applications

Li-ion is a low-maintenance battery that is not possible with other battery types. This battery does not suffer from memory loss, and its self-discharge rate is less than 50% compared with Ni-Cd batteries. Lithium cells cause little harm when exposed accidently. Because the battery's mechanical structure is delicate, it requires a miniaturized protection circuit to maintain its safe operation. The built-in protection circuit limits the peak voltage of each cell during the charge and prevents the cell from dropping to low ion voltage. It is necessary to monitor the cell voltage to prevent temperature extremes. In this particular application, the charge and discharge current level on the battery pack is limited to between 1 C and 2 C. The battery suppliers recommend storage temperature of 15°C or 59°F, because cold storage of these cells slows the aging process of Li-ion cells, thereby providing longer storage life. Li-ion battery manufacturers recommend a 40% charge level for storage level. According to the maintenance experts, no periodic charge is needed. Li-ion cells are capable of providing very high current levels with no compromise in battery reliability and mechanical integrity.

8.8.2 Limitations of Li-Ion Batteries

Li-ion batteries require a microelectronic protection circuit to maintain voltage and current within safe limits without compromising the reliability and electrical performance of the battery. Aging has been observed even when this battery is not in actual use, but storage in a cool place at 40% charge reduces this particular effect. Shipping of these batteries in large quantities is subject to strict regulatory control. These batteries are expensive to manufacture. Li-ion battery costs are roughly 40% higher than the Ni-Cd batteries.

8.8.3 Cell-Balancing Requirements for Li-Ion Rechargeable Battery Packs

Cell balancing of a lithium-based rechargeable battery pack is necessary if the reliability and safety of the battery pack are the principal design requirements [8]. Li-ion battery designers reveal that self-discharge rates of Li-ion rechargeable batteries double for every 10°C rise in temperature. Cell balancing is essential for the safety and longevity of the Li-ion rechargeable battery pack. A number of different cell-balancing technologies are available. But it is up to the battery pack designer to determine how much cell balancing is needed. Studies performed by the author indicate that cell balancing needs to be matched to both the cells and to the way the cells are going to be used [8].

The studies further reveal that thermal mismatch heavily influences the amount of balancing required. Thermal mismatch is due to the fact that the battery pack can experience a difference in surface ambient temperatures that causes the cells near the operating equipment to run much hotter than the cells located on the outside of the battery pack. Self-discharge rises drastically as a cell is heated, as illustrated by the data presented in Table 8.6.

From the tabulated values, it appears that the self-discharge rate doubles for every 10°C rise in cell surface temperature. The amount of balancing required is strictly dependent on the number of cells used by the battery pack and the balancing factors involved [8]. The balancing factors involved include thermal mismatch when the battery is being used, impedance mismatch when the pack is charged and discharged, cell quality dependent on the quality control and quality assurance specifications used by the cell manufacture, rate of cycling dependent on the number of cycles per day or week, desired cycle life dependent on the application requiring a few hundred cycles before wearing out, and pack capacity indicating the number of cells used in the pack. Minimum balancing is required when a few cells are used in the pack. Conversely, more balancing is required when many cells are involved in the pack.

Additional balancing may be required if cheaper cells are used in the battery pack, if operation involving many cycles is desired, if an application wants to get as many cycles as possible out of the pack, and if an application wants to squeeze every milliampere-hour out of the battery pack. Cell balancing is a complex and costly

Table 8.6   Self-Discharge per Month as a Function of Lithium-Ion Cell Temperature

Cell Temperature (°C)	Self-Discharge per Month (%)
0	000
10	0.05
20	1.00
30	2.00
40	4.00
43	5.00
50	8.15
60	15.18

procedure. More efforts and costs are involved for more balancing requirements for the battery pack.

Cell balancing can be accomplished either using a passive or an active cell-balancing technique. The passive cell-balancing approach is relatively less costly and complex. The active-balancing technique is very complex, but it offers an added advantage if it is used during the discharge cycle.

Thermal mismatch is the most critical factor that influences the amount of balancing required. The battery pack can experience differences in temperature, which will cause the cells closer to the equipment to run much hotter than those cells located outside the pack, as illustrated in Figure 8.10. Because of this, the individual cells state of charge can be affected by even a short amount of use in such

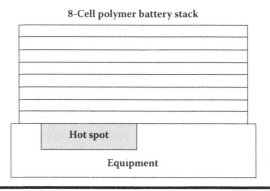

Figure 8.10   Battery pack cells experiencing different temperatures because of their locations.

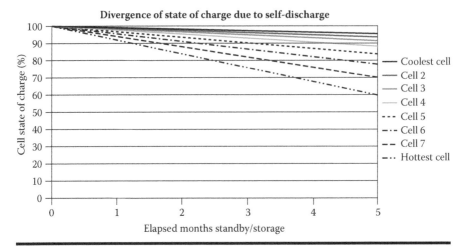

Figure 8.11 State of charge of individual cell as a function of standby or storage duration.

an imbalance situation as illustrated in Figure 8.11. The table that is an integral part of the Figure 8.11 shows that cell number one is the coldest one, whereas cell number eight is the hottest one because it is closest to the equipment. Examination and comprehensive review of Figure 8.11 indicate that some other sources of imbalance must be considered, which will affect the overall cell-balancing requirements. Examination of Figure 8.10 shows that the hottest spot in the battery pack is the lower surface of cell number eight.

The type of balancing required is strictly dependent on the amount of current needed. In addition, how much cell-balancing current does on need to take care of an initial pack imbalance? One should always look at it from a milliampere-hour requirement. The cell-balancing concept can be clear if the following mathematical example of a 10 Ah battery pack capacity with 1 h charge is reviewed using the assumed values of certain parameters:

Parameters assumed

- *Battery pack capacity (C)*: 10 Ah
- *Days required to balance the battery pack*: D
- *Charging duration (t)*: 1 h
- *Cell balancing duration*: 30 days at one cycle per day
- *Charge cycles per day (n)*: one cycle
- *Initial imbalance percentage (p)*: 7%
- *Self-discharge imbalance per month (i)*: <5%
- *The operational duration of the pack*: 5 months with a 20% initial thermal imbalance
- *Self-discharge variation per month*: [20%/5] = 4%

Table 8.7 Amount of Imbalance Drop after Each Month

Months since Beginning	Number of Days	Imbalance Drop after Each Month (%)
1	30	0.36
2	60	0.18
3	90	0.12
4	120	0.09
5	150	0.07

The balance current (I_B) equation can be written as follows:

$$I_B = [C(p + i)]/[(n\ t\ D)] \tag{8.4}$$

Inserting the assumed values of various parameters in this equation, one gets:

$$I_B = [10\ (0.07 + 0.04)]/[1 \times 1 \times 30] = [1.1]/[30]$$
$$I_B = [1.1 \times 1,000]/[30] = (36.66)\ mA \tag{8.5}$$

This balance current of 36.66 mA is 0.3666 or roughly 0.4% balancing of the 10 Ah pack capacity, which will drop after the first month, because the initial imbalance factor, which is 7%, hopefully would be balanced.

Inserting these parameters and number of days in each month into Equation 8.4, the balancing will drop to new values as shown in Table 8.7.

It is evident from the computed values summarized in Table 8.7 that the total imbalance after five months or 150 days is roughly 0.82%. This imbalance can be further reduced using either a passive or active cell-balancing techniques.

If the battery pack is recharged two times a month or twice a month, this translates to only one cycle every 15 days or 0.067. This means the balance current using Equation 8.4 can be written as follows:

$$[10(0.07 + 0.04)]/[0.067 \times 1 \times 30)] = [1.1]/[2.01] = [547]\ mA \tag{8.6}$$

This balance current is roughly (547/10,000) (100)%, which comes to about 5.47% of the 10 Ah battery capacity. This is a very high balance current level that can be handled by using an active-balancing technique.

8.8.4 Active-Balancing Technique

When the balancing current is too high, the active cell-balancing technique appears to be most effective [8]. With active-cell balancing, one gets an added advantage if the technique is used during the discharge phase of the battery. Let us assume another 10 Ah battery pack with eight cells for which one cell has a capacity of 8 Ah. With active balancing during the discharge cycle, the balancer could transfer the charge from the highest cells in the pack to the lowest cell. On the basis of the eight-cell pack concept, this would take about (10 – 8)/8 Ah or (2,000/8) or 250 mAh from the remaining cells and would transfer it to the lowest cell. This means that the battery pack appears to have (10 – 0.25) or 9.75 Ah capacity. This also means that the pack with 9.75 Ah capacity would still be in service, thereby extending the operating life of the battery pack for hundreds of recharge cycles. To achieve longer operating pack life, the active-balancing technique needs to be effective enough to keep up with the electrical load of the battery pack.

These sample calculations do not take into account all the relevant balancing factors. In the absence of other unknown factors, it is advisable to overspecify the amount of balancing current level by 10 to 20%. If too much balancing current is deployed, the cost of the protection circuits could go up slightly, but the safety and reliability of the battery pack will not be compromised, and the balancing circuitry will not operate more often.

To achieve precise balancing current levels, it will be necessary to undertake real-life testing, which is more complex and time-consuming. Once a rate is selected and the battery pack is assembled, the system should not be subjected to the worst conditions and it should be monitored to ensure that the balancing system is keeping up. Additional secondary protection circuits may be required to ensure that when the balancing technique cannot keep up, one must ensure that the battery pack fails gracefully. In summary, real-life testing experts suggest that the cell-balancing level must be selected with the smallest balancing current levels being around 10 mA for an 800 mAh battery pack, which comes to a 1.25% current level.

8.9 Conclusion

This chapter is dedicated to low-power batteries and their applications in various fields. Performance characteristics and major benefits of Li-ion batteries for low-power devices applications are summarized with an emphasis on longevity, reliability, safety, and cost-effectiveness. Materials best suited for anodes, cathodes, and electrolytes used in low-power batteries are briefly discussed. The benefits of solid electrolytes in low-power rechargeable batteries are identified with an emphasis on reliability and safety. The rechargeable batteries most ideal for low-power electrical and electronic device applications for which price and performance are of paramount importance are evaluated. Packaging techniques and materials used in the design and development of low-power rechargeable batteries are described with an emphasis on reliability and cost. A glossary of terms often used to specify rechargeable battery performance parameters is provided. Fabrication requirements of thin-film low-power batteries best suited for low-power electronic devices are briefly discussed. Performance capabilities and limitations of primary and secondary (rechargeable) batteries for low-power applications are highlighted. Critical performance parameters of miniaturized versions of $LiMn_2O_4$, $Li-I_2$, $Li-CF_x$, $Li-FeS_2$, $LiSO_2$, and $Li-SOCl_2$ batteries are summarized with an emphasis on gravimetric and volumetric energy density level, shelf life, operating temperature, and reliability. Performance characteristics of small-size rechargeable batteries such as sealed-lead acid, sealed Li-ion, S-Ni-MH, and lithium polymer electrolyte batteries are summarized with an emphasis on reliability, safety, and price-to-performance ratio. Performance capabilities of thin-film rechargeable microbatteries are summarized with an emphasis on reliability and longevity. Benefits of thin-film metal-oxide electronics used in lithium-based rechargeable microbatteries are identified. Important performance parameters of microbatteries using MEMS technology are discussed with an emphasis on a significant reduction in weight, size, and form-factor. Electrical performance characteristics and benefits of nanobatteries using nanotechnology are summarized with an emphasis on microampere current capability in 10 to 20 mA. Benefits of nanomaterials such as CNTs and the CNT arrays most ideal in the fabrication of nanobatteries are identified. Low-power rechargeable batteries for health-related applications are discussed. Performance requirements of rechargeable batteries best suited for cardiac rhythm–detection applications are summarized with an emphasis on portability, reliability, and absence on electrical spikes. Performance capabilities of implantable microbatteries for the detection and treatment of cardiac diseases and other ailments are identified with a particular emphasis on operating life exceeding 10 years, high reliability, and safety. Performance capabilities of rechargeable microbatteries are summarized for applications in consumer applications, such as cellular phones, electrical toothbrushes, electronic toys, low-cost cordless electrical drills, and other electronic devices.

Chemical reaction expressions are derived wherever required for the benefits of readers and to examine the harmful battery by-products, if any. Li-ion rechargeable batteries capable of supplying ultra-low current levels in a microampere range from 10 to 25 mA best suited for implantable pacemakers are described with an emphasis on longevity, reliability, safety, and minimum size and weight.

Performance characteristics of implantable microbatteries, including Li-ion, Li-AgVO, and $Li-I_2$, are summarized with an emphasis on reliability, safety, and longevity. Rechargeable batteries and their critical performance parameters for three distinct diagnostic systems, namely MGS, IGS system, and self-regulating response therapeutic drug delivery system, are described with an emphasis on weight, size, form-factor, safety, and reliability.

Rechargeable microbatteries incorporating thin-film technology and their performance capabilities for endoscopic capsule are briefly described with an emphasis on the deployment of microelectronic technology and mini-infrared cameras vital for endoscopic procedure and patient comfort. Rechargeable battery performance requirements for atrial defibrillators, cardioverter defibrillators, and the TAH are identified with an emphasis on safety, reliability, weight, size, and form-factor.

A battery pack consisting of 4 to 12 lithium-based cells may be required to meet specific power-consumption requirements. In such a battery pack application, cell balancing is of critical importance to preserve the electrical performance, thermal performance, reliability, and mechanical integrity of the pack. Cell balancing becomes more complicated and expensive as the number of cells in the pack increases. Studies performed by the author indicate that balancing can be accomplished either using a passive or active cell-balancing technology. The passive cell-balancing technique may be just right if cells are incorporated in the pack. The author has presented couple of examples to demonstrate the application of cell balancing involving Li-ion cells because these cells are more temperature-sensitive under high current levels. Self-discharge rates for Li-ion cells as a function of temperature are discussed in great detail with an emphasis on imbalance current dropping out after each month of use. Computed values of balancing currents are provided as a function of battery pack capacity in ampere-hours, days required to balance the battery pack, charging duration, cell-balancing duration, charge cycle per day, self-discharge imbalance per month, and self-discharge variation per month. Benefits of the active cell-balancing technique are briefly discussed.

References

1. Pier Paolopro Sim, Rita Mancini et al., *Solid State Ionics*, London: Elsevier (2001), pp. 185–192.
2. Robert Powers, "Batteries for low power electronics," *Proceedings of the IEEE*, 83, no. 4 (April 1995), pp. 687–693.
3. Eric Lind, "Primary lithium batteries," *Electronics Products* (2010), pp. 14–19.
4. A. R. Jha and R. Goel, *Monolithic Microwave Integrated Circuit (MMIC) Technology and Design,* Norwood, MA: Artech House (1989).
5. A. R. Jha, "MEMS and nanotechnology-based sensors and devices for commercial, medical, and aerospace applications," in *Performance Parameters of a 3-D Thin-Film Micro-Battery*, CRC Press: Boca Raton, FL (2008), p. 349.
6. A. R. Jha, "MEMS and nanotechnology-based sensors and devices for commercial, medical, and aerospace applications," in *Performance Parameters of a 3-D Thin-Film Micro-Battery*, CRC Press: Boca Raton, FL (2008), p. 353.
7. S. M. Ventateswara, N. S. Rao et al., "Indian pacing and electrophysiology" (October 2004), pp. 201–212.
8. Steve Carkner, "Lithium-cell balancing: When is enough, enough?" *Electronic Products* (March 2011), pp. 46–50.

Index

Page numbers followed by *f* and *t* refer to figures and tables, respectively.